This brilliant volume, edited by two of the leading
first century and with contributions from dozens of
vides a lucid, accessible, and actionable introductic
that is intellectually and empirically wider, deeper, a
By embracing the field's rich heterogeneity, the editors have produced, perhaps
terintuitively, the clearest and most coherent overview to date of not just what political ecology is, but of how it can be used by those who seek to understand and change the world. It is required reading and an essential resource for researchers, students, and practitioners in this domain.

James McCarthy, *Leo L. and Joan Kraft Laskoff Professor of Economics, Technology and Environment, Clark University*

A tremendous introduction (or reintroduction) to the insights, interpretive lenses and approaches, and shared critical, justice-minded ethos that have drawn so many to political ecology. Lucky newcomers and current practitioners alike will find guides and inspirations here, as the collection's diverse pool of authors welcomes readers along an engaging exploration of longstanding currents and new wellsprings in a rich, growing field. A timely exhortation to all that doing the collective work of political ecology matters now more than ever, and that embracing the field's ongoing pluralization is crucial to its intellectual abundance, effective alliances, and hopeful future-making for an environmentally changed world.

Sarah Knuth, *Associate Professor of Geography, Durham University, UK*

The origins of political ecology are diverse, multi-disciplinary and transnational, and its conceptual and methodological toolkit is wide-ranging and heterogenous resembling something like a big tent. In their outstanding collection of nineteen rich chapters drawing on scholars located across North America, Europe, Oceania, Africa, and Latin America, Gregory Simon and Kelly Kay's *Doing Political Ecology* offers a comprehensive and innovative census of the tent's occupants. In seeing the field as a rich lattice-work of approaches, the book brilliantly illuminates not simply its breadth, scope, and sophistication but how differing lineages offer up powerful and varied ways of "doing political ecology" as a form or critical knowledge production and practice. A book to think with and to savor.

Michael Watts, *Class of '63 Professor Emeritus, University of California, Berkeley and Long Term Fellow Swedish Collegium for Advanced Study, Uppsala*

Doing Political Ecology is a timely and indispensable volume for scholars, students, and practitioners conducting critical environmental research in the twenty-first century. Its great strength is the inclusion of contributors whose diverse voices and methods are enlivening a new wave of scholarship. This outstanding edited collection offers a provocative encounter with the methodological, theoretical, and ontological approaches that are energizing political ecology as a field of inquiry and practice.

Judith Carney, *Distinguished Research Professor of Geography, University of California, Los Angeles*

While much has been written about what political ecology *is*, this volume explores what it means to *do* political ecology. Each chapter charts a unique story about the relationship between political ecology as a unifying field and the specific lenses that inspire individual political ecologists. Written by some of the subfield's brightest new stars, this volume is both global in scope and grounded in compelling case studies. It showcases political ecology in all its manifold glory: a field where many concepts, methods, and outputs are welcome. This should be required reading for new political ecologists as well as those looking for fresh and inspiring new perspectives.

Andrea Marston, *Assistant Professor, Department of Geography, Rutgers University*

An inspiring collection of work by the leading lights and rising stars of political ecology, and a celebration of its continuing refusal of confinement within method or discipline. Political ecology continues to develop as a complex harmony of voices, and remains a pluripotent lens, instinct, approach, and strategy for understanding the mutual constitution of environment and society. Together, these authors weave a framework for research and action that is both generous and durable.

Morgan Robertson, *Professor of Geography, University of Wisconsin–Madison*

Political ecologists are doing innovative theorizations and praxis, drawn from diverse lineages of analyzing nature-society relations, ones that help address many complex global environmental problems globally. This book showcases this field of scholarship brilliantly by bringing together an impressive array of authors and ideas. *Doing Political Ecology* offers critical, emergent, and insightful arguments that are essential readings for political ecologists and interdisciplinary scholars and students interested in environmental politics and resource governance.

Farhana Sultana, *Professor of Geography and the Environment, Syracuse University*

In *Doing Political Ecology*, Kay and Simon have assembled a cast of luminaries in the field who together offer a novel and refreshing addition to recent edited collections on the subject with its innovative approach that political ecology is something that scholars *do*. Throughout its nineteen chapters, authors engage new pressing debates (data science), while also offering innovative takes on traditional concerns (epistemologies of knowing nature). Together the chapters in this edited book offer a timely and fresh contribution to the field of PE that is more than the sum of its parts. It's a must read and is eminently teachable.

Trevor Birkenholtz, *Professor of Geography, The Pennsylvania State University*

With a focus on *doing* political ecology, this book breaks into the "black box": It is an open invitation to consider political ecology's multiple roots, frameworks, and praxis. An important guide for both old and newcomers to the field.

Diana Ojeda, *Professor of Geography, Indiana University Bloomington*

The volume presents a diverse set of accounts, critiques, and authors, and it does so astutely. It is sensitive to the shifts in political ecology of the last decade and it honors the nexus of "new bodies" of thought that have rightfully critiqued and now been assumed or subsumed into political ecology practice, such as the insights from Black Geographies, settler colonial studies, and racial capitalism.

Eric Perramond, *Professor of Environmental Science and Southwest Studies, Colorado College*

Doing Political Ecology offers well-curated, highly readable, and theoretically rich chapters on the stuff that's animating US-based political ecologists today—from the ways that nature is monetized, enumerated, and narrated to the place of race and bodies in political-ecological struggle. This is a vital, go-to resource for graduate students and others intrigued by the frameworks, methods, and insights that political ecologists contribute to how we understand the world.

Kendra McSweeney, *Professor of Geography, The Ohio State University*

By uniting a host of global voices, contexts, and concerns into a single, fresh, and accessible resource, *Doing Political Ecology* offers something new. This spirited collection demonstrates the urgency of critical environmental scholarship for addressing the world's most urgent challenges—climate change, biodiversity collapse, and rampant environmental injustice—and compels students and readers to imagine and visualize ways they can intervene themselves.

Paul Robbins, *Dean of the Nelson Institute of Environmental Studies, University of Wisconsin–Madison*

Reflecting emergent and established voices, and covering both foundational debates and new perspectives, Kay and Simon's book captures the historical development and current state of PE, while offering insights that researchers can operationalize for years to come.

The contributing authors, approaches, and topics reflected in the book are testimony to the commitment of Kay and Simon—and the field of PE—to maintaining space for multiple ways of doing, being, and knowing. The book platforms a wide series of "lenses" or "frameworks"—including feminist political ecology, Science and Technology Studies, political economies of nature, social constructivist approaches, and critical physical geography—while privileging none. In capturing this multiplicity and the diverse ways that researchers approach the field, the book provides particularly useful guidance for early career researchers cementing their own methodology, epistemology, and ontology position.

Emma Colven, *Lecturer in Risk, Environment and Society, Geography Department, King's College London, UK*

This book offers an exquisite collection of chapters that explore how nature is political, what environmental politics is all about, and the lineages of power through which nature is or ought to be politicized. For those who believe nature is inherently political, this book will surely offer key new insights and perspectives.

Erik Swyngedouw, *Professor of Geography, The University of Manchester, UK*

Doing Political Ecology

Since its inception, the field of political ecology has served as a critical hub for inclusive and transformative environmental inquiry. *Doing Political Ecology* offers a distinctive entry point into this ever-growing field and argues that our scholarly "foundations," today more than ever, comprise a cross-cutting latticework of research approaches and concepts.

This volume brings together 28 leading scholars from a range of backgrounds and geographies, with contributions organized into 18 analytical lenses that highlight different approaches to critical environmental research and "ways of seeing" nature-society interactions. The book's contributors engage the breadth and depth of the field, recognizing a variety of roots and genealogies, and give ample voice to these rich and complementary lineages. This inclusive presentation of the field allows diverse theoretical and empirical approaches to intermingle in novel ways. Readers will emerge with a wide-ranging understanding of political ecology and will attain a diverse toolkit for evaluating human–environment interactions.

Each chapter astutely grounds key methodological, theoretical, topical, and conceptual approaches that animate a range of influential, cutting-edge, and complementary ways of "doing" political ecology.

Gregory L. Simon is a Professor in the Department of Geography and Environmental Sciences at the University of Colorado Denver. He has held positions at ETH Zurich, Stanford University, University of Colorado Boulder, and UCLA. His research examines the development and governance of social-environmental risks and vulnerabilities.

Kelly Kay is an Associate Professor in the Department of Geography at the University of California, Los Angeles. She received her PhD from Clark University and has held academic appointments at UC Berkeley and The London School of Economics and Political Science. Her research is concerned with the political economy of the environment.

Doing Political Ecology

Edited by Gregory L. Simon and Kelly Kay

LONDON AND NEW YORK

Designed cover image: Bloomberg Creative / Getty Images

First published 2025
by Routledge
4 Park Square, Milton Park, Abingdon, Oxon OX14 4RN

and by Routledge
605 Third Avenue, New York, NY 10158

Routledge is an imprint of the Taylor & Francis Group, an informa business

British Library Cataloguing-in-Publication Data
A catalogue record for this book is available from the British Library

ISBN: 978-0-367-75531-7 (hbk)
ISBN: 978-0-367-76095-3 (pbk)
ISBN: 978-1-003-16547-7 (ebk)

DOI: 10.4324/9781003165477

Typeset in Sabon
by SPi Technologies India Pvt Ltd (Straive)

Contents

Illustrations

Figures

Tables

Boxes

Contributors

Christine Biermann is an Associate Professor of Geography and Environmental Studies at University of Colorado Colorado Springs. She co-edited *The Palgrave Handbook of Critical Physical Geography*. Spanning human and physical geography, her work considers the nexus of conservation science, management, and politics.

Michaelanne Butler is pursuing a PhD at the University of California, Santa Cruz in Sociology. Holding both an MBA and MSc in human geography from Oxford University, her interests range from entrepreneurship and innovation to critical food studies and the political-ecological connections between the global North and South. Most recently, she was a contributor to the UC AgriFood Tech Research (AFTeR) project, which explored the emerging Silicon Valley-based Food Tech and Ag Tech sectors across a broad range of innovations, seeking to understand the transformative potential of novel agro-food tech products and the visions that underpin them.

Noel Castree's research and teaching focuses on how people relate to the planet. He's author of *What Future for the Earth?* (Routledge, 2025) and co-author of *David Harvey: A Critical Introduction to His Thought* (Routledge, 2023).

Deepti Chatti is Assistant Professor of Climate Justice at the University of California San Diego, in the Department of Urban Studies and Planning, and the Critical Gender Studies Program. An interdisciplinary scholar, Deepti's research critically analyzes sustainable development projects on energy access, air pollution, and climate change. Her research also examines the role of science and technology to advance sustainable development, and the politics of knowledge production in environmental research.

Alice Cohen is a professor in the departments of Earth & Environmental Science and Environmental & Sustainability Studies at Acadia University. Her research focuses on environmental governance. She is the co-author of *Organizing Nature: Turning Canada's Ecosystems into Resources*, which explores the mutually constitutive relationship between peoples and landscapes. She holds a PhD from the University of British Columbia.

Rosemary Collard is a human geographer and political ecologist who draws on feminist political economy and primary research to shed light on the root drivers of extinction and declining wild animal abundance. She is the author of *Animal Traffic: Lively Capital in the Global Exotic Pet Trade* (Duke, 2020) and co-editor of *Critical Animal Geographies: Politics, Intersections and Hierarchies in a Multispecies World* (Routledge, 2015). She is an Associate Professor in Geography at Simon Fraser University.

Joel E. Correia is author of *Disrupting the Patrón: Indigenous Land Rights and the Fight for Environmental Justice in Paraguay's Chaco* and assistant professor in the Human Dimensions of Natural Resources Department at Colorado State University. His research investigates the uneven effects of development-driven environmental change, particularly the destruction of South America's largest forests—the Amazon and the Gran Chaco—and social responses to that change. Correia's work is rooted in qualitative social science and public political ecology but transgresses traditional disciplinary boundaries through diverse mixed-methods and engaged in-situ research with front-line communities, human rights advocates, policy makers, and scientists.

Diana K. Davis, a geographer and veterinarian, is Professor of History and Geography at the University of California, Davis. She is the author of the award-winning *Resurrecting the Granary of Rome: Environmental History and French Colonial Expansion in North Africa* (Ohio), *The Arid Lands: History, Knowledge Power* (MIT), co-editor of *Environmental Imaginaries of the Middle East and North Africa*, and over four dozen articles and book chapters. Her research has been awarded fellowships by the Guggenheim Foundation, the EPA, the ACLS, the SSRC, and the NEH. She is currently working on a book on the sands of San Francisco.

Jessica Dempsey is an Associate Professor in the Department of Geography at the University of British Columbia. Her current research centers on the political economy drivers of biodiversity loss, and seeks to understand escalating ecological decline in a time of unprecedented efforts to arrest them.

Kate Derickson is an Associate Professor in the Department of Geography, Environment and Society at the University of Minnesota. She is a member of the Gullah/Geechee Sustainability Think Tank and the co-Director of the CREATE Initiative, which engages historically marginalized communities in the process of scholarly research on environmental injustice.

Julie Guthman is a geographer and retiring Professor of Community Studies at the University of California, Santa Cruz, where she has investigated various US-based efforts to transform food production and distribution. Her publications include three multi-award winning monographs, an edited collection, and over 60 articles in peer-reviewed journals. Her most recent book, *The Problem with Solutions: Why Silicon Valley Can't Hack the Future of Food* (University of California Press, 2024), doesn't hold back.

Leila M. Harris is Professor with the Institute for Resources, Environment, and Sustainability and with the Institute for Gender, Race, Sexuality and Social Justice at the University of British Columbia. Her work analyzes a range of governance and political considerations from feminist, equity, and sustainability perspectives—with research focused on water politics, governance, and justice in a variety of contexts (from Turkey, to Ghana, South Africa, and Canada). Recent projects consider lived experiences and equity concerns related to the uneven implementation of the human right to water, ongoing water governance shifts in varied contexts, as well as social, affective, and political dimensions of the uneven geography of water access, quality, and infrastructures.

Matthew T. Huber is Professor in the Department of Geography and the Environment at Syracuse University. His work focuses on the relationships between energy, capitalism, and the politics of climate change. He is the author of *Lifeblood: Oil, Freedom and the*

Forces of Capital (University of Minnesota Press, 2013) and *Climate Change as Class War: Building Socialism on a Warming Climate* (Verso Books, 2022).

Kelly Kay is an Associate Professor in the Department of Geography at the University of California, Los Angeles. She is a political ecologist and economic geographer, and her current research is concerned with how the rapid rise of investor-ownership of private timberland is impacting forest-reliant communities in the United States. Kelly is currently writing a book on the topic and has published articles on the political economy of land conservation, energy transitions, and water politics, among other topics.

Lisa C. Kelley is an Assistant Professor in the Department of Geography and Environmental Sciences at the University of Colorado Denver. Her research is situated in the field of critical physical geography, and combines theories and methods from political ecology, critical agrarian studies, and remote sensing to understand ongoing transformations in land and labor relations at the intersection of agrarian and climate change.

Justine Law is an Associate Professor of Ecology and Environmental Studies in the Hutchins School of Liberal Studies at Sonoma State University. She is a human-environment geographer, and her research focuses on the science and politics of forest conservation, the ecological knowledge and management practices of rural North American communities, and the relationships between people and plants.

Alex A. Moulton is an Assistant Professor in the Department of Geography and Environmental Science at Hunter College and a member of the Faculty in the PhD Program in Earth and Environmental Science at the City University of New York Graduate Center. Previously, he was a faculty member in the Department of Sociology and Department of Geography and Sustainability at the University of Tennessee—Knoxville. His research is concerned with Black geographies and ecologies, socio-ecological justice, and the political ecology of climate change.

Tracey Osborne is Associate Professor and Presidential (endowed) Chair in the Management of Complex Systems Department at the University of California, Merced. She is also the founding director of the UC Center for Climate Justice. Tracey's research focuses on the social and political economic dimensions of climate change mitigation in tropical forests and the role of Indigenous Peoples, the politics of climate finance, global environmental governance, and climate equity and justice. She has worked on these issues globally with extensive field experience in Mexico and the Amazon. She received her PhD from the Energy and Resources Group at the University of California, Berkeley.

Zoe Pearson is an Associate Professor of Geography and International Studies in the School of Politics, Public Affairs, and International Studies at the University of Wyoming, USA. Her research examines the politics of illegal drug control and their impacts on people and the environment in Bolivia and Central America; and resource extraction and conflicts in Ecuador surrounding oil, bamboo, and conservation. More broadly, her research and teaching interests include nature–society relations and theory, the geopolitics of illegal drug control, social and environmental justice, and resource conflicts in the Americas.

Maano Ramutsindela is Professor of Geography and former Dean of the Faculty of Science at the University of Cape Town. He has held several positions including the

Mandela Fellow at the W. E. B. Du Bois Institute for African and African American Research (Harvard University), Distinguished Hubert H. Humphrey Visiting Chair at Macalester College, and Inaugural Guest Professor, African Studies Programme of the Global Studies Institute, University of Geneva. He is the author of *Transfrontier Conservation in Africa: At the Confluence of Capital, Politics and Nature*. His latest co-edited book is *The Violence of Conservation in Africa: State, Militarization and Alternatives*.

Fernanda Rojas-Marchini has a PhD in Geography from the University of British Columbia. Her research addresses the political ecology of biodiversity loss in relation to state formation processes and the emergence of market-based conservation in Chile and Latin America. Working at the Pontificia Universidad Católica de Valparaíso, Fernanda is starting a research grant to examine the biodiversity offset market unfolding in the high Andean wetlands in Chilean Altiplano. Her project establishes a dialogue with the larger research initiative on high Andean peatlands, titled Andespeat, which produces scientific research aiming to expand the protection of these fragile ecosystems.

Gabe Schwartzman is an Assistant Professor in the Department of Geography and Sustainability at the University of Tennessee. He studies the political ecology of rural transformations, examining processes of decarbonization and rural landscape change, most recently in the Appalachian coalfields.

Nari Senanayake is an Assistant Professor in the Department of Geography at the University of Kentucky. Her research bridges geographic work on health/disease, agrarian environments, and scholarship on the politics of knowledge, science and expertise. Her current project focuses on everyday encounters with a severe and mysterious form of chronic kidney disease (CKDu) in Sri Lanka's dry zone.

Gregory L. Simon is a Professor in the Department of Geography and Environmental Sciences at the University of Colorado Denver, and Co-Director of the Community Collaborative Research Center. His research draws from political ecology, STS, and critical development and hazards studies to explore societal risks under conditions of social-environmental change. He is author of *Flame and Fortune in the American West* and Co-Editor of *Cities, Nature and Development: The Politics and Production of Urban Vulnerabilities*. He has been Chair of the AAG CAPE specialty group, a United Nations Foundation Advisor, and a member of the US EPA Board of Scientific Councillors.

Jim Sinner is a social scientist with policy and economics background, based in Nelson, Aotearoa New Zealand. Jim's research has focused on values in freshwater management, regulation of cumulative environmental effects, and the practical challenges facing collective management of catchments and rivers.

Marc Tadaki is an environmental geographer at the Cawthron Institute in Nelson, Aotearoa New Zealand. Marc's research looks at people's values and connections to nature, the inner workings of bureaucracies and decision-making processes, and the value-laden construction of environmental science. His current project, Fish Futures, explores people's connections to freshwater fish and how values for fish are embedded in decision-making processes. Working with Indigenous, community, and agency partners, the project aims to rethink and remake the practice of freshwater fish management.

Rebecca Walker is an assistant professor at the University of Illinois in the Department of Urban and Regional Planning. Her work engages political ecology and critical race theory to study racialization, nature, and urban development.

Sophie Webber, PhD, is an ARC DECRA Research Fellow and Senior Lecturer in Geography in the School of Geosciences at the University of Sydney. Her research sits at the intersections of political ecology and economic geography. Drawing from these fields, Sophie studies the economic and financial logics of climate change adaptation as it is implemented in some of the world's most vulnerable cities in Southeast Asia and the Pacific region.

Introduction

Kelly Kay and Gregory L. Simon

Riding in the passenger's seat of Sandra's truck, we pass through forests of loblolly and longleaf pine trees, planted in neat rows as far as the eye can see. Timber is big business in southeast Georgia, USA, and Sandra's 5,000 acres provide multiple sources of income for her family, ranging from wood to carbon credits, pecan orchards to hunting leases, and high-value non-timber products like pine straw. Sandra also owns and manages a wetland mitigation bank on her land, which allows developments within the Satilla River watershed to be "offset" by purchasing credits from her bank, effectively paying Sandra to protect native habitat so it can be destroyed nearby. As we drive deep into the tree farm, Sandra reveals that this land is much more than just a source of income for her, it has been in her family for generations, it is the place she was born—her family *homeplace* in local parlance, and it has undergone multiple landscape transformations, once serving as a tobacco and cotton farm, then as a plantation to provide turpentine and other industrial chemicals for the US Navy during the late nineteenth and early twentieth centuries, and now as a complex patchwork of timber and non-timber forest products.

This tree farm and the land it sits on are part of Sandra's culture, it is a wildlife habitat, and it is a place she insisted on managing, even after being told that women could not manage forests or work in the woods. For a political ecologist, this tree farm—like many forested places around the globe—is a complex socio-natural landscape that can be understood in many ways, all at once. Forests, as Nancy Peluso reminds us, are deeply political (Peluso 1992; Vandergeest and Peluso 2015), raising key questions about conservation and the territories and subjectivities it produces (Chapter 11, this volume), about how we value and evaluate our ties to our environments (Chapter 7, this volume), about gendered relations of work and home in masculine spaces (Chapter 15, this volume), and about wildlife and the many more-than-human actants that share their homes with the places that humans modify to meet their needs (Chapter 18, this volume).

Critically, for this volume, forests provide a perfect example of the many different lenses—or frameworks—through which political ecologists can and have viewed the politicized environment and society's place within it. Southern pine plantations like Sandra's, for example, serve as a key input to paper production, and nearby mills are some of the most prolific producers of cardboard boxes that help to ensure the movement of commodities across the globe, highlighting the ways that resource consumption links people across space (Chapter 13, this volume) and the ways that globalization has wedded the fortunes of small resource-dependent locales to an international political economy and circulation of goods around the globe (Chapters 9 and 10, this volume). These trees are also part of heavily modified monocrop plantations, their genetically improved

DOI: 10.4324/9781003165477-1

seed stock are but one of the myriad examples of how we have drawn on science and technology to produce new ideas about nature (Chapter 5, this volume) and modify our environments into new material natures (Chapter 2, this volume), thereby making these natures more mappable, predictable, and knowable (Chapter 8, this volume). Finally, these natures have distinct histories, they are entwined in the stories local people tell themselves to make sense of their place in the world (Chapter 6, this volume), and are also entwined in legacies of the Transatlantic slave trade. These forest plantations have direct links to older histories of plantation life—Georgia's coastal plains were home to extensive rice and cotton plantations, leading to the forced movement of countless enslaved peoples, with the region's sea islands and Gullah/Geechee heritage reflecting those racialized landscapes, as well as histories of marronage, counter-territoriality, and resistance (Chapters 4 and 16, this volume; Bledsoe 2017).

In what follows, we briefly explain the justification for the particular approach and framing we have taken in this book, opting to let each author narrate the past, present, and future of the field from the places that inspire and inform their work. This unique approach, based around different "lenses" or frameworks rather than a singular unifying starting point, was intentional as a means of highlighting the diversity of empirical, methodological, theoretical, and ontological approaches employed by our contributors. After explaining the significance of this lens-based approach and the centrality of "doing" for this particular volume, we close with an overview of the major parts of the book and the chapters contained within them.

Political Ecology's Multiple Roots

Scholars have always had a difficult time pinpointing political ecology's origins. Some have traced the first usage of the term "political ecology" to anthropologist Eric Wolf in 1972 (Robbins 2012), and others emphasize geographers Harold Brookfield and Piers Blaikie's definition of the field as combining "the concerns of ecology and a broadly defined political economy" (Blaikie and Brookfield 1987, 17, cited in Neumann 2014). Over time, definitions of the field have grown and expanded. Michael Watts (2000), for example, defines political ecology as a field that aims to "understand the complex relations between nature and society through a careful analysis of what one might call the forms of access and control over resources and their implications for environmental health and sustainable livelihoods" (257), highlighting the key role of power over resource systems. In a similar vein, others have understood *political* ecology as a contextual approach—a counterpoint to an *apolitical* ecology—that identifies the key role of "broader systems rather than blaming proximate and local forces" for ecological degradation, and which sees environmental systems as "power-laden rather than politically inert" (Robbins 2012, 13). This interest in contextualization has meant that scholars in this field have long been concerned with linking local and regional environmental change to broader social forces, including capital, the state, and other institutions, by constructing "chains" (Blaikie 1989; Blaikie and Brookfield 1987), "webs" (Rocheleau 2008), or "networks" (Birkenholtz 2012) of explanation that connect a range of actants and social forces across space and scale.

Volumes of this type often begin by rehearsing a long and authoritative genealogy of the field (e.g., Neumann 2014; Perreault et al. 2015; Robbins 2012)—usually beginning in the late 1970s with the collision of several fields, including hazards research, cultural ecology, and critical development studies—in order to demonstrate that political ecology is a vibrant and established field with a clear starting point and traceable contours. Such an approach recognizes multiple origins, or root systems, but often narrates them as eventually coming together to form a field with a common and identifiable core. This book, however, offers a distinctive entry point into this diverse and ever-growing field and argues that our scholarly "foundations," today more than ever, comprise a diverse and cross-cutting latticework of research approaches and concepts. Since its inception, political ecology has always been a relatively wide-ranging field rife with methodological, ontological, topical, and theoretical pluralities amongst its members. This has only increased over time. While this diversity has at times raised concerns amongst its adherents, with for example, concern about how particular approaches may be neglecting the field's stated commitments to "ecology," "policy," or "politics" (Walker 2005, 2006, 2007), we tend to agree with Martin et al.'s description of political ecology as a "big tent" (2019, 228), signaling the field's welcoming nature and diversity of approaches as perhaps its greatest strengths. The decision to refuse to narrate a singular origin story for our sub-field is also aligned with recent critiques of the canon in our field and of the telling and retelling of histories of the field of geography that are overwhelmingly male, white, and Western (Monk 2012; Kinkaid and Fritzsche 2022; Craggs and Neate 2020; Keighren et al. 2012). Thus, rather than seeking a singular and holistic synthesis of the field with this edited volume (e.g., Zimmerer and Bassett 2003), we take two major points of inspiration as our starting point.

First, with a diverse body of scholars "doing" political ecology, this volume asks us to reflect on the following proposition: Does the field define us? Or do we define the field? As a field that spans human geography, cultural anthropology, environmental studies, rural sociology, development studies, environmental history, gender studies (and more), political ecology covers a lot of empirical and theoretical ground. Our first aim with this volume is to embrace both the breadth and depth of the field, recognizing that many scholars attribute their roots to genealogies that are different from those highlighted elsewhere, by giving ample voice to these other rich and complementary genealogies. Indeed, one aspect of political ecology that we find particularly appealing is the field's obstinate inclusiveness, which is linked to the field's core commitment to polyvocality and pluriversal ontologies. Thus, this "big tent" status allows a wide range of theoretical and empirical approaches to intermingle in novel ways. We see this internal diversity and relationality as one of the field's greatest strengths, and it was one of the major motivating factors for organizing this volume around the vast range of approaches in political ecology. A primary goal of this volume is therefore to highlight the multiple and overlapping approaches to conducting research in the field of political ecology.

Given this diversity, how should those who are new to the field approach political ecology and apply key concepts and methods to their own research topics? A second point of inspiration is to reflect this diversity by organizing the book around distinct lineages of thought and ways of "doing" political ecology. To do so, this book brings together 28 leading scholars at a range of career stages, from a range of positionalities and embodied experiences, and from a diversity of geographical contexts. Contributors are located at universities and research institutes across North America, Europe, Oceania, Africa, and Latin America. These authors' contributions are organized into 18 analytical lenses and highlight different approaches to conducting critical environmental research. We view these lenses as particular paradigms or "ways of seeing" nature-society interactions. While there are meaningful overlaps between them, they each have their own shared, yet unique, theoretical-conceptual histories and methodological approaches.

By engaging with the diverse approaches highlighted in each chapter, readers will emerge with a broad understanding of political ecology, and will attain a diverse toolkit for evaluating human-environment interactions in their own work. Accordingly, each chapter examines key methodological, theoretical, topical, and conceptual approaches that animate the authors' particular understanding of the field and approach to their work.

Overview of the Book

The book is divided up into four major parts: "Politicizing Environmental Management" (Part I); "Making Nature Knowable" (Part II); "Capital, Colonialism, and Political Economy" (Part III); and "Political Ecologies of Identities, Difference, and Justice" (Part IV). Each part is animated by a major current of research and thought in political ecology. Below, we outline the major themes of each part and the specific analytical lenses that are grouped under each.

Part I: Politicizing Environmental Management

The chapters in Part I of this volume share an emphasis on the longstanding interest of political ecologists with environmental politics, policy, and governance. Each of the reviews begin with the idea that environmental governance—by state and non-state actors like NGOs, community members, scientists, land managers, Indigenous peoples, and other entities—is complex, contested, and power-laden. While the management of sea level rise, chestnut tree DNA, First Nations water systems, or endangered sea turtles may appear straightforward, scientific, and uncontroversial at the outset, these animating cases demonstrate the importance of accounting for embedded socio-cultural conditions, power dynamics, political economic institutions, and other taken-for-granted assumptions. As Derickson et al. (Chapter 4) put it, a politicized "nature both becomes a site of contestation and is a site through which power is contested." These chapters introduce readers to concepts that are often the subject of political ecology analysis, including formal and informal environmental governance; risks, hazards, and vulnerability; scale; co-production; biopower; resilience and resourcefulness; as well as the importance of community and social movements for enacting political change. Several chapters also examine how environmental change stemming from the emergence of the Anthropocene could present bright spots or opportunities for more innovative and just governance outcomes, such as recuperating the political potentialities inherent in concepts like sustainability, or acknowledging the long-term co-production of people and their environments as a means of taking a more liberatory approach to socio-ecological research and management.

Part I consists of four chapters. Chapter 1, by Sophie Webber, begins by reviewing the risks, hazards, and vulnerability traditions within geography and environmental research. For Webber, managing our environment sustainably means fundamentally understanding how disasters—both ongoing and acute—interface with human societies in ways that are deeply conditioned by histories of investment and disinvestment, as well as politically motivated approaches to risk management. Through an extended consideration of the Central Pacific nation of Kiribati, Webber also demonstrates the need to critically examine narratives of risk and sustainability. Chapter insights will be particularly helpful for researchers interested in climate change, adaptation, and critical approaches to the study of risks and hazards. Next, in Chapter 2, Christine Biermann, Justine Law, and Zoe

Pearson consider the concept of co-production, tracing the long lineage of ways that human societies and environments make one another through techniques of management and the production of ecological knowledge. This framework highlights how resulting human and non-human entanglements are often complex, messy, politicized, and always changing. The management of the environment is also deeply biopolitical, as humans make critical interventions that allow some species, environments, or genes to thrive, and others to die. This chapter will be useful for researchers interested in novel ecologies, critical physical geography and the critical environmental sciences, and science and technology studies.

Alice Cohen and Leila Harris lay out many foundational concepts in environmental politics and governance in Chapter 3, focusing in particular on the topics of power, scale, and inequality. Through their extended engagement with water politics, the authors demonstrate the critical roles that formal and informal institutions play in environmental decision making, as well as the critical need to foreground ontological questions and knowledge claims in environmental management. The chapter has a strong engagement with water, and will be of interest to scholars of water, as well as those interested in environmental governance and government, more generally. Finally, in Chapter 4, Kate Derickson, Gabe Schwartzman, and Rebecca Walker focus on community-based responses to changing political and environmental conditions. The authors begin with the idea of nature as a field of power to illustrate how environments articulate with various forms of difference to become a key site of struggle, oppression, and resistance. Using a range of examples, including neighborhood movements for resourcefulness in Glasgow UK and urban social movements opposing green gentrification in St Louis USA, the authors highlight a range of ways that communities and activists engage with always-already politicized natures, and how movements might be scaled up or reworked toward more equitable human-environment relations into the future. Chapter insights will be useful for those interested in community and social movements and resistance.

Part II: Making Nature Knowable

The four chapters in Part II explore how ideas about nature emerge and inform contemporary approaches to knowing, valuing, studying, and evaluating the environment. These chapters draw on recurring constructivist impulses in political ecology that have explored the politics of knowledge production and its implications for environmental management and policy. They also highlight how political ecology has productively contributed to debates in fields such as geography, anthropology, science and technology studies, and the critical environmental sciences to explain how, why, and for whom we make sense of nature. Collectively, the chapters underscore key concepts to help readers engage critically with the embedded assumptions and logics that shape environmental narratives and values; the influence of enduring colonial, State, and other powerful forces shaping nature imaginaries and environmental governance; and the politics of social-environmental data and scientific ways of knowing and counting nature. They also underscore not only the prevailing influence of epistemic critique in political ecology but also the importance of committing to pluriversal ontologies as we search for inclusive and just approaches to valuing and managing nature for all. Finally, chapters highlight the limits to truth claims, particularly those emanating from sources of power, and also efforts to challenge such authoritative knowledge by political ecologists given our own subjective and embedded scholarly positions.

In Chapter 5, Noel Castree describes the long-standing and still prominent influence of "social constructionism" in shaping how political ecologists view and study nature-society interactions. The chapter highlights how a constructivist lens usefully underscores how societies and individuals "come to perceive reality in various ways that become 'sedimented' and institutionalized over time." To further unpack how nature is socially constructed, Castree offers three value- and power-laden dimensions of this process: "construal," "constitution," and "composition." The chapter reminds us that answering questions about why society uses and misuses our environments must begin by understanding ourselves and the nature of our ideas. Observations from this chapter will be helpful to those seeking foundational insights connecting environmental knowledge to social practice and the world of our (un)making. In Chapter 6, Diana Davis describes the importance of storytelling as method and subject of analysis in critical environmental research. In particular, Davis focuses on environmental narratives and describes how these stories and imaginaries often emerge and become widely adopted and engrained over time. Using arborocentrism (the privileging of trees as a way of improving both landscapes and societies) as an example, Davis describes how robust historical contextualization allows researchers to highlight the ideational origins of influential explanations about nature, society, and their interactions—ideas and beliefs that continue to shape policies and management orthodoxies today. Doing so allows us to counter and retell the apolitical (and often ahistorical) storylines found in popularized environmental accounts. Chapter insights will be particularly helpful for researchers interested in the virtues and tactics of doing historical data collection and analysis in political ecology.

Chapter 7 by Marc Tadaki and Jim Sinner begins by acknowledging that environmental values underscore almost every aspect of environmental decision making. Using the case of trout in Aotearoa New Zealand, the authors highlight the variegated and power-infused perspectives and values that shape how nature matters to people. The authors describe how political ecologists have rendered environmental values and processes of environmental valuation political through critiques associated with commensurability metrics, such as monetary value, and the power-laden ways values are prioritized and actualized (and also marginalized and suppressed) in environmental governance. Tadaki and Sinner underscore how the valuation process is often presented as merely "technical" or "administrative," despite actually being subjective, non-neutral, and severely constrained in both its development and application. This chapter will be useful to researchers interested in the politics and practices of values-based environmental decision making. Chapter 8, by Lisa Kelley and Gregory Simon, examines the data collection process as a site and source of politics. The authors note the post-positivist orientation of research in political ecology and highlight the politics of scientific knowledge production. The authors review several recurring themes in political ecology used to deconstruct the enumeration process such as the "intuitive conceptual obviousness" of existing science practices, the role of knowledge co-production in shaping how we study the environment, and the political work in/of knowledge blind spots. Kelley and Simon go on to suggest that spaces of scientific knowledge production are not just shaped by political forces but are also sites of political action that may produce alternative, justice-based, and transdisciplinary approaches, allowing researchers to re-imagine and counter hegemonic scientific practices. The Enumerating Nature chapter will aid research that explores and seeks to counter the "colonial and imperial legacy of scientific knowledge production about the environment, the positivistic orientations of related datasets, and the instrumentalization of scientific practices to advance state and corporate interests."

Part III: Capital, Colonialism, and Political Economy

The chapters in Part III are anchored by a broad perspective on political economy—a perspective that focuses on capitalism and markets, the historic and ongoing role of colonialism and extractivism in turning nature into both property and global commodities, and the role of states and other global political economic institutions in (re)making environments that are more legible (Scott 1999) and commodifiable. These are topics that have animated decades of research in political ecology. For many scholars, grounding research in political economy has long provided a way of speaking across diverse case studies. As Matt Huber puts it in Chapter 12: "whether we are looking at mining (Arboleda 2020), farmland investors (Fairbairn 2020), or fishery conglomerates (Campling and Colás 2021), the root cause of much ecological degradation is often capital seeking a return (M-C-M') via a production process that turns nature into surplus value." A shared commitment to placing nature-society relations within the *longue durée* of the capitalist world economy underscores how the appropriation and enclosure of land and resources is neither new nor incidental to capitalist political economy, but is instead foundational to it (Moore 2015). In addition to a broader historical and political perspective on capital's engagements with nature, the chapters also recognize the importance of thinking about the political economy of the environment through approaches that center both consumption and production. The chapters in this part provide extended consideration of a number of key concepts, including green neoliberalism and neoliberal natures, (neo-)extractivism, rent and financialization, environmental diplomacy, green violence, philanthrocapitalism and environmental altruism, Marx's hidden abode of production, valorization and metabolism, boycotts and buycotts, commodity fetishism, and visceral politics.

Part III of the book consists of five chapters, with each taking a distinct approach to the topics of Capital, Colonialism, and Political Economy. In Chapter 9, Fernanda Rojas-Marchini and Jessica Dempsey draw on the case study of state reregulation of forestry in Chile to highlight the critical role that states play—and have always played—in the making of resources and in reducing friction for the free movement of capital. The authors expand our analysis of globalization and nature by harkening back to 1492 and the Columbian Exchange to situate modern-day approaches to globalization within longer histories of colonialism, imperialism, and ecological exchanges. The authors compellingly place the process of re-regulation, that is a hallmark of neoliberalism, into a longer genealogy that begins and ends—in many respects—with extractivism, which they understand as "not only an economic structure but also an ecological-political regime on the fabric of life, territories, flows of energy and matter, bodies, institutions, and cultural webs" (Terán Mantovani 2016, 257, as cited in Chapter 9). Insights from this chapter will be particularly valuable for those researchers working on topics related to current and historical flows of capital and resources, extraction, and global governance.

Kelly Kay, in Chapter 10, argues that understanding twenty-first-century approaches to generating profit from nature requires both the centering of economic theories of value production and a close engagement with law and state power. The chapter reviews a range of key debates within the political economy of nature, including discussions of neoliberalization and the "green economy," property and enclosures, rent and rentiership, and the financialization of nature. This chapter will be useful for researchers interested in the political economy of the environment and in understanding the evolving ways that value is produced in and through "capitalist natures." Next, in Chapter 11,

Maano Ramutsindela focuses on the political ecology of environmental conservation, and uses transboundary protected areas like peace parks as a means of highlighting the importance of situating political economies of conservation in colonial histories and political geographies. For Ramutsindela, protected areas—so critical to both colonial era plunder and the imperialist remaking of the world—constitute "the colonial metanarrative of human hegemony over nature" (Coates 1998, 18, as cited in Chapter 11). The chapter carries the geopolitical role of protected areas forward into the present, through consideration of environmental diplomacy, conservation and development initiatives, philanthrocapitalism and environmental altruism—all ways of making nature pay political and economic dividends. Researchers interested in political geographies of conservation, colonialism, philanthropy, and development will gain insights from this chapter.

Matt Huber, in Chapter 12, introduces the reader to the field's long-standing commitment to Marxian political economy in environmental research. Huber explores foundational debates on the role of the land manager and on ecosocialist thought to make a case for more sustained engagement with what he terms, borrowing from Nancy Fraser, the "hidden abode" of production. After a review of key literature on Marxian political economy and the valorization process, Huber's chapter traces what a political ecology of industrial production might look like, with equal emphasis on theoretical and methodological considerations. In particular, the chapter is notable for its extended consideration of research approaches, and for its recognition that core considerations of a more "orthodox" Marxism—like capital/labor relations—are just as relevant for the study of ecological Marxism. Insights from this chapter will be of interest for researchers concerned with Marxian political economy, and to those who are looking to conduct research on topics related to industrial production. Finally, in Chapter 13, Julie Guthman and Michaelanne Butler draw on a sustained engagement with food as a means of capturing the breadth of ways that scholars can engage with questions of consumption. Their review ranges from more economic approaches centered around commodity supply chains and purchasing decisions to more embodied approaches centered around the politics of ingestion. For the authors, food provides a means of drawing upon consumption—one of the most commonplace ways that humans engage with resources and the more-than-human world—as a critical means of bridging and broadening the political ecology literature toward more culturally rooted approaches to politics. This chapter will be of interest to food and agriculture scholars, as well as those who are interested in consumption—broadly defined.

Part IV: Political Ecologies of Identities, Difference, and Justice

The five chapters in Part IV shed light on the large and diverse body of literature informing questions of justice and difference in political ecological research. With increasing precision and acuity, scholars have examined how different social identities (e.g., race, ethnicity, gender, age, sex) influence how groups experience social-environmental change and participate in both mundane and formal management solutions. These chapters demonstrate active engagement with scholarship that is (increasingly) central to political ecology research, including fields such as African Studies, Black Studies, Gender Studies, Indigenous Studies, Latinx Studies, and Subaltern Studies. In their own way, each chapter points to the importance of genealogical approaches to research for highlighting the origins of racialized and gendered knowledges of, and experiences with, nature. The chapters also reveal the valuable insights of intersectional research, showing how changing human-environment interactions illuminate forms of social difference, and how knotted identities combine and interact to shape social-ecological injustices. As several of the chapters in Part IV highlight, political ecology research has also seen a shift toward the study of the intimate and visceral in order to explore embodiments of nature and the way bodies are a critical source, target, and conduit of power. Extending our examination of representation and agency to the ecological realm, the final chapter by Rosemary Collard reminds us of nature's disruptive influence, thus underscoring the need to understand questions of representation and justice across the nature-society divide. As Joel Correia and Tracey Osborne point out in Chapter 14, these political ecological commitments have challenged scholars to incorporate, and in some cases revive, critical methodological practices that decenter the researcher in favor of ethnographic and community-based participatory research methodologies which foreground the insights, priorities, and embodied experiences of our research partners.

Correia and Osborne address the transformative potential of political ecology as a tool for developing tangible environmental governance pathways and just solutions. In short, the authors discuss political ecology as a form of political action as society is confronted with a series of environmental crises. Focusing on climate justice, the authors outline a process of engaged political ecology scholarship, or "public political ecology" that is structured around relational, decolonial, and participatory research methodologies. For Correia and Osborne, public political ecology highlights the need for scholars of the environment to convert our critical toolkit into public facing, grounded, and inclusive action in pursuit of environmental justice (i.e., by "doing" political ecology in the most literal sense). This chapter will be of particular interest to those who are curious about key strategies and challenges associated with enacting political ecology research through public-facing and justice-based engagements.

Chapter 15, by Deepti Chatti, details a wide range of feminist scholarship used to assess dominant discourses on environmental change and the social constructs of gender and nature more broadly, including "the interlinkages between (and consequences of) the feminization of nature and the naturalization of gender." Chatti highlights the internal heterogeneity of this feminist environmental scholarship and draws on diverse epistemic traditions found in political ecology, environmental anthropology, feminist science studies, environmental humanities, and development studies. Chatti describes how these diverse lineages and differences of opinion, rather than splintering work in political ecology, have generated more nuanced, innovative, and complementary ways of thinking about gender-environment relations. This chapter will be particularly useful to those interested in exploring the gendered nature of environment-society interactions, environmental impacts and hazards, and knowledge about the environment.

In Chapter 16, Alex Moulton highlights approaches to prioritizing race and forms of racialization in our conceptions of nature-society relationships. This innovative review highlights research prioritizing questions of race and indigeneity related to "historical geographies of racialized nature, racial dispossession of natural resources, and racialized regimes of governing nature." Moulton describes how these threads of scholarship inform contemporary political ecological research endeavors. Moulton goes on to describe the contours of a varied and increasingly refined body of political ecology research addressing race-nature relationships. This chapter will be helpful to researchers exploring, among other topics, how the intersection of race and nature shape forms of social discipline, land and financial dispossession, control over natural resources and amenities, and visions for environmental justice and equitable resource management. In Chapter 17, Nari Senanayake turns our attention to the body in political ecology research. The chapter examines how the body as socio-nature comes into contact with the environment and is made, modified, and remade under capitalism and the Anthropocene. Building on these insights, Senanayake highlights key concepts in political ecology used to connect patterns of environmental change to bodily health and disease while also detailing how "emotional, affective, and biochemical dimensions of bodies...materialize in everyday life." Insights from this chapter will be of particular interest to those engaging with the "relational, intersectional, and uneven entanglements of bodies and nature."

While the chapters thus far primarily focus on how society acts upon nature, Chapter 18 by Rosemary Collard is about "nature acting back." Collard examines various forms of nature's unruliness including nature's spatial movements and infringements, nature's incompatibility with capitalist "improvement," and the intractability of nature's dynamism with common scientific and resource management practices. Using a number of examples, the chapter highlights a recurring theme in political ecology: that "efforts to master nature have largely encountered nature's agency as a problem." Collard provides an important counterpoint by also noting the forms of hope, possibility, and survival that such unruliness also provides for humans and more-than-human life. Given its attention to physical environmental geographies, this chapter will be valuable to those wishing to center nature's agency within complex social-environmental assemblages.

References

Arboleda, M. (2020). *Planetary Mine: Territories of Extraction Under Late Capitalism.* London: Verso.

Birkenholtz, T. (2012). Network political ecology: Method and theory in climate change vulnerability and adaptation research. *Progress in Human Geography*, 36(3), 295–315.

Blaikie, P. (1989). Explanation and policy in land degradation and rehabilitation for developing countries. *Land Degradation and Development*, 1(1), 23–37.

Blaikie, P., and Brookfield, H. (1987). *Land Degradation and Society.* London: Methuen.

Bledsoe, A. (2017). Marronage as a past and present geography in the Americas. *Southeastern Geographer*, 57(1), 30–50.

Campling, L., and Colás, A. (2021). *Capitalism and the Sea: The Maritime Factor in the Making of the Modern World.* London: Verso.

Coates, P. (1998). *Nature.* Oxford: Blackwell.

Craggs, R., and Neate, H. (2020). What happens if we start from Nigeria? Diversifying histories of geography. *Annals of the American Association of Geographers*, 110(3), 899–916.

Fairbairn, M. (2020). *Fields of Gold: Financing the Global Land Rush.* Ithaca, NY: Cornell University Press.

Keighren, I. M., Abrahamsson, C., and Della Dora, V. (2012). On canonical geographies. *Dialogues in Human Geography*, 2(3), 296–312.

Kinkaid, E., and Fritzsche, L. (2022). The stories we tell: Challenging exclusionary histories of geography in US graduate curriculum. *Annals of the American Association of Geographers*, 112(8), 2469–2485.

Martin, J. V., Epstein, K., Bergmann, N., Kroepsch, A. C., Gosnell, H., and Robbins, P. (2019). Revisiting and revitalizing political ecology in the American West. *Geoforum*, 107, 227–230.

Monk, J., 2012. Canons, classics, and inclusion in the histories of geography. *Dialogues in Human Geography*, 2(3), 328–331.

Moore, J. W. (2015). *Capitalism in the Web of Life: Ecology and the Accumulation of Capital.* London: Verso Books.

Neumann, R. (2014). *Making Political Ecology.* Abingdon: Routledge.

Peluso, N. L. (1992). *Rich Forests, Poor People: Resource Control and Resistance in Java.* Berkeley, CA: Univ. of California Press.

Perreault, T. A., Bridge, G., and McCarthy, J. P. (Eds.). (2015). *The Routledge Handbook of Political Ecology.* Abingdon: Routledge.

Robbins, P. (2012). *Political Ecology: A Critical Introduction.* Chichester: John Wiley & Sons.
Rocheleau, D. E. (2008). Political ecology in the key of policy: From chains of explanation to webs of relation. *Geoforum*, 39(2), 716–727.
Scott, J. C. (1999). *Seeing Like a State: How Certain Schemes to Improve the Human Condition Have Failed.* New Haven, CT: Yale University Press.
Terán Mantovani, E. (2016). Las nuevas fronteras de las commodities en Venezuela: Extractivismo, crisis histórica y disputas territoriales. *Ciencia Política*, 11(21). doi: 10.15446/cp.v11n21.60296
Vandergeest, P., and Peluso, N. L. (2015). Political forests. In R. L. Bryant (Ed.), *The International Handbook of Political Ecology*, 162–175. Cheltenham: Edward Elgar Publishing.
Walker, P. A. (2005). Political ecology: Where is the ecology? *Progress in Human Geography*, 29(1), 73–82.
Walker, P. A. (2006). Political ecology: Where is the policy? *Progress in Human Geography*, 30(3), 382–395.
Walker, P. A. (2007). Political ecology: Where is the politics? *Progress in Human Geography*, 31(3), 363–369.
Watts, M. J. (2000). Political ecology. In E. Sheppard and T. Barnes (Eds.), *A Companion to Economic Geography*, 257–274. Oxford: Blackwell.
Zimmerer, K. S., and Bassett, T. J. (2003). Approaching political ecology. In K. S. Zimmerer and T. J. Bassett (Eds.), *Political Ecology: An Integrative Approach to Geography and Environment-Development Studies*, 1–25. New York: Guilford Press.

Part I

Politicizing Environmental Management

1 Sustaining Nature

Climate Change and the Catastrophe to Come in Kiribati

Sophie Webber

In newspaper articles, blog posts, and documentaries, just as in scientific articles and policy reports, the Central Pacific nation of Kiribati is often described as one of the most vulnerable places on earth. In these reports, Kiribati—like other small island developing states—is depicted as remote, low-lying, and lacking in resources and capacity. Physical and scientific projections also contribute to this image of vulnerability, predicting future sea level rise, more frequent and intense disaster events, ocean warming, and associated destruction to lives, livelihoods, ecosystems, and infrastructures. Images circulate alongside descriptions, capturing the intimacy of water lapping at permeable housing, children playing in floodwaters, or, zooming out to scattered islands amidst a vast ocean. Together, they depict an existential climate catastrophe to come, where the seas and storms swallow up the islands of Kiribati and the i-Kiribati peoples.

Of course, Kiribati is currently experiencing and will continue to experience the impacts of climate change. In line with global trends, temperatures and sea levels have steadily increased in South Tarawa, the capital of Kiribati, influenced by regional climate phenomenon including the El-Niño Southern Oscillation (ENSO; Australian Bureau of Meteorology and CSIRO 2014). Attempts to project warming and its impacts in the archipelagos are confounded by the limited spatial resolution of climate models. Nonetheless, warming is projected to continue at slightly below global average, reaching approximately three degrees by the 2090s (World Bank 2021). A majority of climate models project increases in monthly and annual average rainfall. Extreme weather events are anticipated to increase in frequency and severity, with impacts on ecosystems and livelihoods, particularly reefs and fisheries.

Sea level rise is among the most uncertain of anticipated climate impacts, with local experiences mediated by global trends, regional climate phenomenon, and proximate dynamics (Donner and Webber 2014). Sea level rise is often associated with slow-onset changes, but it will be felt through extreme sea level events such as storms, high tides, and waves, which may produce increased water levels at the shoreline, overwash, and flooding. The spectre of this slow-slow-fast sea level change is often portrayed as an existential threat. Kiribati consists of 32 coral atolls and one limestone island, with a maximum elevation of around three meters and more than two-thirds of land below two meters (Woodroffe 2008). As a result, some scholars have suggested that Kiribati will—like most atoll countries—become uninhabitable by the middle of the century due to destructive flooding, economic and infrastructural impacts, and salinisation of the freshwater lens (Storlazzi et al. 2018; but see Webb and Kench 2010).

According to this "extinction narrative of the sinking island state" (Weatherill 2022, 1), the future of Kiribati is constrained due to its physical vulnerabilities, an unfortunate

DOI: 10.4324/9781003165477-3

geographic coincidence of low elevation and high exposure to climate extremes. Other calculations of vulnerability introduce social and economic dimensions as determinants of adaptive capacity in the face of such threats (Dean, Green, and Nunn 2016); accordingly, Kiribati is limited by high levels of poverty, water and food insecurity, and a small, remote and precarious economy (e.g., Storey and Hunter 2010). But, few researchers highlight the long historical and ongoing processes through which these adaptive capacities have been underdeveloped, including through colonial occupations, dispossessive and extractive regimes, forced labour migration and relocation, and nuclearism (Weatherill 2022; Germano 2022; Bordner, Ferguson, and Ortolano 2020).

What are the consequences of these catastrophic projections of contemporary vulnerability and the emergent climate crisis? In a famous essay, Pacific scholar Hau'ofa (1993) argues for an alternative reckoning than the dominant understanding of the region as comprised of "pitiful microstates" that are "too small, too poorly endowed with resources and too isolated from the centres of economic growth." These externally established and imposed perceptions are founded on colonial relations of "dominance and subordination" which reproduce hopelessness, just as in framings of Kiribati as a climate hotspot in race-to-the-bottom adjudications of extreme vulnerability (Webber 2013). An overwhelming and universalising understanding of the climate-catastrophe-to come undermines efforts to nurture and facilitate sustainable ecologies and livelihoods in the islands (Barnett 2017). The existing catalogue of globally produced adaptation and resilience projects in Kiribati is not optimistic, showing failed interventions and maladaptive strategies (Piggott-McKellar, McNamara, and Nunn 2020). In contrast, Hau'ofa, and other Pacific scholars, articulate the "large world" of a "sea of islands" which foregrounds self-determination and the resilience of "surviving for thousands of years in a challenging oceanic environment" (Teaiwa 2018, 69).

This story of climate change in Kiribati hinges on our changing ideas about hazards, vulnerability, adaptation, resilience, and risk, and the ways these ideas are translated onto multiple scales and sites of action. Concepts like hazards, vulnerability, and so on, each model a different way of diagnosing more, or less, sustainable relations between societies and natures. And they each also invite particular kinds of remediation—from big infrastructural solutions (Kumar and Taylor 2015), to Indigenous, self-determined social protection measures (Ratuva 2014)—in the face of anticipated and experienced environmental threats. Each of these ideas has an intellectual and institutional history and set of political commitments. As key concepts in political ecology, they have contested meanings and utility. While characterising the sustainability of human-environment relations has clearly evolved, defining the relative contribution of social versus natural processes and individual versus structural determinants, are enduring questions for political ecology. In the sections that follow, key diagnostic and programmatic concepts are reviewed in roughly chronological order such that their historical context, critiques, and inter-relations are evident.

For Whom Is a Hazard a Hazard?

Hazards and disasters have been animating concepts of political ecological inquiry for much of the last century. Natural hazards researchers have studied volcanic eruptions, earthquakes, cyclones, and hurricanes, and more recently the perilous impacts of climate change. These events are all hazards, which is to say both experienced disaster events as well as potential future catastrophic phenomenon.

The "hazard tradition" of research investigates the causes of disasters, their uneven outcomes, and appropriate responses towards sustainability. Emerging with the influential work of Gilbert White in the 1940s, this "human ecology" approach centres social processes as determinants of hazard events. Where the physical sciences model extreme events with a goal to inform physical and engineered mitigation measures, White (e.g., 1974) and his human ecology collaborators (Burton, Kates, and White 1993) instead consider public perceptions of and behaviours in response to hazard risks. Their surveys of perceptions of risk in the face of hazards reveal differentiated "envelopes" of opportunity and a variety of adjustment opportunities, while highlighting how engineered solutions such as levees encourage blithe overdevelopment despite ongoing hazard risks.

While influential for its socio-natural interactionist approach, the individualistic and behaviouralist tendencies of this hazards research have been excoriated by political economists. In his famous essay on "The Poverty of Theory," Watts (1983) argues that the hazards paradigm "erod[es] the irreducible social character of human life to atomised individuals or organisms" and lacks any consideration of the "complex social production of material life" (234, 235). Watts flings a long list of political ecological insults at the approach: it is functionalist, utilitarian, deterministic, naturalising, and empiricist. Moreover, as a research paradigm built up in the context of the United States floodplains, the expansion of their survey approach to "cross cultural investigation" is "parochial," "rigid," "extraordinarily naïve," and "crude scientism" with "little credibility" for understanding the occurrence and effects of hazard events (239, 240). On the basis of these critiques, and drawing from more explicitly Marxian and post-colonial theoretical traditions, Watts and his interlocuters instead focus on identifying, explaining, and responding to the root causes of vulnerability to disaster as structurally produced (Emel and Peet 1989).

Rather than surveying, or blaming, individual behaviours, this political economic approach—often termed the vulnerability paradigm—sees vulnerability as secreted over time, produced through social relations and colonial and postcolonial processes. Principally focused in the Global South, the vulnerability paradigm seeks to build explanations capable of accounting for the uneven distributions of extreme hunger and famine, particularly in sub-Saharan Africa in the 1980s. The overarching research question is: why are some people more vulnerable than others in the face of disasters? While being structural in its explanatory orientation, class, or poverty, are not the only markers of vulnerability; as Watts and Bohle (1993, 44) describe, there are a "multiplicity of [co-determinant] factors" that produce vulnerability, including gender and sexuality, race and ethnicity, and age. These vectors of social and political economic marginality are not only found to make certain people more exposed to disaster they also reproduce physical vulnerabilities through ecological and environmental feedback loops.

These are foundational debates for political ecology as a practice and critique. Nonetheless, hazards research has remained vibrant, introducing new modes of research (see Box 1.1) and animating ideas. For instance, Mustafa has developed the concept of a "hazardscape" that explains hazards and their effects through both "material geographies of vulnerability" and "how those hazardous geographies are viewed, constructed, and reproduced" (2005, 566; see also Yamane 2009; Saguin 2017). With reference to flooding in the twin cities of Islamabad and Rawalpindi, Mustafa conceptualises hazardscapes as landscapes onto which differentiated vulnerability is produced and reproduced—materially, politically, and discursively. Causal factors producing hazardscapes include migration onto marginal, low-lying land, crony capitalism in real-estate markets, policymakers' tendencies towards individuating rather than integrating risks, and externally developed, high-modern flood-mitigation measures. The idea of hazardscapes, then, builds on natural hazards research in identifying the "envelope" of choices for constituents in pursuit of sustainable socio-natural watersheds, while recognising that these remain "circumscribed by social structures and discourses" (Mustafa 2005, 583).

Box 1.1 A methodological question: the future of vulnerability and adaptation?

The methodologies associated with approaches to sustaining socio-ecological relations in the face of disturbances are diverse, both in formal method and epistemology. The early hazards tradition often combined the physical geographies of the hazards themselves with surveys of affected communities. For instance, in her vulnerability science approach, Cutter (2003) integrates social, natural, and engineering risks to identify vulnerable people and places. In assessing vulnerability in South Carolina USA, Cutter et al. (2000) model historic hazard occurrence, map spatial impacts, and characterise populations by "age, race/ethnicity, income levels, gender, building quality, [and] public infrastructure" (726). These data combine to produce composite maps of hazard vulnerability. Other researchers collaborate with research participants to co-produce maps and models of risk and recovery (Brun 2009). In coastal Masantol, Philippines, Cadag and Gaillard (2012) use participatory 3-Dimensional Mapping to produce hazard risk assessments and planning tools, alongside school communities, local organisations and governments, NGOs, and academics. Participants in workshops plotted places, people, and assets they perceived to be vulnerable, and identified strengths and capacities. The resulting participatory maps led communities to identify adaptive actions.

While both these studies are integrative, multi-scalar, spatially differentiated, and normatively focused on reducing the impacts of disasters, the methodological approaches are diametric. Where Cutter et al. (2000) have produced hotspots of vulnerability from afar, drawing from expert scientific knowledge and disembedded statistics, Cadag and Gaillard (2012) have mobilised local knowledge and experiences, dependent on social relations of trust and reciprocity built up over extended periods of time.

Which methodologies will enliven political ecological research about sustainability in the current conjuncture? Political ecologists have long analysed vulnerability and other key concepts through in-depth, case-study based research, "in the field." But, what of fieldwork? Knowledge production on sustaining nature faces a multifaceted crises: a place-bound pandemic (Oliver 2022), constrained carbon emissions associated with travel and other climate-impacted research (Rickards and Watson 2020), and a demand to produce more abundant (Collard, Dempsey, and Sundberg 2015), decolonial nature-society futures (Osborne 2021). There are no easy answers to chart the methodological way forward; understanding and conceptualising the political ecology of sustaining nature attends to globalising differentiations and responsibilities that stretch socially and spatially. Future research about vulnerability, adaptation, and emerging paradigms must grapple with the contradictory imperatives of being attentive to both here and there.

From Famines to Floods: What Causes Vulnerability?

As the form, cause, and effects of environmental risks proliferate, so too has research about vulnerability. Directly inspired research about hazards, scholars that define vulnerability to climate change seek to incorporate both socioeconomic and biophysical measures. As Adger (2006, 269) summarises: "in all formulations, the key parameters of vulnerability are the stress to which a system is exposed, its sensitivity, and its adaptive capacity." This echoes Watts and Bohle's (1993; see also Bohle, Downing, and Watts 1994) definition of vulnerability to hunger and famine as being the culmination of the risks and consequences of exposure and capacities to cope. Definitions of vulnerability aim to "denaturalise" natural disasters and climate changes, and instead emphasise underlying social, political, and economic factors that produce vulnerability (for example Thomas et al. 2019). The causal commitment is that identifying where, how, and why people and places are vulnerable might lead to ameliorative actions to reduce that vulnerability.

Over decades of debates, important distinctions have emerged in how vulnerability to climate impacts is understood and operationalised. O'Brien et al. (2007) distinguish between "outcome" and "contextual" vulnerability. The former is the "linear result of the projected impacts of climate change on a particular exposure unit ... offset by adaptation measures" (75). In contrast, contextual vulnerability considers that climate changes occur in the context of overlapping social, political, and economic structures and processes "which interact dynamically with contextual conditions associated with a particular 'exposure unit'" (76). These competing understandings of vulnerability are not merely different interpretations of the term or concept, but rather reflect entirely different ideologies, political commitments, and moral imperatives about the problem of climate change itself (O'Brien et al. 2007; see also Fussel and Klein 2006; Kelly and Adger 2000).

Different approaches to vulnerability ask different kinds of questions, use different methodologies, and invite different kinds of responses. Outcome vulnerability is often associated with scenario-based assessment tools, which combine Global Circulation Model projections of biophysical changes with socioeconomic system changes over time (Burton et al. 2002). As an example, Rosenzweig and Parry (1994) combine models of crop growth under a variety of climate scenarios with global trade models to assess the impact of climate change on world food supplies. Drawing from the outputs of these models, which include predicted world food prices and risks of hunger, the authors assess agricultural vulnerability, finding wide global disparities. The result of these "first generation" vulnerability assessments, therefore, is to focus on future vulnerabilities driven by anticipated climate impacts (Fussel and Klein 2006). In contrast, contextual vulnerability is typified by case study and household survey research, often seeking to understand how multiple climatic and non-climatic stressors interact in a specific location to produce differentiated vulnerabilities (O'Brien et al. 2007). Liverman (1990), for example, combines physical geography, meteorological and census data, and land tenure regimes to compare vulnerability to drought in two regions of Mexico, finding that reported losses from droughts are not predicted by rainfall observations. Vulnerability to drought losses is instead driven, in the Sonora and Puebla cases, to the marginality of land and access to technology. This "second generation" vulnerability assessment approach is focused on current vulnerabilities and adaptive capacities as a latent condition in the context of political economic structures and affordances, livelihood coping strategies, and resources (Fussel and Klein 2006). Competing definitions and methodologies for vulnerability research reach well beyond scholarly journals, effecting global scientific and policy arenas including changing definitions within the Intergovernmental Panel on Climate Change (IPCC).

Alongside disagreements about how to conceptualise vulnerability, there have been fundamental critiques of the term as a guiding principle in climate change research and policy. Vulnerability is operationalised and measured through indices, indicators, models, and frameworks for allocative purposes (Barnett, Lambert, and Fry 2008; Hinkel 2011). But these measures often fail to capture cross scalar interactions, tend to be clumsily additive with climatic and non-climatic processes, and can miss the dynamism and complexity at the heart of vulnerability (Brown 2016) driven by "exploitation, exclusion, marginalization [and] socially stratified societies" (Ribot 2011, 1160). In practice, vulnerability profiles and rankings have invited external interventions that may not align with community priorities or needs. Not only are the systemic and structural causes of vulnerability, such as ongoing colonialism, often overlooked, such external interventions risk perpetuating these relations (Cameron 2012). As a pejorative descriptor, vulnerability has been dismissed as disempowering, categorising communities and places as victims.

Adaptation Regimes: Transformation or Business as Usual?

Different conceptualisations of adaptation mirror debates about vulnerability and hazard risk: is it supposed to be transformative or reformist, or address proximate or contextual impacts (Bassett and Fogelman 2013; Pelling 2011)? In the natural hazards tradition, adaptation is conceived as purposeful adjustments to biophysical phenomenon, which are principally technocratic and largely seek to maintain the "status quo" (Bassett and Fogelman 2013; Head 2010). In contrast, political economic formulations of adaptation see it as potentially transformative of structures that produce vulnerability. Despite being described as slippery and vague, adaptation commonly refers to adjustments to experienced or anticipated climate impacts.

Adaptation has often been differentiated as either planned or autonomous, anticipatory or reactive (Smit and Wandel 2006). For instance, Brown (2016) differentiates between adaptation planning at the national scale, addressed through climate risk mapping, identification of adaptation gaps and vulnerabilities, and assumptions about how to address these, and actual adaptation actions that individuals are undertaking. Adaptation can also be either incremental or transformative, where the latter can be defined as "fundamental alterations to political, economic, and socionatural relations, practices, values, and meaning-making" (Nightingale, Gonda, and Eriksen 2022, 2; see also O'Brien 2012). Rather than addressing measurable vulnerabilities emerging from climate impacts in isolation, transformative adaptation works with wider social relations as the basis from which individuals and collectives might navigate change (Eriksen, Nightingale, and Eakin 2015).

However, while political ecologists have long argued the need for transformative, sustainable adaptations attuned to uneven power relations across scales, examples of these remain few and far between. Instead, catalogues of existing adaptations show persistent limits, barriers, maladaptations, and failures. Adaptation is maladaptive when it has perverse impacts: inadvertently increasing greenhouse gas emissions, reproducing vulnerability somewhere else, or reducing future opportunities for transformative adaptation (Barnett and O'Neil 2010). It is easy to imagine examples where actions taken to reduce individual risks create new risks for others: changed water consumption patterns, new flood protection measures, or intensified agricultural practices. Indeed, Atteridge and Remling (2018) argue that adaptation often redistributes risks between peoples and places rather than reducing them. The teleconnected effects of adaptation are greatly concerning if reducing vulnerability is to be a collective achievement.

Uneven effects are not incidental or accidental, rather they are constitutive of adaptation as a sociopolitical process that involves trade-offs and feedbacks across space and time (Paprocki 2018). In Khulna, southwestern Bangladesh—a deltaic region often proclaimed to be among the most vulnerable to climate impacts in the world—adaptation must contend with rising seas. Global collectives of scientists, financiers, development planners, and consultants propose that shrimp aquaculture can protect against the "inevitable ruination" of climate change in Khulna as it is adapted to watery futures and is economically productive (Paprocki 2019). But these proposals ignore the social and ecological destruction shrimp farming causes: chemical effluents, mangrove deforestation, water logged and saline soils, alongside land-grabbing and restricted labour opportunities. Adaptation through shrimp farming in Khulna is founded upon the dispossession of agrarian livelihoods and migration from the region. Meanwhile, local activists and farmers resist shrimp futures.

From Countryside to Town: Resilience

Although adaptation remains one of the two overarching policy arenas of the United Nations Framework Convention on Climate Change (UNFCCC), some scholars regard it as reactive and clunky. Enter resilience, the seemingly more flexible and dynamic idea, referencing emergent adaptability in the face of unpredictable and complex "radical uncertainties" (Berkes 2007). There are both ecological and engineering perspectives on resilience, ambiguously meaning both the ability to "retain or rapidly return to desired functions" or to "quickly transform" socioecological and sociotechnical systems when faced with uncertainties (Meerow, Newell, and Stults 2016). Where adaptation is largely associated with political ecological concerns of environment and development in agrarian contexts of the Global South, resilience tends to map onto an explicitly urban climate focus. Accordingly, urban resilience sees cities as preeminent in governing climate change, celebrated for their entrepreneurial zeal, unencumbered by flabby nation states and their bureaucratic politics (Long and Rice 2019).

Like the other keywords introduced here, resilience is slippery and vague—a malleable concept that means something to everyone but nothing precisely (Davoudi 2012). For activist and academic critics its biggest shortcoming is its affinity with neoliberalisation. An emphasis on resilience in the face of climate impacts demands communities and individuals that are able to withstand various crises, without redistributive assistance from states or political or ecological alternatives (MacKinnon and Derickson 2013). Kaika describes: resilience prepares citizens and environments for "larger doses of inequality and degradation in the future" (2017, 89). As a paradigm that seeks to cultivate responsiveness in self-organising systems, resilience is also internalist, ignoring broader processes and social relations that produce vulnerabilities.

However, resilience, reformulated, can be attentive to power and agency, and a foundation for transformative political agency. As Harris et al. (2018) show, if thought of as a negotiated and improvised process, resilience is always situated and contested rather than a thing or outcome that inheres to specific systems. While often associated with technological and financial prowess of global cities, the urban poor of cities of the Global

South have centred the politics of resilience amidst social and ecological crises. In Jakarta, for instance, the emblematic 100 Resilient Cities program facilitated negotiations over global norms to address broad and encompassing risks and local political and ecological contexts, what constitutes vulnerability, risk and resilience, and how these might be addressed through existing programs and policies (Webber, Leitner, and Sheppard 2021). At the same time, resilience is also negotiated by residents of Jakarta's kampungs, or informal settlements (Betteridge and Webber 2019; Colven and Tri Irawaty 2019). Everyday resilience is a practice that draws on and produces social and material relations of endurance in the face of floods and evictions. And, these relations also provide a basis to foster collective, if differentiated and sometimes uncoordinated, resistance to political and ecological threats.

Risky Objects and Subjects

Alongside resilience, risk has reoriented the analytical, policy, and even individual response to climate change, emphasising contingency and dynamism (Derickson 2018). Rather than static measures of vulnerability and adaptation, risk and resilience are entrepreneurial governance strategies that emphasise management through crisis and uncertainty. For sociologists of risk, climate change is the ultimate "manufactured" risk of our contemporary "risk society" (Beck 2006). Rather than a "natural" phenomenon predicted with probabilities of hazards, these manufactured risks are globalised, unpredictable, and socio-technically mediated. Identifying, classifying, measuring, and managing risk is an increasing preoccupation of climate policy. Accordingly, climate politics is expressed through the norms of financial governance that, instead of mitigating risks, model and distribute them in search of the highest returns (Christophers 2018).

Risk is a common meta-framework of both climate change science and financial markets. This is evident in the adaptive technique of insurance, the increasingly prevalent financial tool that, in pricing climate risks, seeks to facilitate adaptive behaviours across scales, systems, and sectors (Collier, Elliott, and Lehtonen 2021). Insurance mediates climate risk in a number of ways with uneven distributional consequences. It collectivises risk across social, spatial, and temporal scales through premiums. With the understanding that different publics have different exposures, insurance collectivises loss and thereby seeks to reduce its overall cost. For instance, the World Bank has sought to facilitate more financially sustainable insurance mechanisms to manage the impacts of frequent catastrophes in the Pacific region by creating a sovereign risk pool, the Pacific Catastrophe Risk Insurance Company (PCRIC; see Christophers et al. 2020). In many Pacific Island Countries, there is high exposure to extreme events and high average annual losses from these disasters as a proportion of GDP. As a risk pooling insurance tool, the PCRIC manages these catastrophe risks by collectivising pay-outs to member countries once a pre-specific parametric trigger—a predetermined level of loss or damage—is met. But there is high "basis risk": divergences between expected losses predicted by the parametric model that determines the trigger, and losses experienced during hazard events. In the PCRIC and other similar regional tools, member countries are disappointed with financial pay-outs—in the Solomon Islands, one disaster that claimed losses of almost 10% of GDP was not covered—and several have withdrawn from schemes (Christophers, Bigger, and Johnson 2020).

Because it seeks to price risk through premiums, marketeers assume that insurance incentivises risk-subjects to take risk-mitigating actions. These "price signals" are far more complex, however. For example, in the United States insurance premiums are determined by local regulators, who balance their constituents desires for low insurance premiums, their own dependence on maintaining high property values as the basis for their tax revenues, and ongoing property development in vulnerable locations that low insurance premiums encourage (Elliott 2021). The mitigation action to reduce risk for homeowners or regulators is unclear in this context. Even more concerning, globally, a growing numbers of places and hazards are now "uninsurable," such that insurance companies will not sell insurance products to cover them, and many people are unable or unwilling to buy formal insurance products (Lucas and Booth 2020; Müller, Johnson, and Kreuer 2017). This suggests that the proliferation of new and experimental risk pricing tools may be undesirable for supposed beneficiaries and perhaps even ultimately maladaptive. And yet, as a measure that reaches across time and space, risk and insurance create new publics and solidarities (Christophers, Bigger, and Johnson 2020). Like the other frameworks identified above, risk provides an invitation to produce new collective accountabilities and responsibilities to each other and towards sustainability.

The Politics of Competing Global Environmental Regimes

The ideas of hazards, vulnerability, adaptation, resilience, and risk have been applied in Kiribati in response to climate impacts by a wide variety of international actors. These external interventions in search of sustainability range from micro-infrastructure grants in the "outer islands," innovations in the education curriculum to include climate change, to regional catastrophe insurance funds to pool risk across Pacific Islands. The Kiribati Joint Implementation Plan (KJIP; Government of Kiribati 2014) seeks to integrate the diversity of these climate-responsive investments, suturing together competing problematisations of climate change, impacts, and responses. The KJIP works across disasters, hazards, and climate impacts, across resilience, risk, and adaptation, and therefore tries to balance the competing demands of low-carbon sustainable economic development, among many others. As global environmental regimes, hazards, risk, and so on, each have their own inherited academic hang-ups, international policy architectures, models and frameworks, and innovative project designs. These matter for how projects are imagined, designed, and implemented, and their outcomes. There was, for instance, a highlighted shift from the imperative to adapt to climate change to building local resilience in Kiribati, away from a slogan of "adapt or perish" towards protection "for OUR future." The existence of the KJIP also signals a desire by the Government of Kiribati to gain some coordinating control of—the more than 100 identified—climate change projects largely conceived by external actors and imposed locally.

There are few redeemable characteristics of the proliferating family of climate change projects that seek to sustain collective life and livelihoods in Kiribati—whether focused on resilience, adaptation, or another buzzword altogether. Each of the critiques identified over the decades of political ecological research on these global environmental regimes lives on: the terms are slippery and vague, they bracket out structural social processes, and there are failed projects littered across landscapes. And yet, the demands articulated by i-Kiribati people through these global environmental regimes are political ones. The promise of responding to climate change has become more and more significant, as the efforts to mitigate greenhouse gas emission stagnate and the socio-spatially differentiated impacts of climate change rapidly multiply. The practical implementation of ideas like adaptation and resilience remains contested and subject to myriad failures. Because, these regimes are a domain of politics in which so-called vulnerable countries demand recognition and recompense—maybe even climate reparations (Táíwò 2022)—and the fundamental inequalities of climate change and its impacts are laid bare. Future political ecological research about sustaining natures in the face of multiple crises might now fruitfully turn to recuperating its progressive political and ecological potentials.

References

Adger, WN. 2006. "Vulnerability." *Global Environmental Change* 16 (3): 268–281.

Atteridge, A, and E Remling. 2018. "Is Adaptation Reducing Vulnerability or Redistributing It?" *WIREs Climate Change* 9: e500.

Australian Bureau of Meteorology and CSIRO. 2014. "Climate Variability, Extremes and Change in the Western Tropical Pacific: New Science and Updated Country Reports." Pacific-Australia Climate Change Science and Adaptation Planning Program Technical Report. Melbourne, Australia: Australian Bureau of Meteorology and Commonwealth Scientific and Industrial Research Organization.

Barnett, J, and S O'Neil. 2010. "Maladaptation." *Global Environmental Change* 20 (2): 211–213.

Barnett, J. 2017. "Don't Give up on Pacific Island Nations Yet." *The Conversation* 2017. https://theconversation.com/dont-give-up-on-pacific-island-nations-yet-83300

Barnett, J, S Lambert, and I Fry. 2008. "The Hazards of Indicators: Insights from the Environmental Vulnerability Index." *Annals of the Association of American Geographers* 98 (1): 102–119.

Bassett, TJ, and C Fogelman. 2013. "Deja vu or Something New? The Adaptation Concept in the Climate Change Literature." *Geoforum* 48: 42–53.

Beck, U. 2006. "Living in the World Risk Society: A Hobhouse Memorial Public Lecture given on Wednesday 15 February 2006 at the London School of Economics." *Economy and Society* 35 (3): 329–345.

Berkes, F. 2007. "Understanding Uncertainty and Reducing Vulnerability: Lessons from Resilience Thinking." *Natural Hazards* 41 (2): 283–295.

Betteridge, B, and S Webber. 2019. "Everyday Resilience, Reworking and Resistance in North Jakarta's Kampungs." *Environment and Planning E: Nature and Space* 2 (4): 944–966.

Bohle, HG, TE Downing, and MJ Watts. 1994. "Climate Change and Social Vulnerability: Toward a Sociology and Geography of Food Insecurity." *Global Environmental Change* 4 (1): 37–48.

Bordner, AS, CE Ferguson, and L Ortolano. 2020. "Colonial Dynamics Limit Climate Adaptation in Oceania: Perspectives from the Marshall Islands." *Global Environmental Change* 61 (March): 102054.

Brown, K. 2016. *Resilience, Development and Global Change*. London and New York: Routledge. www.routledge.com/Resilience-Development-and-Global-Change/Brown/p/book/9780415663472

Brun, C. 2009. "A Geographers' Imperative? Research and Action in the Aftermath of Disaster." *The Geographical Journal* 175 (3): 196–207.

Burton, I, RW Kates, and GF White. 1993. *The Environment as Hazard*, 2nd edition. New York: Guilford University Press.

Burton, I, S Huq, B Lim, O Pilifosova, and Li Schipper. 2002. "From Impacts Assessment to Adaptation Priorities: The Shaping of Adaptation Policy." *Climate Policy* 2 (2–3): 145–159.

Cadag, JRD, and JC Gaillard. 2012. "Integrating Knowledge and Actions in Disaster Risk Reduction: The Contribution of Participatory Mapping." *Area* 44 (1): 100–109.

Cameron, ES. 2012. "Securing Indigenous Politics: A Critique of the Vulnerability and Adaptation Approach to Human Dimensions of Climate Change in the Canadian Arctic." *Global Environmental Change* 22 (1): 103–114.

Christophers, B. 2018. "Risk Capital: Urban Political Ecology and Entanglements of Financial and Environmental Risk in Washington, DC." *Environment and Planning E: Nature and Space* 1 (1–2): 144–164.

Christophers, B, P Bigger, and L Johnson. 2020. "Stretching Scales? Risk and Sociality in Climate Finance." *Environment and Planning A: Economy and Space* 52 (1): 88–110.

Collard, R-C, J Dempsey, and J Sundberg. 2015. "A Manifesto for Abundant Futures." *Annals of the Association of American Geographers* 105 (2): 322–330.

Collier, SJ, R Elliott, and TK Lehtonen. 2021. "Climate Change and Insurance." *Economy and Society* 50 (2): 158–172.

Colven, E, and D Tri Irawaty. 2019. "Critical Spatial Practice And Urban Poor Politics: (Re)Imagining Housing In A Flood-Prone Jakarta." *Society + Space* (blog). 2019. www.societyandspace.org/articles/critical-spatial-practice-and-urban-poor-politics-re-imagining-housing-in-a-flood-prone-jakarta

Cutter, S, JK Mitchell, and MS Scott. 2000. "Revealing the Vulnerability to People and Places: A Case Study of Georgetown County, South Carolina." *Annals of the Association of American Geographers* 90: 713–737.

Cutter, SL. 2003. "The Vulnerability of Science and the Science of Vulnerability." *Annals of the Association of American Geographers* 93 (1): 1–12.

Davoudi, S. 2012. "Resilience: A Bridging Concept or a Dead End?" *Planning Theory and Practice* 13 (2): 299–307.

Dean, A, D Green, and P Nunn. 2016. "Too Much Sail for a Small Craft? Donor Requirements, Scale, and Capacity Discourses in Kiribati." In *Island Geographies: Essays and Conversations*, edited by E Stratford. New York and London: Routledge.

Derickson, K. 2018. "Urban Geography III: Anthropocene Urbanism." *Progress in Human Geography* 42 (3): 425–435.

Donner, S, and S Webber. 2014. "Obstacles to Climate Change Adaptation Decisions: A Case Study of Sea-Level Rise and Coastal Protection Measures in Kiribati." *Sustainability Science* 9: 331–345.

Elliott, R. 2021. *Underwater: Loss, Flood Insurance, and the Moral Economy of Climate Change in the United States*. New York: Columbia University Press.

Emel, J, and R Peet. 1989. "Resource Management and Natural Hazards." In *New Models in Geography*, by R Peet and N Thrift. Routledge.

Eriksen, S, A Nightingale, and H Eakin. 2015. "Reframing Adaptation: The Political Nature of Climate Change Adaptation." *Global Environmental Change* 35: 523–533.

Fussel, HM, and RJT Klein. 2006. "Climate Change Vulnerability Assessments: An Evolution of Conceptual Thinking." *Climatic Change* 75: 301–329.

Germano, M. 2022. "'Neutral' Representations of Pacific Islands in the IPCC Special Report of 1.5°C Global Warming." *Australian Geographer* 53 (1): 23–39.

Government of Kiribati. 2014. "Kiribati Joint Implementation Plan for Climate Change and Disaster Risk Management 2014-2023." Plan. Tarawa, Kiribati: Government of Kiribati and Secretariat of the Pacific Community. www.spc.int/images/news/20140901/KJIP-BOOK.pdf

Harris, L, E Chu, and G Ziervogel. 2018. "Negotiated Resilience." *Resilience* 6 (3): 196–214.

Hau'ofa, E. 1993. "Our Sea of Islands." In *A New Oceania: Rediscovering Our Sea of Islands*. Suva, Fiji: University of the South Pacific.

Head, L. 2010. "Cultural Ecology: Adaptation—Retrofitting a Concept?" *Progress in Human Geography* 34 (2): 234–242.

Hinkel, J. 2011. "'Indicators of Vulnerability and Adaptive Capacity': Towards a Clarification of the Science–Policy Interface." *Global Environmental Change* 21 (1): 198–208.

Kaika, N. 2017. "'Don't Call Me Resilient Again!': The New Urban Agenda as Immunology … or What Happens When Communities Refuse to Be Vaccinated with 'Smart Cities' and Indicators." *Environment and Urbanization* 29 (1): 89–102.

Kelly, PM, and WN Adger. 2000. "Theory and Practice in Assessing Vulnerability to Climate Change and Facilitating Adaptation." *Climatic Change* 47 (4): 325–352.

Kumar, L, and S Taylor. 2015. "Exposure of Coastal Built Assets in the South Pacific to Climate Risks." *Nature Climate Change* 5 (11): 992–996.

Liverman, D. 1990. "Vulnerability to Global Environmental Change." In *Understanding Global Environmental Change: The Contributions of Risk Analysis and Management*, by RE Kasperson et al., 27–44. Worcester: Clark University Press.

Long, J, and J Rice. 2019. "From Sustainable Urbanism to Climate Urbanism." *Urban Studies* 56 (5): 992–1008.

Lucas, CH, and KI Booth. 2020. "Privatizing Climate Adaptation: How Insurance Weakens Solidaristic and Collective Disaster Recovery." *WIREs Climate Change* 11 (6): e676.

MacKinnon, D, and K Derickson. 2013. "From Resilience to Resourcefulness: A Critique of Resilience Policy and Activism." *Progress in Human Geography* 37 (2): 253–270.
Meerow, S, J Newell, and M Stults. 2016. "Defining Urban Resilience: A Review." *Landscape and Urban Planning* 147: 38–49.
Müller, B, L Johnson, and D Kreuer. 2017. "Maladaptive Outcomes of Climate Insurance in Agriculture." *Global Environmental Change* 46 (September): 23–33.
Mustafa, D. 2005. "The Production of Urban Hazardscape in Pakistan: Modernity, Vulnerability and the Range of Choice." *Annals of the Association of American Geographers* 95 (3): 566–586.
Nightingale, AJ, N Gonda, and SH Eriksen. 2022. "Affective Adaptation = Effective Transformation? Shifting the Politics of Climate Change Adaptation and Transformation from the Status Quo." *WIREs Climate Change* 13 (1): e740. https://doi.org/10.1002/wcc.740
O'Brien, K. 2012. "Global Environmental Change II: From Adaptation to Deliberate Transformation." *Progress in Human Geography* 36 (5): 667–676.
O'Brien, K, S Eriksen, L Nygaard, and A Schjolden. 2007. "Why Different Interpretations of Vulnerability Matter in Climate Change Discourses." *Climate Policy* 7 (1): 73–88.
Oliver, C. 2022. "'Significant Nothingness' in Geographical Fieldwork." *Geoforum* 134: 82–85. https://doi.org/10.1016/j.geoforum.2022.06.010
Osborne, T. 2021. "Decolonizing Methodologies for Climate Justice Research." *Environment and Planning E: Nature and Space* 4 (2): 405–408.
Paprocki, K. 2019. "All That Is Solid Melts into the Bay: Anticipatory Ruination and Climate Change Adaptation." *Antipode* 51 (1): 295–315.
Paprocki, K. 2018. "Threatening Dystopias: Development and Adaptation Regimes in Bangladesh." *Annals of the American Association of Geographers* 108 (4): 955–973.
Pelling, M. 2011. *Adaptation to Climate Change: From Resilience to Transformation*. New York: Routledge.
Piggott-McKellar, AE, KE McNamara, and PD Nunn. 2020. "Who Defines 'Good' Climate Change Adaptation and Why It Matters: A Case Study from Abaiang Island, Kiribati." *Regional Environmental Change* 20 (2): 43.
Ratuva, S. 2014. "'Failed' or Resilient Subaltern Communities?: Pacific Indigenous Social Protection Systems in a Neoliberal World." *Pacific Journalism Review* 20 (2): 40–58.
Ribot, J. 2011. "Vulnerability before Adaptation: Toward Transformative Climate Action." *Global Environmental Change* 21 (4): 1160–1162.
Rickards, L, and JEM Watson. 2020. "Research Is Not Immune to Climate Change." *Nature Climate Change* 10 (3): 180–183.
Rosenzweig, C, and ML Parry. 1994. "Potential Impact of Climate Change on World Food Supply." *Nature* 367 (6459): 133–138.
Saguin, K. 2017. "Producing an Urban Hazardscape beyond the City." *Environment and Planning A: Economy and Space* 49 (9): 1968–1985.
Smit, B, and J Wandel. 2006. "Adaptation, Adaptive Capacity and Vulnerability." *Global Environmental Change* 16: 282–292.
Storey, D, and S Hunter. 2010. "Kiribati: An Environmental 'Perfect Storm'." *Australian Geographer* 41 (2): 167–181.
Storlazzi, CD, SB Gingerich, A van Dongeren, OM Cheriton, PW Swarzenski, E Quataert, CI Voss, et al. 2018. "Most Atolls Will Be Uninhabitable by the Mid-21st Century Because of Sea-Level Rise Exacerbating Wave-Driven Flooding." *Science Advances* 4 (4): eaap9741.
Táíwò, OO. 2022. *Reconsidering Reparations*. New York: Oxford University Press.
Teaiwa, K. 2018. "Moving People, Moving Islands in Oceania." In *Paradigm Shift: People Movement*, 63–69. Canberra: Australian National University.
Thomas, K, RD Hardy, H Lazrus, M Mendez, B Orlove, I Rivera-Collazo, JT Roberts, M Rockman, BP Warner, and R Winthrop. 2019. "Explaining Differential Vulnerability to Climate Change: A Social Science Review." *WIREs Climate Change* 10 (2): e565.
Watts, M. 1983. "On the Poverty of Theory: Natural Hazards Research in Context." In *Interpretations of Calamity*, edited by K Hewitt, 231–262. Boston: Allen and Unwin.
Watts, MJ, and HG Bohle. 1993. "The Space of Vulnerability: The Causal Structure of Hunger and Famine." *Progress in Human Geography* 17 (1): 43–67.

Weatherill, K. 2022. "*Sinking Paradise? Climate Change Vulnerability and Pacific Island Extinction Narratives*." Geoforum, May.
Webb, AP, and PS Kench. 2010. "The Dynamic Response of Reef Islands to Sea-Level Rise: Evidence from Multi-Decadal Analysis of Island Change in the Central Pacific." *Global and Planetary Change* 72 (3): 234–246.
Webber, S. 2013. "Performative Vulnerability: Climate Change Adaptation Policies and Financing in Kiribati." *Environment and Planning A* 45: 2717–2733.
Webber, S, H Leitner, and E Sheppard. 2021. "Wheeling Out Urban Resilience: Philanthrocapitalism, Marketization, and Local Practice." *Annals of the American Association of Geographers* 111 (2): 343–363. https://doi.org/10.1080/24694452.2020.1774349
White, GF. 1974. *Natural Hazards, Local, National, Global.* New York: Oxford University Press.
Woodroffe, C. 2008. "Reef-Island Topography and the Vulnerability of Atolls to Sea-Level Rise." *Global and Planetary Change* 62 (1–2): 77–96.
World Bank. 2021. "Climate Risk Country Profile: Kiribati." Washington DC: World Bank Group. https://climateknowledgeportal.worldbank.org/sites/default/files/country-profiles/15816-WB_Kiribati%20Country%20Profile-WEB.pdf
Yamane, A. 2009. "Climate Change and Hazardscape of Sri Lanka." *Environment and Planning A: Economy and Space* 41 (10): 2396–2416.

2 Producing Nature

Where Biophysical Materialities Meet Social Dynamics

Christine Biermann, Justine Law, and Zoe Pearson

Introduction: Planting Trees, Producing Nature

It is April 2012, and I (Christine) am standing on a Pennsylvania hilltop, looking out at the site where United Airlines Flight 93 crashed on September 11, 2001. The site has been repeatedly scoured and strip-mined to extract bituminous coal. Now designated the Flight 93 National Memorial, the US National Park Service is leading a massive reforestation project on the landscape, and I'm here to plant American chestnut (*Castanea dentata*) trees. The goals of the plantings are many: to turn a strip mine into forest, to cultivate a living memorial, to create a wildlife habitat, and to bring back the American chestnut tree. For me, there is another goal: to conduct participant observation research for my doctoral dissertation about the restoration of the American chestnut, a tree that once dominated eastern North American upland forests.

Now, the species is facing functional extinction. At the turn of the twentieth century, the chestnut blight fungus (*Cryphonectria parasitica*) was inadvertently imported to New York City from Asia. It spread rapidly throughout eastern North America, infecting and killing virtually all mature American chestnuts by the mid-twentieth century. Today, American chestnuts occasionally hide in the understory, re-sprouted from the roots of trees killed by blight. But as these sprouts grow, they almost always become reinfected and die before reproducing (Paillet 2002). Gridlocked by blight, the American chestnut still lives but cannot make new life.

Over the past century, chestnuts have been divided, pollinated, crossed, backcrossed, genotyped, phenotyped, inoculated, cloned, and engineered in various efforts to ward off extinction. The American Chestnut Foundation has been working for four decades to produce a blight-resistant tree by backcrossing the American chestnut with the Chinese chestnut (*Castanea mollissima*). Meanwhile, other researchers have produced a transgenic American chestnut tree that can withstand the blight (Newhouse and Powell 2021). In 2020, the transgenic chestnut became the first genetically engineered organism developed specifically for restoration in the wild to be considered for federal deregulation in the US (Barnes and Delborne 2021). Many involved in chestnut restoration hope to cross transgenic trees with wild-type and backcrossed chestnuts, like those at the Flight 93 Memorial, to produce a blight-resistant population that is genetically diverse and adaptable to changing environmental conditions.

The American chestnut is at once both novel and ancient, social and natural. For political ecologists, it is a story that can be understood through the lens of co-production. Rather than emphasizing either the *destruction* or *construction* of environmental systems or landscapes, co-production foregrounds interconnections among ideas about nature,

DOI: 10.4324/9781003165477-4

techniques for managing nature, and material nature (Robbins 2012). In other words, the scientists creating ecological knowledge, the applications of that knowledge, and the ecosystems themselves actively shape each other. Co-production suggests, for example, that we see a forest as a vast, messy web of trees, foresters, restorationists, deer, mushroom hunters, woodpeckers, lichen, loggers, and so on—all of whom can tug on, and thereby change, that web. This approach raises questions like: How do political economic motivations, diverse cultural meanings, and interactions among humans and nonhumans shape the production of a blight-resistant chestnut? What are the ecological ramifications of a blight-resistant chestnut? Is it possible for the tree to both be enrolled in and resist new systems of environmental management?

Before we consider these questions further, let's examine how human activities, land use practices, economic and environmental policies, and emerging technologies co-produce new material ecologies, and how political ecologists make sense of these productions. As we do so, we emphasize recent research in political ecology, critical physical geography, and cognate fields. We start by considering a major theme of political ecology since the field's outset: ecologies of long-term stewardship. Later, we attend to ecologies shaped through conservation, neoliberal economies, molecular technologies, and the Anthropocene.

Ecologies of Long-term Stewardship

Are humans an inevitable blight on ecosystems? For political ecologists, the answer to that question is a resounding "no." Indeed, political ecology research regularly spotlights traditional ecological knowledge and environmental stewardship practices of peasants, Indigenous peoples, and other communities who derive at least part of their livelihood from the physical environment. This research intentionally combats popular narratives that claim—or at least suggest—that nature would be better off without us pestilent humans. At the same time, this research demonstrates how seemingly "wild" landscapes have been produced through intricate webs of human–nonhuman interaction that stretch across generations. This means, to put it more plainly, that the very natures we exalt, from biodiverse rainforests to sweeping savannas filled with charismatic megafauna, look and function the way they do *because of* humans.

Take the rainforests of Central and South America. Over 30 years ago, Denevan (1992) published "The Pristine Myth," a groundbreaking account of how pre-Columbian populations numbering in the tens of millions regularly cultivated these rainforests, thereby shaping their soils, species composition, and landscape diversity. For example, a combination of swidden burns, lightning fires, and buried pottery, organic matter, and animal bones created the charcoal-studded *terra preta* soils of the Amazon basin (Denevan 1992). Subsequent ecological and archeological research has strengthened Denevan's argument by revealing, for instance, that domesticated trees such as Brazil nut (*Bertholletia excelsa*) and maripa palm (*Attalea maripa*) are more likely to be dominant species in Amazonian rainforests than non-domesticated trees, especially in the transitional forest zones with seasonal rainfall patterns that many communities inhabited (Levis et al. 2017). The rainforest's biodiversity is clearly a product of biophysical processes *and* the people who have inhabited it for thousands of years.

And it is not just the rainforest. Humans have stewarded forest landscapes across the world. Fairhead and Leach (1995), for example, dismantled nearly a century of colonial science by showing how Guinean communities established and tended forest "islands" around their villages. Rather than the threat to forests that colonial narratives painted them to be, these communities were a force of greening on the landscape. Other political ecology research from southeast Ohio (e.g., Law and McSweeney 2013; Mansfield et al. 2015) to southeast Asia (e.g., Brookfield et al. 2002; Rerkasem et al. 2009) similarly demonstrates how established user communities can help dictate forest cover, structure, and function. In fact, even populations of understory plants (Law 2022) and fungi (Barron 2015) owe much to the guardianship of people. American ginseng (*Panax quinquefolius*) seeds, for instance, germinate best at a ¾-inch soil depth, but they are more likely to reach that depth via human hands than they are via animal dispersal or fallen berries (Law 2022).

The same is true for non-forested ecosystems that are so often cast as "untouched." Research in the African savanna—including Butt's highly-interdisciplinary research that draws on ethnographic methods, GPS data from tracking collars, and satellite-derived vegetation data (2014)—has shown how the population levels and migration patterns of megafauna such as wildebeest are produced through a delicate balance of human and nonhuman forces (e.g., Neumann 2002; Butt 2014; Goldman 2018). In fact, some scholars suggest that we wield terms like "savanna" carefully, as they have been used to obscure (and eliminate) the role of humans on the landscape (Duvall, Butt, and Neely 2018).

In sum, one overriding point of this work is to say: "see, people *can* help produce vibrant, biodiverse, resilient ecologies—especially when they have longstanding knowledge of these ecologies and especially when their livelihoods depend, at least in part, on their survival." Effective ecological management, then, may require us to maintain human–nonhuman networks where and when they produce sustainable, equitable natures.

Ecologies of Biodiversity Conservation

New ecologies are also brought into being through human attempts to conserve or manage biodiversity. While conservation shapes material ecologies in many ways, here we focus on how it aims to control individual organisms, promotes desirable species (for example, through legal protection or assisted migration and reintroduction projects), and kills or diminishes populations seen a threat (for example, through invasive species eradication) (Collard 2012; Hodgetts 2016; Biermann and Anderson 2017; Margulies 2019; Perkins 2020). Responding to these efforts, political ecologists have asked: what are the material eco-social impacts of biodiversity conservation agendas, techniques, and metrics? This question not only foregrounds interactions among landscapes, species, ideas, science, and human activities, but also recognizes that any effort to protect nature inevitably alters and re-shapes the very landscape, ecosystem, or species that it aims to safeguard.

For example, in the Western US, Anderson and colleagues have documented how grey wolf (*Canis lupus*) recovery and management efforts rely upon tools such as guard dogs, lights, sirens, and non-lethal munitions in an attempt to influence wolf pack behavior by cultivating and sustaining wolves' fear of people (Anderson et al. 2022). In this management scenario, protecting and promoting wolf populations involves disciplining wolves who do not appear sufficiently afraid of people in order to change how they behave, move, and interact. In other words, wolf conservation is not merely a project of *protecting* wolves as they are but is a project of *producing* desirable wolf behaviors and ecologies. Even wolves—the ultimate symbol of wildness and independence from human activity—are co-produced.

While wolves represent conservation's drive to protect certain populations, white-tailed deer management illustrates how conservation produces some species as "killable" and some spaces as "sanctioned sites for killing" (Connors and Short Gianotti 2021). In many suburban communities across North America, deer populations have flourished over the past several decades due to lack of predators, abundant food sources, and a favorable patchy habitat. As a result, white-tailed deer are now commonly labeled "overabundant," and their heavy browsing on vegetation is considered a threat to ecosystem health. This new ecology of "overabundant" deer is itself a product of linked environmental-political economic processes such as suburbanization, agricultural abandonment, establishment of urban green spaces, and prohibition of hunting in residential areas. The resurgence of deer has led many wildlife managers and municipalities to enact lethal control measures to reduce deer populations, to maintain a desired forest structure, species composition, and aesthetic (Connors and Short Gianotti 2021). Not surprisingly, such measures are often subject to intense public scrutiny and debate.

On a more conceptual level, the killing of deer because their population size poses a threat to forest ecologies is an example of the biopolitics of conservation (e.g., Srinivasan 2017; Cavanagh 2018; Bluwstein 2018; Wynne-Jones et al. 2020; Kiggell 2021). Extending Foucault's concept of biopower (summarized as the power to "make live and let die"; Foucault 2003, 241), conservation biopolitics encompasses the intertwined logics of care and killing at work in biodiversity science and management. In the case of deer, caring for the forest becomes synonymous with reducing deer populations through lethal means. In ecologies of biodiversity conservation, it is the population as a whole that is the primary target of management interventions that aim to protect ("make live") desirable taxa or diminish ("let die") those that serve as a threat to other life. Crucially, however, not all conservation strategies lead to the same eco-social outcomes, even when they similarly aim to manage life at the level of the population or species. Instead, the resultant ecologies are determined not only by the human logics and actors involved, but also through the agency of nonhuman others and their interactions with people and one another. Producing richly detailed accounts of these ecologies requires researchers to embrace a wider suite of research methods and partnerships not only across academic disciplines but also beyond the academy. This is exemplified by the work of the Applied Conservation Science Lab in the University of Victoria's Department of Geography. For example, through a recent community-academic partnership with the Kitasoo Xai'xais (KX) First Nation, lab members evaluated the conservation status of mountain goats in British Columbia using aerial surveys of goats, interviews with KX Knowledge Holders, and long-term hunter kill data (Jessen et al. 2022). For political ecologists interested in producing conservation research that informs environmental management decisions, this integration of methods, knowledge systems, and cultures provides a useful model.

Ecologies of Neoliberal Globalization

We turn now to the current "neoliberal" era of capitalism, highlighting a few of the ways in which neoliberalism manifests as a project of environmental governance with material outcomes for nature. While Chapters 9 and 10 of this book discuss neoliberalism in greater detail, here we showcase mixed-methods research that examines its material eco-social impacts. Following McCarthy and Prudham (2004), we can divide research on ecologies of neoliberalism into two veins. The first focuses on the environmental impacts of prototypical (i.e., not explicitly "green") global neoliberal development projects. The second examines natures produced through the incorporation of environmentalism into neoliberal approaches, and neoliberal logics into environmental efforts (i.e., "green neoliberalism").

In the first vein, infrastructural mega-development projects carried out over large geographic areas are highly visible political economic strategies that drive material environmental change on vast scales. Coastal reclamation, for example, creates land and water spaces out of shallow coastal areas for agriculture, fish farming, urban development, port development, and more (Choi 2014). Choi writes that such coastal land reclamation was and is an East Asian development strategy of the last several decades, with the dual political economic benefit of creating new means of production while—following Harvey (2003)—also providing a "spatial fix" for capital over-accumulation crises (Choi 2014). Building upon and incorporating the existing material world, these projects have involved the creation of several hundred square kilometers of new land or water expanses—quite literal and visible changes to material ecosystems, with enormous implications for coastal and marine habitats and a myriad of eco-social effects. Techniques such as mapping and remote sensing analysis, biodiversity surveys, and ecological field methods have allowed researchers to begin to document spatial and temporal patterns of coastal land reclamation. Analysis of remote sensing imagery from China, for example, allowed Sengupta et al. (2019) to identify multiple trajectories of reclamation in varying urban contexts. But rather than being satisfied with identifying patterns, political ecologists also aim to explain the processes that undergird them. Here, political economic concepts and theories are put into conversation with results from remote sensing analysis, allowing researchers to describe how novel landscape patterns arise from national policy reforms, privatization of property rights, flows of capital, and contradictions among economic growth, social justice, and environmental protection (Choi 2014; Sengupta et al. 2019).

In the second vein, green neoliberalism describes the incorporation of environmentalism into neoliberal approaches, and neoliberal logics into environmental efforts. Examples of this trend abound, from the buying and selling of carbon offsets in a global marketplace to the rebranding of ecosystems as "natural capital" (Benabou and Harms 2021). Critics contend that green neoliberalism is nothing more than "green washing"—efforts to accumulate profit through the appropriation of commons and resources (McAfee 1999; Bakker 2010). But political ecological critiques of green neoliberalism have rarely examined the specific material ecological outcomes of neoliberal and market-based environmental strategies, focusing instead on social dynamics of land/resource theft, acts of enclosure, privatization, and commodification. Here, integrative methods are needed to clarify whether the *ecological* outcomes of green neoliberalism are as problematic as critics suggest.

One example of research on green neoliberalism and its material ecologies assesses stream mitigation banking, a form of market-based management in which stream damage associated with development is "offset" by restoration elsewhere. To understand this market-based strategy and the streams it produces, an interdisciplinary team of political ecologists and stream geomorphologists performed geomorphological surveys of stream channels along with social science methods such as interviews, policy analysis, and participant observation (Doyle et al. 2015; Lave, Doyle, Robertson, and Singh 2018b). This integrative critical physical geography approach (Box 2.1) found that the locations, morphologies, and ecologies of restored streams are shaped by economic motives and regulatory requirements. Because mitigation credits are allocated by linear foot of stream, restoration has focused more heavily on small headwater channels that cost less to restore than larger streams (Doyle et al. 2015). Similarly, streams restored for mitigation credits tend to be wide, shallow, and more homogenous, probably because such channels are less likely to erode and stream stability is necessary to sell mitigation credits (Doyle et al. 2015). This research suggests that the conservation potential of market-based approaches such as stream mitigation banking is severely limited by tensions between the simplicity and stability required by the market and the inherent dynamism and complexity of ecosystems (Lave, Biermann, and Lane 2018a). Other prime examples of the use of market logics to solve environmental problems under neoliberalism, and objects of study for political ecologists taking seriously the physical materiality of nature, include payments for ecosystem services (PES) (Etchart et al. 2020; Tuijnman et al. 2020; vonHedemann 2023); conservation and eco-tourism (Brondo 2013); and carbon emissions trading (Turhan and Gündoğan 2019), to name a few.

Box 2.1 Critical physical geography

In order to understand produced natures—including how they work and the work they do in the world—a combination of methods is often required. Some early political ecologists frequently integrated biophysical and social data and methods in their work (e.g., Blaikie and Brookfield 1987; Leach and Mearns 1996), but in the late 1990s and 2000s this integrative approach became somewhat less common as post-structural ideas about discourse, language, and cultural formations of nature began to dominate the field.

Critical physical geography is a new field of environmental research which emphasizes the need for mixed social and biophysical methods to understand interconnections among material ecologies, knowledge politics, and social power relations (Lave et al. 2014; Lave, Biermann, and Lane 2018a). Beginning as a series of discussions in the early 2010s, this field shares much in common with political ecology. Like political ecology, it brings together deep knowledge of ecological processes and attention to social dynamics and structural relations of power. Where critical physical geography differs from political ecology, however, is in its core commitment to the integration of biophysical and social methods. While political ecology recognizes that the matter of nature matters (Bakker and Bridge 2006), critical physical geography takes this a step further to argue that if the worlds we inhabit are fundamentally eco-social, then our research methods, theories, and concepts must be as well (Biermann, Kelley, and Lave 2020).

There are three ideas at the core of critical physical geography. First, most landscapes are now deeply shaped by human actions and structural inequalities around race, gender, and class. Rather than existing outside of "nature," these power relations draw upon and are impacted by the materiality of nature (Urban 2018). Second, the same power relations that shape the landscapes and systems we study also shape who studies them and how we study them (King and Tadaki 2018). Third, the knowledge we produce has deep impacts on the people and systems we study (Law 2018).

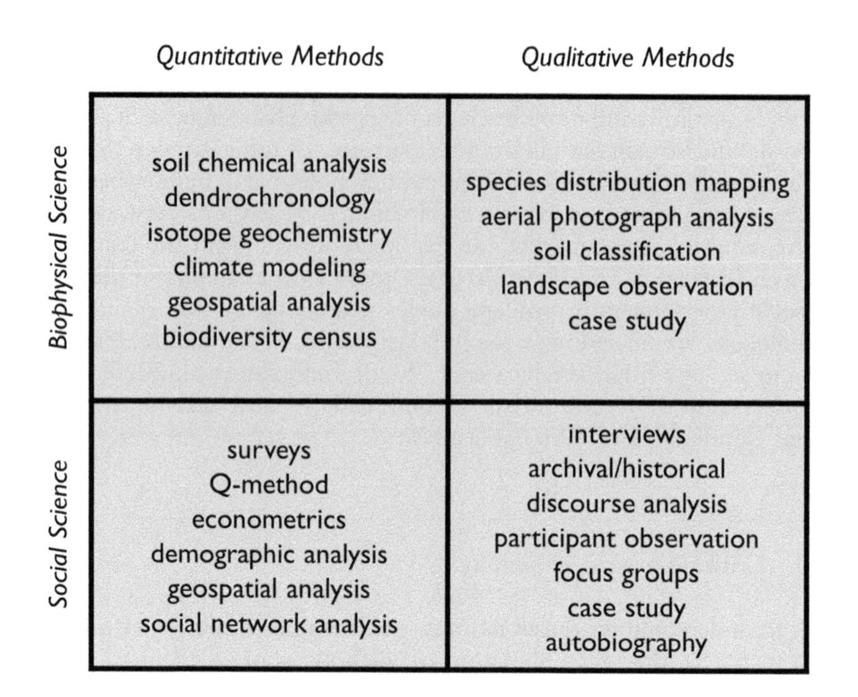

Figure 2.1 The diversity of methods used by critical physical geographers. This methods four-square depicts methods used by critical physical geographers and political ecologists, distinguishing between methods common in the social science and biophysical sciences, and quantitative and qualitative methods.

Source: Altered from Lave, Biermann, and Lane (2018a).

But how might we as political ecologists go about expanding our conceptual and methodological ranges? Figure 2.1 presents a framework for thinking about methodological integration: a four-square that distinguishes between biophysical science and social science methods, and quantitative and qualitative methods, recognizing that these categories themselves often work to create and maintain siloed approaches to environmental problem-solving. Critical physical geography aims to span multiple boxes within the four-square, and particularly emphasizes a reach across the social-biophysical divide. This integration allows researchers to triangulate results across disparate data sources and identify frictions that add nuance to our knowledge.

For a few examples of critical physical geography in action, see Kelley (2018) on cacao growers and agricultural expansion in Sulawesi, Indonesia, Colucci et al. (2021) on genocide and landscape transformation in Cambodia, and Malone and McClintock (2022) on no-till agriculture and conservation policies in Oregon, USA.

Ecologies of the Molecular Turn

A further development of the late twentieth and early twenty-first century is the application of genetic and genomic tools in biodiversity conservation. From the use of environmental DNA (eDNA) in detecting species invasions to the proposed genetic engineering of extinct taxa, what counts as conservation is being radically reimagined. While biodiversity has long been defined as encompassing ecosystem, species, and genetic diversity, this molecular turn in conservation has shifted the emphasis away from the level of ecosystem or species (Hennessy 2015). It is not that ecosystems or species are no longer significant, but rather that ecosystems and species are increasingly understood as outcomes of processes occurring at the molecular scale.

Hennessy has detailed how this scalar shift in conservation has played out on the Galápagos Islands, where advancements in mitochondrial DNA and nuclear microsatellite data have enabled new phylogenetic reconstructions of the evolutionary dispersal of the Galápagos' famed giant tortoise populations. Genetic studies have fundamentally altered tortoise taxonomy, bringing new species into being by revealing striking genetic differences between tortoise populations previously understood as the same species based on their common morphological traits. According to Hennessy (2020), these "new" tortoise species are not discoveries *per se* but are outcomes of the co-production of genetic science, conservation, and nature. These findings were achieved through decades of attention and care from conservationists working to protect the Galápagos as a natural laboratory of evolution. Yet this evolving tortoise taxonomy has also challenged the conservation status quo by, for example, generating debate about how to best manage hybrid tortoises that are crosses of multiple species. In this way, conservation and genetic science continue to shape one another, co-producing the materiality and ecologies of tortoises themselves through conservation breeding, relocation efforts, and prioritization of certain genetics.

This scalar shift in conservation, political ecologists have noted, relies on a reductionist view of genes as discrete, transferable bits of information that produce controllable and predictable traits (McAfee 2003; Valve 2011). Critical examinations have also shown how genetic studies become the basis for geopolitical decisions about the demarcation of conservation territories (Campbell and Godfrey 2010) and life-and-death decisions about wildlife populations (Fredriksen 2016; Peltola and Heikkilä 2018), even as notions of what genetics are "pure" for a given species or subspecies are constantly changing (Biermann and Havlick 2021).

On one hand, genetics and genomics provide a new avenue through which to imagine and protect a pure, pristine nature that exists outside of human activity. The human influence on ecosystems overall may not be able to be erased, but genomes can be envisioned and produced to match an imagined level of purity (Hennessy 2015; Havlick and Biermann 2020). Yet this same molecular turn is also giving rise to conservation visions and goals that embrace hybridity and boundary-crossing between nature and society (e.g., Lorimer and Driessen 2013; Redford and Adams 2021). Political ecologists working at the science-society interface have explored the ethical and philosophical dimensions of genetic and genomic approaches to conservation, focusing on spectacularly controversial forms of conservation (e.g., de-extinction) made possible through biotechnology and synthetic biology (Adams 2016; Cohen 2014; Jørgensen 2013; Rossi 2013). Metaphors of genes as programmable information or code have generated optimism among some conservationists that techniques such as gene editing might provide simple fixes to protect beleaguered taxa, reverse extinction, or eradicate pests and pathogens. Political ecologists have responded to possible conservation applications of gene editing and synthetic biology with a mix of both cautious optimism and warnings that genome alteration may produce un-intended outcomes, both through off-target mutations and through the deployment of technologies in ways that are unjust, exploitative, or otherwise problematic (Rossi 2017; Adams 2020; Redford and Adams 2021).

Ecologies of the Anthropocene

The final set of ecologies we examine here are the ecologies of the Anthropocene, the ecologies of today and of the future. The Anthropocene—sometimes called the "capitalocene" to shift the focus from human populations to the capitalist economy (Moore 2017)—refers to the current, and unofficial, epoch in which human processes dominate geologic, chemical, and biological processes (Crutzen 2006). And, certainly, with global temperatures expected to be 2–4°C higher by 2100 as a result of human-induced climate change (IPCC 2021), material nature is bound to be different in the twenty-first century than it was in recent centuries.

Several scholars have defined some of this differentness as "novel" ecology. More precisely, novel ecosystems are characterized by "abiotic, biotic and social components that, by virtue of human influence, differ from those that prevailed historically, having a tendency to self-organize and manifest novel qualities without intensive human management" (Hobbs, Higgs, and Hall 2013). In other words, they are self-sustaining ecosystems that exist because of human forcings. And "novel ecosystem" is not a neutral term; while some use it to deride "trash" ecosystems (Simberloff et al. 2015), most scholars and science writers who use the term do so sympathetically (e.g., Hobbs, Higgs, and Harris 2009; Marris 2011; Lorimer 2015; Pearce 2016). We have fundamentally changed this planet, their argument goes, so it does not make sense to turn back the clock to a nonexistent and irrelevant baseline condition. Instead, we should embrace the sometimes-surprising novel species assemblages that *do* thrive in twenty-first century ecosystems and take the ecological functions they provide (e.g., water filtration, soil stabilization, carbon sequestration, habitat) seriously. To this end, scientists are researching the value of forests filled with nonnative species (Mascaro, Hughes, and Schnitzer 2012), reefs filled with previously uncommon but comparatively-resilient corals and seaweeds (Graham et al. 2014), and artificial ponds filled with diverse insects (Simaika, Samways, and Frenzel 2016). Meanwhile, practitioners such as landscape architects are starting to use "spontaneous urban plants"—otherwise known as weeds—to help create beautiful urban spaces where plants can survive without much maintenance (Sack 2013; Del Tredici 2020).

This novel ecology turn connects to an argument that political ecologists have been making for decades: nonhuman entities can be resilient, can be hard to control, and can do unexpected things when they interact with humans, especially when they are under stress. It is, in a sense, the "life finds a way" argument from *Jurassic Park*. Early theoretical antecedents of this vein of work include actor-network theory, which was developed in the 1980s by science and technology studies (STS) scholars and which posits that human and nonhuman "actants" alike are continually produced through webs of relation with each other (Callon, Law, and Rip 1986; Latour 2005), as well as Haraway's writings on companion species (2007) and hybrid "cyborgs" that confuse the boundaries of natural and artificial (1991). Meanwhile, empirical political ecology research has demonstrated how the material characteristics of everything from otters (Goedeke and Rikoon 2008) to copper (Bridge 2000) can "act back" and transform environmental management regimes, sometimes in emancipatory ways. Novel ecosystems, similarly, are ecosystems that find a way in degraded landscapes and in the face of rapid environmental change—and *could* help provide a path forward.

All of this is not to say that ecologies of the Anthropocene are fundamentally, or even mostly, "novel" or hopeful. We should expect that many species will be extirpated or go extinct, particularly as climate change continues to transform the planet over the twenty-first century (Bellard et al. 2012) and beyond (Lyon et al. 2022). Some ecosystems will vanish entirely. And the people who depend on threatened species and ecosystems will bear the brunt of these losses. A further threat is the emergence of new rounds of enclosure and exploitation as various actors attempt to tackle the challenges of the Anthropocene. For example, while experimentation will surely be necessary for maintaining human and nonhuman nature, we should be wary of narratives that cast the Anthropocene as an era *ripe for* experimentation, as such narratives tend to originate from centers of socioeconomic power (e.g., technology firms, think tanks) and tend to focus on globalized feats of hubris, such as solar geoengineering or the creation of Half-Earth—a system of "untouched" preserves covering half of the planet (Wilson 2017).

So what are ecologies of the Anthropocene? Well, they may look like biodiverse assemblages of native and nonnative plants, fungi, and animals in brownfield sites next to freeways, on top of reclaimed strip mines, or in melting Arctic landscapes. Or they may look like kelp plantations that are owned by Elon Musk, cover vast swaths of the shallow ocean, and are genetically engineered to rapidly sequester carbon. Our job, as political ecologists, is to tell the difference between scenarios such as these, and to assess the social and biophysical consequences.

The American Chestnut, Once More

Let's turn back to the case that we introduced at the outset of this chapter. What might it look like to *do* a political ecology of the American chestnut? How can we understand this production of nature using some of the theories and concepts discussed here? We offer a few brief ideas.

First, and most broadly, a political ecology of the chestnut might start by recognizing that the restoration of the species is not merely a project of "bringing back" the tree or turning the clock back to pre-blight forests. Instead, its restoration represents the continued co-production of the species and the landscapes of which it is a part. Crucially, this process of co-production is not itself new. Today, the American chestnut is very obviously eco-social: a poster child of the Anthropocene, shaped by dynamics ranging from natural selection to genetic engineering (Brister and Newhouse 2020). But political ecologies of stewardship remind us that we should not assume pre-blight chestnut forests to have existed apart from human interactions and values.

This might prompt us to trace the historical co-production of the American chestnut and chestnut forests. To do so, we might consider using a critical physical geography approach that integrates methods or data from the biophysical sciences with those from the social sciences or humanities. Such an approach could use a variety of biophysical methods/data, such as forest structure and composition data or pollen analysis from sediment or soil cores.

A first step might involve analyzing existing historical ecology research. Biogeographic studies of pollen abundance suggest that the dominance of the chestnut in turn-of-the-twentieth-century forests was most likely a product of Native American burning and cultivation and Euro-American settler land uses (Delcourt and Delcourt 1997). Indeed, the chestnut increased in density and dominance following European conquest and settlement, likely due to trees re-sprouting in cut-over forests (Brugham 1978; Paillet 2002) and settlers planting and tending orchards in the forest (Zon 1904; Ashe 1911; Russell 1987; Lutts 2004).

Still thinking historically, we might next consider how the species' cultural and sociopolitical histories shaped the materiality of the tree and the ecologies of which it is part.

Here, we might reckon with the idea that chestnut restoration shares a genealogy with the eugenics movements, anti-immigration sentiment and policy, and racist and masculinist expressions of American nationalism (Biermann 2016). These conditions crucially shaped how the blight was understood and managed, and as such, remain entangled with efforts to bring back the species.

Turning toward the present, a political ecology of chestnut restoration might also analyze relationships among science and technology, neoliberal capitalism, and nature. The chestnut in its transgenic form is a material product of the molecular turn and an example of the opportunities and risks posed by genetic engineering applications for conservation and restoration. While genetic biotechnologies have been critiqued for deepening neoliberal patterns of commodification and privatization of the environment, some political ecologists have cautiously refuted the inevitability of this outcome (Robbins 2020; Barnes and Delborne 2021). In the case of the transgenic chestnut, researchers have resisted pressure to patent or protect the organism as intellectual property, leading some critical scholars to suggest that in this context genetic technologies might challenge rather than extend logics of neoliberal capitalism (Barnes and Delborne 2021).

As the tree's restoration proceeds, there is a continued need to interrogate the process as it unfolds. Numerous questions remain for political ecologists and fellow travelers from related fields: Who or what is involved in decision making around the deregulation of the transgenic chestnut, and what forms of evidence are used in this decision making? At what sites and in what landscapes have chestnuts been planted, and what are the long-term impacts of such plantings? Is the chestnut fulfilling the promises that restorationists have made regarding its ecological benefits, social impacts, and economic possibilities? These questions signal back to an overarching idea at the core of political ecology and critical studies of nature's production: It is not sufficient to ask *if* nature is produced. Instead, we must not shy away from the complex questions of *how*, *why*, *for whom*, and *with what effects*.

Conclusion

In this chapter we have explored how different natures are materially shaped through interconnected biophysical, social, scientific, technological, and political economic forces. In doing so, however, we engage only a small handful of the broad range of empirical topics that political ecologists have studied as produced natures. To conclude, we provide here a "tasting menu" that represents the wide variety of material landscapes and processes under study:

- Examining soils and agriculture, political ecologists have used mixed social and biophysical methods to consider urban soil contamination (McClintock 2015), soil carbon farming (Brockett et al. 2019), soil degradation and soil fertility dynamics (Turner 2018; Engel-DiMauro 2018), and no-till agriculture (Malone and McClintock 2022), among other topics. These works extend from the long history of research on agricultural landscapes in political ecology (e.g., Blaikie and Brookfield 1987; Bell and Roberts 1991).
- Around water and hydrology-related topics, mixed social-biophysical research has examined stream restoration and floodplain recovery (Wölfle Hazard 2022), dams and dam removal (Van Dyke 2015; Dufour et al. 2017; Fox, Magilligan, and Sneddon 2016), deltas and salt marsh dynamics (Hatvany, Cayer, and Parent 2015; Jensen and Morita 2020), water quality, availability, and justice (Arce-Nazario 2018; Correia

2022); flood management (Gillett et al. 2018; Warner, Vogel, and Hatch 2022), and recession of glaciers (La Frenierre and Mark 2017).
- Regarding the interlinkages among landscape, climate, and livelihoods, the range of topics includes deforestation and forest regrowth (McSweeney et al. 2014; Beitl et al. 2019; Ahmed et al. 2019; Tellman et al. 2020), neoliberal strategies such as payments for ecosystem services, forest carbon offsetting, REDD+, and company-community partnerships (Fent et al. 2019; vonHedemann 2023), historical timber use and trade (Greer et al. 2018; Simon and Peterson 2019), invasive species (Kull 2018), climate-related displacements (Kelley, Shattuck, and Thomas 2022), pests, pathogens, and disease (Biehler et al. 2018; Ferring and Hausermann 2019), and urban climate dynamics (Beray-Armond 2022).

Crucially, knowledge of one produced nature cannot be directly translated to other contexts and materialities. Each empirical object requires distinct literatures, methods, and areas of expertise. We can, however, identify processes that are common across multiple produced natures and thereby serve as unifying themes.

Summarizing the sections above, we offer five broad points related to the co-production of material nature, ideas about nature, and techniques for managing nature. First, many species, landscapes, and processes commonly referred to as natural or wild—from Amazonian rainforests to wildlife migration patterns—are the products of webs of human–nonhuman interaction that stretch across generations. Second, even when we as humans aim to protect nonhuman nature (through conservation initiatives, for example), we are inevitably materially altering the very natures we seek to safeguard. At the same time our own human subjectivity is re-shaped as well. Third, political economic forces associated with neoliberal globalization and neoliberal environmental governance are yielding natures with varying material outcomes—for example, through efforts to establish environmental markets (e.g., for carbon, restoration credits, biodiversity offsets, etc.) Fourth, genetic and genomic tools have enabled molecular-scale interventions that reshape both the material genetic composition of nonhuman taxa and how we understand, know, and classify nonhuman nature. Fifth and finally, as a result of anthropogenic climate change and related processes (e.g., pollution, urbanization, etc.), novel ecosystems are proliferating across our planet. These novel natures, we believe, should neither be embraced nor cast off with a broad brush, but rather investigated for the mixed social and biophysical functions they provide.

References

Adams, W. 2016. Geographies of conservation I: De-extinction and precision conservation. *Progress in Human Geography* 41 (4): 534–545.

Adams, W. 2020. Gene editing for climate: Terraforming and biodiversity. *Scottish Geographical Journal* 136 (1–4): 24–30.

Ahmed, N., M. Islam, M. Hasan, T. Motahar, and M. Sujauddin. 2019. Understanding the political ecology of forced migration and deforestation through a multi-algorithm classification approach: The case of Rohingya displacement in the southeastern border region of Bangladesh. *Geology, Ecology, and Landscapes* 3 (4): 282–294.

Anderson, R., S. Charnley, K. Epstein, K. Gaynor, J. Martin, and A. McInturff. 2022. The socio-ecology of fear: A critical geographical consideration of human–wolf–livestock conflict. *The Canadian Geographer/Le Géographe canadien* 67 (1): 17–34.

Arce-Nazario, J. 2018. The science and politics of water quality. In *The Palgrave handbook of critical physical geography*, ed. R. Lave, C. Biermann, and S. Lane, 465–483. London: Palgrave.

Ashe, W. 1911. *Chestnut in Tennessee*. Nashville, TN: Baird-Ward Printing Company.
Bakker, K. 2010. The limits of "neoliberal natures": Debating green neoliberalism. *Progress in Human Geography* 34 (6): 715–735.
Bakker, K., and G. Bridge. 2006. Material worlds? Resource geographies and the matter of nature. *Progress in Human Geography* 30 (1): 5–27.
Barnes, J., and J. Delborne. 2021. The politics of genetic technoscience for conservation: The case of the blight-resistant American chestnut. *Environment and Planning E: Nature and Space* 5 (3): 1518–1540.
Barron, E. 2015. Situating wild product gathering in a diverse economy: Negotiating ethical interactions with natural resources. In *Making other worlds possible: Performing diverse economies*, ed. G. Roelvink. K. St. Martin, and J. K. Gibson-Graham, 173–193. Minneapolis: University of Minnesota Press.
Beitl, C., P. Rahimzadeh-Bajgiran, M. Bravo, D. Ortega-Pacheco, and K. Bird. 2019. New valuation for defying degradation: Visualizing mangrove forest dynamics and local stewardship with remote sensing in coastal Ecuador. *Geoforum* 98: 123–132.
Bell, M., and N. Roberts. 1991. The political ecology of dambo soil and water resources in Zimbabwe. *Transactions of the Institute of British Geographers* 16 (3): 301–318.
Bellard, C., C. Bertelsmeier, P. Leadley, W. Thuiller, and F. Courchamp. 2012. Impacts of climate change on the future of biodiversity. *Ecology Letters* 15 (4): 365–377.
Benabou, S., and A. Harms. 2021. Revisiting green neoliberalism in India through the lens of market-based restoration and reforestation projects. *Journal of South Asian Development* 16 (3): 327–341.
Beray-Armond, N. 2022. A call for a critical urban climatology: Lessons from critical physical geography. *Wiley Interdisciplinary Reviews: Climate Change* 13 (4): e773.
Biehler, D., J. Baker, J. Pitas, Y. Bode-George, R. Jordan, A. Sorensen, S. Wilson, H. Goodman, M. Saunders, D. Bodner, and P. Leisnham. 2018. Beyond "the mosquito people": The challenges of engaging community for environmental justice in infested urban spaces. In *The Palgrave handbook of critical physical geography*, ed. R. Lave, C. Biermann, and S. Lane, 295–318. London: Palgrave.
Biermann, C. 2016. Securing forests from the scourge of blight: The biopolitics of nature and nation. *Geoforum* 75: 210–219.
Biermann, C., and R. Anderson. 2017. Conservation, biopolitics, and the governance of life and death. *Geography Compass* 11 (10): e12329.
Biermann, C., L. Kelley, and R. Lave. 2020. Putting the Anthropocene into practice: Methodological implications. *Annals of the American Association of Geographers* 111 (3): 808–818.
Biermann, C., and D. Havlick. 2021. Genetics and the question of purity in cutthroat trout restoration. *Restoration Ecology* 29 (8): e13516.
Blaikie, P., and H. Brookfield. 1987. *Land degradation and society*. London: Methuen.
Bluwstein, J. 2018. From colonial fortresses to neoliberal landscapes in Northern Tanzania: A biopolitical ecology of wildlife conservation. *Journal of Political Ecology* 25 (1): 144–168.
Bridge, G. 2000. The social regulation of resource access and environmental impact: Production, nature and contradiction in the US copper industry. *Geoforum* 31 (2): 237–256.
Brister, E., and A. Newhouse. 2020. Not the same old chestnut: Rewilding forests with biotechnology. *Environmental Ethics* 42 (2): 149–167.
Brockett, B., A. Browne, A. Beanleand, M. Whitfield, N. Watson, G. Blackburn, and R. Bardgett. 2019. Guiding carbon farming using interdisciplinary mixed methods mapping. *People and Nature* 1 (2): 191–203.
Brondo, K. 2013. *Land grab: Green neoliberalism, gender, and Garifuna resistance in Honduras*. Tucson, AZ: University of Arizona Press.
Brookfield, H., C. Padoch, H. Parsons, and M. Stocking. 2002. *Cultivating biodiversity: Understanding, analysing and using agricultural diversity*. New York: United Nations Environmental Program, Columbia University Press.
Brugham, R. 1978. Pollen indicators of land-use change in southern Connecticut. *Quaternary Research* 9: 349–362.
Butt, B. 2014. The political ecology of "incursions": Livestock, protected areas and socio-ecological dynamics in the Mara region of Kenya. *Africa* 84 (4): 614–637.
Callon, M., J. Law, and A. Rip. 1986. *Mapping the dynamics of science and technology: Sociology of science in the real world*. London: Palgrave.

Campbell, L., and M. Godfrey. 2010. Geo-political genetics: Claiming the commons through species mapping. *Geoforum* 41 (6): 897–907.
Cavanagh, C. 2018. Political ecologies of biopower: Diversity, debates, and new frontiers of inquiry. *Journal of Political Ecology* 25 (1): 402–425.
Choi, Y. 2014. Modernization, development and underdevelopment: Reclamation of Korean tidal flats, 1950s–2000s. *Ocean & Coastal Management* 102: 426–436.
Cohen, S. 2014. The ethics of de-extinction. *NanoEthics* 8 (2): 165–178.
Collard, R. 2012. Cougar-human entanglements and the biopolitical un/making of safe space. *Environment and Planning D: Society and Space* 30 (1): 23–42.
Colucci, A., J. Tyner, M. Munro-Stasiuk, S. Rice, S. Kimsroy, C. Chhay, and C. Coakley. 2021. Critical physical geography and the study of genocide: Lessons from Cambodia. *Transactions of the Institute of British Geographers* 46 (3): 780–793.
Connors, J., and A. Short Gianotti. 2021. Becoming killable: White-tailed deer management and the production of overabundance in the Blue Hills. *Urban Geography* 44 (10): 2121–2143.
Correia, J. 2022. Between flood and drought: Environmental racism, settler waterscapes, and indigenous water justice in South America's Chaco. *Annals of the Association of American Geographers* 112 (7): 1890–1910.
Crutzen, P. J. 2006. The "Anthropocene." In *Earth system science in the Anthropocene*, eds. E. Ehlers and T. Krafft, 13–18. Berlin, Heidelberg: Springer.
Del Tredici, P. 2020. *Wild urban plants of the northeast: A field guide*. Ithaca: Cornell University Press.
Delcourt, H., and P. Delcourt. 1997. Pre-Columbian Native American use of fire on southern Appalachian landscapes. *Conservation Biology* 11: 1010–1014.
Denevan, W. M. 1992. The pristine myth: The landscape of the Americas in 1492. *Annals of the Association of American Geographers* 82 (3): 369–385.
Doyle, M., J. Singh, R. Lave, and M. Robertson. 2015. The morphology of streams restored for market and nonmarket purposes: Insights from a mixed natural-social science approach. *Water Resources Research* 51 (7): 5603–5622.
Dufour, S., A. Rollet, M. Chapuis, M. Provansal, and R. Capanni. 2017. On the political roles of freshwater science in studying dam and weir removal policies: A critical physical geography approach. *Water Alternatives* 10 (3): 853–869.
Duvall, C., B. Butt, and A. Neely. 2018. The trouble with savanna and other environmental categories, especially in Africa. In *The Palgrave handbook of critical physical geography*, ed. R. Lave, C. Biermann, and S. Lane, 107–127. London: Palgrave.
Engel-DiMauro, S. 2018. Soils in eco-social context: Soil pH and social relations of power in a northern Drava River floodplain agricultural area. In *The Palgrave handbook of critical physical geography*, ed. R. Lave, C. Biermann, and S. Lane, 393–419. London: Palgrave.
Etchart, N., J. Freire, M. Holland, K. Jones, and L. Naughton-Treves. 2020. What happens when the money runs out? Forest outcomes and equity concerns following Ecuador's suspension of conservation payments. *World Development* 136: 105124.
Fairhead, J., and M. Leach. 1995. False forest history, complicit social analysis: Rethinking some West African environmental narratives. *World Development* 23 (6): 1023–1035.
Fent, A., R. Bardou, J. Carney, and K. Cavanaugh. 2019. Transborder political ecology of mangroves in Senegal and The Gambia. *Global Environmental Change* 54: 214–226.
Ferring, D., and H. Hausermann. 2019. The political ecology of landscape change, malaria, and cumulative vulnerability in central Ghana's gold mining country. *Annals of the Association of American Geographers* 109 (4): 1074–1091.
Foucault, M. 2003. *"Society must be defended": Lectures at the Collège de France, 1975–1976*. London: Allen Lane.
Fox, C., F. Magilligan, and C. Sneddon. 2016. "You kill the dam, you are killing a part of me": Dam removal and the environmental politics of river restoration. *Geoforum* 70: 93–104.
Fredriksen, A. 2016. Of wildcats and wild cats: Troubling species-based conservation in the Anthropocene. *Environment and Planning D: Society and Space* 34 (4): 689–705.
Gillett, N., E. Vogel, N. Slovin, and C. Hatch. 2018. Proliferating a new generation of critical physical geographers: Graduate education in UMass's RiverSmart Communities Project. In *The Palgrave handbook of critical physical geography*, ed. R. Lave, C. Biermann, and S. Lane, 515–536. London: Palgrave.
Goedeke, T., and S. Rikoon. 2008. Otters as actors: Scientific controversy, dynamism of networks, and the implications of power in ecological restoration. *Social Studies of Science* 38 (1): 111–132.

Goldman, M. 2018. Circulating wildlife: Capturing the complexity of wildlife movements in the Tarangire ecosystem in Northern Tanzania from a mixed method, multiply situated perspective. In *The Palgrave handbook of critical physical geography*, ed. R. Lave, C. Biermann, and S. Lane, 319–338. London: Palgrave.

Graham, N., J. Cinner, A. Norström, and M. Nyström. 2014. Coral reefs as novel ecosystems: Embracing new futures. *Current Opinion in Environmental Sustainability* 7: 9–14.

Greer, K., K. Hemsworth, A. Csank, and K. Calvert. 2018. Interdisciplinary research on past environments through the lens of historical-critical physical geographies. *Historical Geography* 46 (1): 32–47.

Haraway, D. 1991. *Simians, cyborgs, and women: The reinvention of nature*. New York: Routledge.

Haraway, D. 2007. *When species meet*. Minneapolis, MN: University of Minnesota Press.

Harvey, D. 2003. *The new imperialism*. Oxford, New York: Oxford University Press.

Hatvany, M., D. Cayer, and A. Parent. 2015. Interpreting salt marsh dynamics: Challenging scientific paradigms. *Annals of the Association of American Geographers* 105 (5): 1041–1060.

Havlick, D., and C. Biermann. 2020. Wild, native, or pure: Trout as genetic bodies. *Science, Technology, and Human Values* 46 (6): 1201–1229.

Hennessy, E. 2015. The molecular turn in conservation: Genetics, pristine nature, and the rediscovery of an extinct species of Galápagos giant tortoise. *Annals of the Association of American Geographers* 105 (1): 87–104.

Hennessy, E. 2020. Saving species: The co-evolution of tortoise taxonomy and conservation in the Galápagos Islands. *Environmental History* 25 (2): 263–286.

Hobbs, R., E. Higgs, and J. Harris. 2009. Novel ecosystems: Implications for conservation and restoration. *Trends in Ecology & Evolution* 24 (11): 599–605.

Hobbs, R., E. Higgs, and C. Hall. 2013. Defining novel ecosystems. In *Novel ecosystems: Intervening in the new ecological world order*, ed. R. Hobbs, E. Higgs, and C. Hall, 58–60. London: John Wiley & Sons.

Hodgetts, T. 2016. Wildlife conservation, multiple biopolitics and animal subjectification: Three mammals' tales. *Geoforum* 79: 17–25.

IPCC. 2021. Summary for policymakers. In *Climate change 2021: The physical science basis. Contribution of working group I to the sixth assessment report of the Intergovernmental Panel on Climate Change*, ed. V. Masson-Delmotte, P. Zhai, A. Pirani, S. Connors, C. Péan, S. Berger, N. Caud, Y. Chen, L. Goldfarb, M. Gomis, M. Huang, K. Leitzell, E. Lonnoy, J. Matthews, T. Maycock, T. Waterfield, O. Yelekçi, R. Yu, and B. Zhou, 3–34. Cambridge: Cambridge University Press.

Jensen, C., and A. Morita. 2020. Deltas in crisis: From systems to sophisticated conjunctions. *Sustainability* 12 (4): 1322.

Jessen, T., C. Service, K. Poole, A. Burton, A. Bateman, P. Paquet, and C. Darimont. 2022. Indigenous peoples as sentinels of change in human-wildlife relationships: Conservation status of mountain goats in Kitasoo Xai'xais territory and beyond. *Conservation Science and Practice* 4 (4): e12662.

Jørgensen, D. 2013. Reintroduction and de-extinction. *BioScience* 63 (9): 719–720.

Kelley, L. 2018. The politics of uneven smallholder cacao expansion: A critical physical geography of agricultural transformation in Southeast Sulawesi, Indonesia. *Geoforum* 97: 22–34.

Kelley, L., A. Shattuck, and K. Thomas. 2022. Cumulative socionatural displacements: Reconceptualizing climate displacements in a world already on the move. *Annals of the American Association of Geographers* 112 (3): 664–673.

Kiggell, T. 2021. Monitoring extinction: Defaunation, technology and the biopolitics of conservation in the Atlantic Forest, Brazil. *Journal of Political Ecology* 28 (1): 845–863.

King, L., and M. Tadaki. 2018. A framework for understanding the politics of science (core tenet # 2). In *The Palgrave handbook of critical physical geography*, ed. R. Lave, C. Biermann, and S. Lane, 67–88. London: Palgrave.

Kull, C. 2018. Critical invasion science: Weeds, pests, and aliens. In *The Palgrave handbook of critical physical geography*, ed. R. Lave, C. Biermann, and S. Lane, 249–272. London: Palgrave.

La Frenierre, J., and B. Mark. 2017. Detecting patterns of climate change at Volcán Chimborazo, Ecuador, by integrating instrumental data, public observations, and glacier change analysis. *Annals of the American Association of Geographers* 107 (4): 979–997.

Latour, B. 2005. *Reassembling the social: An introduction to actor-network-theory*. Oxford: Oxford University Press.

Lave, R., M. Wilson, E. Barron, C. Biermann, M. Carey, C. Duvall, L. Johnson, K. Lane, N. McClintock, D. Munroe, and R. Pain. 2014. Intervention: Critical physical geography. *The Canadian Geographer/Le Géographe canadien* 58 (1): 1–10.

Lave, R., C. Biermann, and S. Lane. 2018a. *The Palgrave handbook of critical physical geography*. London: Palgrave.

Lave, R., M. Doyle, M. Robertson, and J. Singh. 2018b. Commodifying streams: A CPG approach to stream mitigation banking in the US. In *The Palgrave handbook of critical physical geography*, ed. R. Lave, C. Biermann, and S. Lane, 443–464. London: Palgrave.

Law, J. 2018. The impacts of doing environmental research (core tenet #3). In *The Palgrave handbook of critical physical geography*, ed. R. Lave, C. Biermann, and S. Lane, 89–103. London: Palgrave.

Law, J. 2022. The knotty politics of ginseng conservation and management in Appalachia. *Journal of Political Ecology* 29 (1): 36–50.

Law, J., and K. McSweeney. 2013. Looking under the canopy: Rural smallholders and forest recovery in Appalachian Ohio. *Geoforum* 44: 182–192.

Leach, M., and R. Mearns. 1996. *The lie of the land: Challenging received wisdom on the African environment*. Portsmouth, NH: Heinemann.

Levis, C., F. Costa, F. Bongers, M. Peña-Claros, C. Clement, A. Junqueira, E. Neves, E. Tamanaha, F. Figueiredo, R. Salomão, C. Castilho, W. Magnusson, O. Phillips, J. Guevara Andino, D. Sabatier, J.-F. Molino, D. López, A. Mendoza, N. Pitman, and H. ter Steege. 2017. Persistent effects of pre-Columbian plant domestication on Amazonian forest composition. *Science* 355: 925–931.

Lorimer, J. 2015. *Wildlife in the Anthropocene: Conservation after nature*. Minneapolis: University of Minnesota Press.

Lorimer, J., and C. Driessen. 2013. Bovine biopolitics and the promise of monsters in the rewilding of Heck cattle. *Geoforum* 48: 249–259.

Lutts, R. 2004. Like manna from God: The American chestnut trade in southwestern Virginia. *Journal of Environmental History* 9 (3): 497–525.

Lyon, C., E. Saupe, C. Smith, D. Hill, A. Beckerman, L. Stringer, R. Marchant, J. McKay, A. Burke, and P. O'Higgins. 2022. Climate change research and action must look beyond 2100. *Global Change Biology* 28 (2): 349–361.

Malone, M., and N. McClintock. 2022. A critical physical geography of no-till agriculture: Linking degraded environmental quality to conservation policies in an Oregon watershed. *The Canadian Geographer/Le Géographe canadien* 67 (1): 74–91.

Mansfield, B., C. Biermann, K. McSweeney, J. Law, C. Gallemore, L. Horner, and D. K. Munroe. 2015. Environmental politics after nature: Conflicting socioecological futures. *Annals of the Association of American Geographers* 105 (2): 284–293.

Margulies, J. 2019. Making the 'man-eater': Tiger conservation as necropolitics. *Political Geography* 69: 150–161.

Marris, E. 2011. *Rambunctious garden: Saving nature in a post-wild world*. New York: Bloomsbury Publishing USA.

Mascaro, J., R. Hughes, and S. Schnitzer. 2012. Novel forests maintain ecosystem processes after the decline of native tree species. *Ecological Monographs* 82 (2): 221–228.

McAfee, K. 1999. Selling nature to save it? Biodiversity and green developmentalism. *Environment and Planning D: Society and Space* 17 (2): 133–154.

McAfee, K. 2003. Neoliberalism on the molecular scale: Economic and genetic reductionism in biotechnology battles. *Geoforum* 34 (2): 203–219.

McCarthy, J., and S. Prudham. 2004. Neoliberal nature and the nature of neoliberalism. *Geoforum* 35 (3): 275–283.

McClintock, N. 2015. A critical physical geography of urban soil contamination. *Geoforum* 65: 69–85.

McSweeney, K., E. Nielsen, M. Taylor, D. Wrathall, Z. Pearson, O. Wang, and S. Plumb. 2014. Drug policy as conservation policy: Narco-deforestation. *Science* 343 (6170): 489–490.

Moore, J. W. 2017. The Capitalocene, Part I: On the nature and origins of our ecological crisis. *The Journal of Peasant Studies* 44 (3): 594–630.

Neumann, R. 2002. *Imposing wilderness: Struggles over livelihood and nature preservation in Africa.* Berkeley: University of California Press.

Newhouse, A., and W. Powell. 2021. Intentional introgression of a blight tolerance transgene to rescue the remnant population of American chestnut. *Conservation Science and Practice* 3: e348.

Paillet, F. 2002. Chestnut: History and ecology of a transformed species. *Journal of Biogeography* 29 (10–11): 1517–1530.

Pearce, F. 2016. *The new wild: Why invasive species will be nature's salvation.* Boston: Beacon Press.

Peltola, T., and J. Heikkilä. 2018. Outlaws or protected? DNA, hybrids, and biopolitics in a Finnish wolf-poaching case. *Society & Animals* 26 (2): 197–216.

Perkins, H. 2020. Killing one trout to save another: A hegemonic political ecology with its biopolitical basis in Yellowstone's native fish conservation plan. *Annals of the American Association of Geographers* 110 (5): 1559–1576.

Redford, K., and W. Adams. 2021. *Strange natures: Conservation in the era of synthetic biology.* New Haven: Yale University Press.

Rerkasem, K., D. Lawrence, C. Padoch, D. Schmidt-Vogt, A. Ziegler, and T. Bruun. 2009. Consequences of swidden transitions for crop and fallow biodiversity in Southeast Asia. *Human Ecology* 37 (3): 347–360.

Robbins, P. 2012. *Political ecology: A critical introduction.* Malden, MA: John Wiley & Sons.

Robbins, P. 2020. Is less more... or is more less? Scaling the political ecologies of the future. *Political Geography* 76: 102018.

Rossi, J. 2013. The socionatural engineering of reductionist metaphors: A political ecology of synthetic biology. *Environment and Planning A* 45 (5): 1127–1143.

Rossi, J. 2017. Synthetic biology: A political economy of molecular futures. In *The Routledge handbook of the political economy of science*, ed. D. Tyfield, R. Lave, S. Randalls, and C. Thorpe, 289–302. London and New York: Routledge.

Russell, E. 1987. Pre-blight distribution of *Castanea dentata* (Marsh.) Borkh. *Bulletin of the Torrey Botanical Club* 114 (2): 183–190.

Sack, C. 2013. Landscape architecture and novel ecosystems: Ecological restoration in an expanded field. *Ecological Processes* 2 (1): 35.

Sengupta, D., R. Chen, M. Meadows, Y. R. Choi, A. Banerjee, and X. Zilong. 2019. Mapping trajectories of coastal land reclamation in nine deltaic megacities using Google Earth Engine. *Remote Sensing* 11 (22): 2621.

Simaika, J., M. Samways, and P. Frenzel. 2016. Artificial ponds increase local dragonfly diversity in a global biodiversity hotspot. *Biodiversity and Conservation* 25 (10): 19211935.

Simberloff, D., C. Murcia, and J. Aronson. 2015. "Novel ecosystems" are a Trojan horse for conservation. https://ensia.com/voices/novel-ecosystems-are-a-trojan-horse-for-conservation/

Simon, G., and C. Peterson. 2019. Disingenuous forests: A historical political ecology of fuelwood collection in South India. *Journal of Historical Geography* 63: 34–47.

Srinivasan, K. 2017. Conservation biopolitics and the sustainability episteme. *Environment and Planning A* 49 (7): 1458–1476.

Tellman, B., S. Sesnie, N. Magliocca, E. Nielsen, J. Devine, K. McSweeney, M. Jain, D. Wrathall, A. Dávila, K. Benessaiah, and B. Aguilar-Gonzalez. 2020. Illicit drivers of land use change: Narcotrafficking and forest loss in Central America. *Global Environmental Change* 63: 102092.

Tuijnman, W., M. M. Bayrak, P. X. Hung, and B. D. Tinh. 2020. Payments for environmental services, gendered livelihoods and forest management in Vietnam: A feminist political ecology perspective. *Journal of Political Ecology* 27 (1).

Turhan, E., and A. Gündoğan. 2019. Price and prejudice: The politics of carbon market establishment in Turkey. *Turkish Studies* 20 (4): 512–540.

Turner, M. D. 2018. Questions of imbalance: Agronomic science and sustainability assessment in dryland West Africa. In *The Palgrave handbook of critical physical geography* ed. R. Lave, C. Biermann, and S. Lane, 421–441. London: Palgrave.

Urban, M. 2018. In defense of crappy landscapes (core tenet # 1). In *The Palgrave handbook of critical physical geography*, ed. R. Lave, C. Biermann, and S. Lane, 49–66. London: Palgrave.

Valve, H. 2011. GM trees on trial in a field: Reductionism, risks and intractable biological objects. *Geoforum*, 42 (2): 222–230.

Van Dyke, C. 2015. Boxing daze: Using state-and-transition models to explore the evolution of socio-biophysical landscapes. *Progress in Physical Geography* 39 (5): 594–621.

vonHedemann, N. 2023. The importance of communal forests in carbon storage: Using and destabilizing carbon measurement in understanding Guatemala's payments for ecosystem services. *The Canadian Geographer/Le Géographe canadien* 67 (1): 106–123.

Warner, B., E. Vogel, and C. Hatch. 2022. Exactly where does the river need space to move? Seeking participatory translation of fluvial geomorphology into flood management. *Journal of the American Water Resources Association* 58 (6): 1454–1469.

Wilson, E. O. 2017. *Half-earth: Our planet's fight for life.* New York: W.W. Norton & Company.

Wölfle Hazard, C. 2022. *Underflows: Queer trans ecologies and river justice.* Seattle: University of Washington Press.

Wynne-Jones, S., C. Clancy, G. Holmes, K. O'Mahony, and K. Ward. 2020. Feral political ecologies? *Conservation & Society* 18 (2): 71–76.

Zon, R. 1904. *Chestnut in Southern Maryland, US Department of Agriculture Bureau of Forestry Bulletin 53*. Washington, DC: Government Printing Office.

3 Governing Nature

Political Ecology and the Knotty Tangle of Environmental Governance

Alice Cohen and Leila M. Harris

Water, Water, Everywhere? Political Ecology and the Knotty Tangle of Governing Natures

In the months leading up to the 2015 Canadian federal election, leadership hopeful Justin Trudeau proudly declared that providing clean water to Indigenous communities was a cornerstone of his electoral platform—promising clean drinking water on all reserves within five years. In a town hall event during the campaign, he stated, "We have 93 different communities under 133 different boil-water advisories ... A Canadian government led by me will address this as a top priority because it's not right in a country like Canada. This has gone on for far too long" (Canadian Broadcasting Corporation 2015). Trudeau was ultimately elected, but improvement on these issues remains slow. As of Fall 2021, 45 advisories remain (Government of Canada 2021).

The current situation arises from a knotty tangle of issues, including: historic and ongoing colonialism including forced displacement with the creation of the reserve system and chronic underfunding of services, governance processes that maintain an unhelpful binary between drinking and "other" water, and decades of decentralization of water institutions. That some first nation reserves face an ongoing water crisis is a direct result of differential governance—Indigenous communities are a federal responsibility under the Canadian constitution, yet drinking water provision falls under the "health" umbrella, which is a provincial responsibility. As such, there is a divide where reserves are subject to federal government intervention and oversight, while other Canadians enjoy water provided by provincial and municipal governments. Tackling this problem can be challenging. Infrastructurally, many communities have out-of-date treatment equipment or inadequate piping. In terms of human resources, there can be a shortage of certified plant operators, and trained individuals are often recruited by industry for jobs elsewhere. Environmental contamination and extractive practices such as forestry and mining also make water more difficult to treat due to the presence of sediment, heavy metals, or other contaminants.

In surveying these challenges, an engineer might look at technical or treatment solutions; governments might look at questions around training; environmental scientists might look at contaminant levels; political scientists or legal scholars might study the constitution and other regulatory frameworks. Adding to this, political ecologists might ask: What are the historical and governance roots of this challenge, and how do drinking water concerns fit with other key inequalities in contemporary Canada? What political, economic, historic, cultural, ecological, or scalar issues play a role in this situation and possible remedies? What can we learn from cases such as this about the complex challenge of governing water, forests, fisheries, or other resources?

DOI: 10.4324/9781003165477-5

In this brief chapter, we will address a range of complex issues regarding formal and informal institutions, socio-cultural dynamics, as well as broader political and cultural considerations important for understanding how decisions are made about diverse "natures"—in a word, we will address the subject of "governance." While we accent water challenges and the situation in Canada, the themes apply to other locales, sectors, and environmental conditions.

In what follows, we first unpack what it means to understand governance through a political ecology (PE) lens. As part of this discussion, we identify and explore the meaning of the term "governance," and how a political ecological lens offers important insights for understanding relationships between people, place, and varied "natures," notably through focus on issues of power, scale, and inequity. We then turn to governance institutions, from formal to informal, to highlight the myriad ways the governance landscape is imbued with socio-cultural norms, knowledge politics, and power relations that can be revealed through political ecological analyses. Turning then to issues of scale—including the spatial dynamics of governance—we examine recent shifts from centralized to decentralized and local governance, processes that have been further fueled by attention to local and global contexts as well as by neoliberalism and similar processes. Following this, and building on the discussions on power and scale, we delve into key inequities and exclusions in resource governance (e.g., gender and Indigeneity), as well as the role of ontological and epistemological assumptions (that is, different worldings, as well as varied ways of knowing) that underpin various governance imperatives. Significant manifestation of these power dynamics includes the centering of science in environmental decision making, and the types of experts and knowledges that are privileged through those processes. In particular, we explore the ways in which units of environmental governance are defined and managed in ways that reflect particular—often techno-scientific, and/or colonial—worldviews. We conclude by suggesting that political ecology analyses are not only critical to understanding these issues, but indeed that such an understanding is foundational for meaningful change towards more equitable and sustainable nature–society relations over the long term.

Theorizing Political Ecologies of Resource Governance: Foregrounding Power and Scale

What does it mean to understand governance through a political ecology lens? A shorthand to understand political ecology is the foregrounding of "politics" as key to any aspect of "ecology" (see Introduction, this volume). A decade ago, Paul Robbins (2012) offered that political ecology is the opposite of *a*political ecology. That is, that PE is characterized by "the difference between identifying broader systems rather than blaming proximate and local forces; between viewing ecological systems as power-laden rather than politically inert; and between taking an explicitly normative approach rather than one that claims the objectivity of disinterest" (p. 13). At the core, *apological* ecology—or ecology without politics—sees ecosystem functions and outcomes as existing *apart* from political and economic systems, rather than as *a part* of them. In the world of water governance, for example, "apolitical ecology" appears in many forms (e.g., techno-managerial approaches), and "doing political ecology" often involves pulling back the curtain to expose and explore connections between water bodies, aquatic ecosystems, and social, political and economic dynamics across scales.

Understanding ecosystem functions and outcomes as *a part* of political processes opens new ways of understanding environmental challenges. We focus below on broad definitions of governance, as well as on three connected concepts—power, scale, and inequality—to explore how politics is fundamentally part and parcel of environmental decision making and outcomes. Indeed, themes of power, scale, and inequality are hallmarks of the political ecology approach (Bryant 1992). Attending to power dynamics, it is common to analyze institutions, individuals, events, and places as embedded in historical, contemporary, and place-based inequalities. Political ecology analysis often also includes focus on participation and decision making, asking questions such as: who is at the table and who is not, or, more fundamentally, who built the table? Taking climate governance as an example, the UNFCCC framework is a platform for nation states, with Indigenous groups reduced to "non state observers" despite a growing consensus that Indigenous Peoples' knowledges and rights should be central to climate governance (Belfer et al. 2019).

The political ecology of "governing nature" hinges on nuanced and varied understandings of "governance" generally defined as "the set of regulatory processes, mechanisms and organizations through which political actors influence environmental actions and outcomes" (Lemos and Agrawal 2006, 298). This includes government (i.e., formal rules and regulations), but also the broad range of actors and issues that affect how decisions are made about the use and management of "natures." As such, governance necessarily includes attention to non-governmental organizations (NGOs), industry, communities, diverse publics, or others working to mediate the relationships between people, place, and natures. With this broad framing, examining forest or water governance entails focus on formal government and institutions (e.g., municipalities, regulations, legal frameworks), as well as other political, economic, cultural, or historical considerations that affect what uses and management practices entail (e.g., cultural practices or values, see section below on informal norms and related considerations).

Political ecology also learns from the concept of "governmentality" first theorized by Michel Foucault. Foucault argued that power is diffuse; that it does not only come from the Government or in the form of repressive power often associated with state power, but rather, that it is infused into practices that manifest in less overt ways. To that end, many PE studies have shifted focus away from the centers of power, to the diverse ways that infrastructures and populations are managed, and how resources work to help shape important everyday power dynamics—including how people "manage themselves" as key to the governmentality approach (e.g., work on water and sanitation in Maputo, Mozambique by Biza et al. 2021).

Work on political ecology of resources and state theory have also highlighted how governing natures is often central to state- and nation-building projects, including consolidation of power in particular institutions (Bridge 2014). Among other examples, work on dam removal in the US has revealed the ways that states exert power in relation to nature and society in strategic and contradictory ways (Sneddon et al. 2022). As well, work on infrastructure projects in Mozambique has shown how these efforts are central to different stages of state-building (Menga et al. 2019), while other examples show the consolidation of power in key state institutions—e.g., ways that water scarcity helped to justify and consolidate power and territory in newly established Israel (Alatout 2008). Political ecologies of governing nature also necessarily entail consideration of key inequities, such as questions around who is able to participate in decisions (e.g., public engagement or consultation), or how different populations might be differentially affected by key environmental and resource uses or practices (e.g., inundation of culturally valuable lands with dam building, or gender differentiated effects of livelihood and landscape shifts). Linked with inequities, we can also consider a number of varied cultural, ideological, or economic considerations, such as what types of forests are considered valuable or mappable, spiritual relationships to resources, or trade and economic growth imperatives that might drive pollution or resource exploitation.

Indeed, political ecology analyses often emphasize and question the causes and effects of pairing particular ecological phenomena with specific levels of governance, or by insisting that any treatment of the "local" is necessarily embedded and situated in relation to multiple broader scales of analysis which might impinge on those outcomes (e.g., trade relations, broader governance entities, etc.). As one clear example, because governments operate at specific scales (i.e, municipal, state/provincial, national) that are almost always distinct from ecological ones (i.e., migratory paths, watersheds, forest regions), the question of how to govern ecological phenomena is often at the core a question about scale. The "fit" between the scale of a problem and the scale of its solution thus emerged as a prominent feature in the environmental conversations of the early 2000s (Cumming et al. 2006; Moss 2004). From Blaikie's (1985) early contributions on soil erosion onwards, there has been a keen interest in embedding observed "local" patterns and outcomes in broader dynamics and circuits—whether they be international prices for a particular crop, connecting deforestation in a particular locale to policies that govern the international timber trade (e.g., certification schemes), or trade networks whereby forests are logged in certain regions to serve consumption in faraway markets where there is a demand for palm oil or wood products (e.g., Dauvergne 1997). Again, political ecology approaches tend to understand these diverse and embedded scales not as pre-given, but as dynamically linked with politics.

As summarized by Brown and Purcell (2005), work in political ecology had previously too often "treat(ed) scale as pregiven and inherent rather than socially produced through political struggle" (p. 612). In addition to treating scales increasingly as emergent rather than given, debates regarding the most appropriate scales for governance of particular issues is ongoing. For instance, some suggest that local governance is likely the most effective, while others have emphasized the importance of polycentricity—the idea that governance is most likely to be effective when local institutions and responses are nested in, and supported by, broader scalar institutions and processes (Pahl-Wostl et al. 2012). Discussions of scale, and the social and political production of scale, have remained key themes of the past several decades. Among other concerns, the emergence of global change science and similar understandings demonstrated the need to engage with dynamics and processes at the global or planetary scale (Taylor and Buttel 1992; Jasanoff and Martello 2004). As Dietz et al. (2003) argued, such phenomena are caused by translocal and transboundary factors and can therefore only be resolved at international or global scales. Similar ideas apply to ocean pollution or climate change, which are problems that emerge in part due to the nature of the global commons—shared spaces (e.g., oceans or the atmosphere) over which no governance entity has exclusive control (e.g., Steinberg 2001). Note here that collective solutions need not take place at an inter-*national* level; international networks of interconnected local actors can also be useful here, as Harriet Bulkeley (2010) and Jennifer Rice (2014) point out in work on cities' responses to climate change. The simultaneous pull towards "global" and "local" has been coined "glocalization" by geographer Erik Swyngedouw, who describes it as "(1) the contested restructuring of the institutional level from the national scale both upwards to supra-national or global scales and downwards to the scale of the individual body or the local, urban or regional configurations and (2) the strategies of global localisation of key forms of industrial, service and financial capital" (Swyngedouw 2004, 37). Undoubtedly shifts and tensions across scale will remain a key theme of political ecologies of governance moving forward.

Governance: From Formal Institutions, Regulations, and Policies to Informal Norms and Dynamics

Formal governance structures and institutions often provide a key starting point for PE analyses building on insights from institutional economists, legal scholars, and others. Consider for instance, research on the power dynamics and inequalities associated with the human right to water and United Nations (UN) frameworks (Sultana and Loftus 2013), or institutional analyses that open up the workings behind key institutions such as the World Bank (see Bakker 2013, which investigates the types of water projects funded by the World Bank during different historical periods and leadership), and recent land claim agreements in the settler colonial context of Canada (Wilson 2019).

Beyond the water realm, other examples of work centered on formal institutions and policies might include attention to multilateral agreements and international governance institutions (e.g., the Convention on International Trade in Endangered Species for biodiversity, or the Intergovernmental Panel on Climate Change for climate change; e.g., Goldman et al. 2018), specific policies and regulations (e.g., wetlands policy in the US or as part of African conservation efforts, e.g., Robertson 2000; King et al. 2018), or critical evaluations of carbon pricing schemes or payments for ecosystem services (McElwee 2012; Chalifour 2010). Recent work by Collard, Dempsey, and Holmberg (2020) offers an excellent example of the power of analyses of such formal governance frameworks. As they show, while the stated purpose of environmental assessments is to mitigate socio-ecological concerns, the ways that approvals work in practice means that very few projects are prevented from moving forward, even in cases that clearly negatively affect First Nations and habitat of caribou as a key "species at risk" ostensibly protected by other regulations in the context of Canada. Here we see failure of formal legislation and stated goals of habitat and species protection, calling attention to the *why* of these gaps—in this example, highlighting economic imperatives, corporate influence, and broad notions of "public or national interest."

While some of the more obvious aspects of governance relate to formal laws, politics, regulatory frameworks, and institutions, political ecology also necessarily involves consideration of informal norms, culture, livelihoods, and historical trajectories that affect governance beyond what is written, visible, or generally acknowledged. Among early contributions to these studies of formal and informal networks were studies on common property resources. Here, institutional theory provided clear blueprints for the need to examine *informal norms and institutions*—including community-level norms or monitoring practices that might be important to prevent overfishing, for instance (e.g., Ostrom 1990). More recently, Liboiron's (2021) work, *Pollution is Colonialism*, highlights the ways in which particular assumptions about science and land access within research institutions have facilitated environmental destruction and marginalization of Indigenous knowledges and governance practices. This trajectory, and associated degradation of ecosystems and governance practices, is made possible by colonial assumptions about access to land and about what constitutes "good science.' As another example, the concept of bricolage highlights the ways that actors cobble together practices and knowledges depending on their context to manage and use resources, creating hybrid arrangements (Cleaver et al. 2013). As such, PE analysis often goes beyond the formal rules of who can participate in decisions of community managed conservation to also include the subtle and unspoken dynamics of how gender, caste, class, or other exclusions might work to allow some to be involved meaningfully, while excluding others (examples include exclusions of black farmers in water governance in the new democratic South Africa—Goldin 2010; or gendered dynamics of institutions involved with payment for ecosystem services and REDD+—Bee 2017).

Other examples bridge formal and everyday spaces and engagements. As examples, critical feminist scholars have highlighted connections between global and local climate justice (Sultana 2021), while other works have sought to bring local situated knowledge into conversation with international perspectives, as Chakraborty and Sherpa (2021) do in their research addressing how climate knowledge is produced in the Nepalese Himalaya. In this last example, the authors summarize many of the points we touch on here in their description of knowledge-framing in writing that "the plurality of Himalayan voices is marginalized both by the political aspirations of the nation-state which fails to represent the aspirations and experiences of its diverse citizens, especially when they threaten its developmental and geopolitical goals, and by the hegemonic narrative of the Anthropocene that erases historical inequities and injustices and replaces them with an overarching "risk society" which afflicts all individuals and communities and can only be resolved through expert-driven initiatives" (Chakraborty and Sherpa 2021, 3). Still other PE works attempt to weave household and individual political economic considerations into understanding ecological changes in a broader regional context, focusing centrally on livelihoods, as is the case in King's work on the Okovango (King et al. 2018) or Budds' work in the La Ligua River Basin in Chile (2016). Such examples help to ascertain the interplay of institutions and multi-level governance (from international financial institutions, to formal state institutions, to everyday lived realities and livelihood dynamics at the household scale). Taken together, the interplay between these institutions and spaces are all likely to be important for conditioning environmental changes, or effects for diverse communities, locales, and individuals.

The complex interplay between formal and informal rules is also well demonstrated through attention to the issue of property ownership. Beyond legal frameworks, there are ideological and cultural dynamics important for notions of property—for instance, a shared understanding that property owners hold decision-making power over the land that they own that at times exceeds formal recognition and practices (see special issue, Ranganathan and Bonds 2022). Inquiry into issues such as water pollution or endangered species from a political ecology perspective thus might involve evaluation of historical land stewardship, ideals and norms of land ownership and associated ideological underpinnings, or historical legacies that might unpack the role of racism, sexism, or discourses of individual rights in creating or exacerbating dynamics of interest (see also Ranganathan 2022 for an example of environmental logics that underpin caste-based slum evictions in India, Jacobs 2003 or McCarthy 2002 for property and land rights movements in the United States, or Deere and Léon 2002 for analysis of collective versus individual land rights in Latin America). In such examples, formal institutions and legal frameworks are clearly important in terms of consolidating property relations and particular types of landscapes and their associated risks, but so are informal expectations and norms related to racialized patterns of suburbanization (Pulido 2000), liberal notions of individual freedom (Ranganathan 2022), or heteronormative ideals around nuclear families (cf. Robbins 2007). As noted above, governmentality has also been a key theme in PE studies that highlights diffuse operations and effects of power (see special issue on "environmentality," Fletcher and Cortes-Vazquez 2020), or examples of how water and sanitation infrastructures produce and naturalize key social and spatial differences (along racial lines in the colonial context of Maputo, see Biza et al. 2021).

Local Scales, Neoliberalism, and Shifting Political Economies of Governance

Significant PE attention has also been paid to critiques of neoliberalism. Often emphasizing deregulation, privatization, decentralization, and market approaches alongside shifting state practices, the neoliberalization of nature has been a central theme in political ecology studies (e.g., Batterbury and Fernando 2006; Castree 2008; Fletcher 2010; Larson and Soto 2008; Perreault 2005; Penning-Rowsell and Johnson 2015). As such, much has been written about the merits and drawbacks of payments for ecosystem services (e.g., Dempsey and Robertson 2010; Kull et al. 2015), privatization of water services (Bakker 2013), or market-based solutions like avoided deforestation schemes (see Heynen et al. 2007 for a range of examples).

The implications of shifts towards neoliberalism have been profound, from dismantling of regulatory structures, to the devolution of governance to reveal a patchwork approach. Among other things, these practices provide new frontiers and opportunities for extraction and profit (e.g., markets for carbon credits (Bumpus 2011)), and new frontiers for the involvement of market trading or global corporations (e.g., Bauer 1997). Given the toolkit of political ecologists, the lens of neoliberal critique has led to research exploring the policy implications of market-based tools, and to ideological and discursive analysis regarding ideas that underpin these shifts (e.g., claims of bloated governance, market efficiency, or individual liberties—recall the example of property rights above). Works that intersect with theories of the state have also attended to the ways that state institutions and practices condition particular uses and management of "natures" (Loftus 2018). As well, works attentive to green governmentality have highlighted shifting subjectivities and power dynamics, for instance, as ways that new instituions and scalar dynamics affect senses of the self or risk associated with certain operations of power (Clarke-Sather 2017; see Fletcher and Cortes-Vazquez 2020 for an overview of recent work).

A PE focus on neoliberalism loops back into the scalar discussion. Consider for instance the movement from centralized to decentralized governance, where downscaling responsibilities to lower levels of government is often seen as validating local knowledge and enabling cost saving. Yet, at once, the off-loading of responsibility can also result in uneven governance frameworks that reward those jurisdictions with more financial and human resources, or perhaps might exacerbate environmental justice challenges (e.g., creating hot zones for pollution). Political ecology has thus often provided key insights to reveal these complex and nuanced shifts and the varied effects they are having on the ground in diverse sites (given the social, ecological, and equity implications). These processes and associated effects are often enacted through inequality and exclusion, which we discuss next.

Inequalities and Exclusions in Decision Making and Lived Experiences

In many of the examples noted, we see that PE emphasizes power dynamics and exclusions. As these instances make clear, a key consideration relates to socio-political dynamics around who is able to participate in deliberations, whose voices are heard, whose knowledge counts as valid, and whose knowledge and preferences are left aside (often as linked with gender, race, or other key axes of difference and inequality). As one example, analyzing water user groups in Egypt, anthropologist Jessica Barnes (2014) showed that women were formally included in institutions, but often were relegated to only speak about domestic water needs due to gendered expectations (and not irrigation issues, productive uses, or water concerns more generally). On forest issues in India or Nepal, Agarwal (2001) and Nightingale (2011) have highlighted the ways that gender, caste, or class affect who is able to be involved in the management and governance of forests. Highlighting the lived experiences of resource use, Heather O'Leary's (2019) work on class and caste in urban Delhi is another compelling illustration of how household level water uses flow from gendered and caste-based experiences (and expectations) of hygiene and class mobility.

Such detailed ethnographic and household level work has been a hallmark of many political ecological and allied anthropological and geographic works, highlighting the everyday lived effects and experiences of resource use and governance. Often, political ecology invites us to consider the lived experience of these exclusions and decision-making processes, including how it might feel to be excluded, or have one's knowledge discounted. Such examples convey the need to speak to these diverse, interconnected, and often hidden dimensions, including the emotional and affective dimensions of resource issues and inequalities (Sultana 2015). All in all, political ecology work on governance is most compelling when it speaks the knotty tangle of issues always at once political, economic, social, cultural, and ecological—weaving together narratives that simultaneously attend to power, inequality, and scale, as well discursive, material, ecological, infrastructural, emotional, or biophysical dimensions that make such issues all the more complex or important (see Bakker and Bridge 2006 on the material and biophysical dimensions).

Box 3.1 Dimensions of power in resource governance

Ahlborg and Nightingale (2018) draw on the case of an electrical project in Tanzania to identify four sites of power in resource governance projects. First is *knowledge*—the prioritization of some epistemologies and ontologies over others. A key point here is that when it comes to understanding how the non-human world works, some ways of knowing and being in the world—notably Western science—are often accorded more power and salience than others. A second site of power includes the *things* that support power hierarchies—structures such as decision-making mechanisms, and physical things including roads, electrical lines, or landscapes. The third site of power identified is *access and entitlements*—that is, the modes of access to benefits derived from the resources or infrastructures at play. Who will access and benefit from the electricity, clean water, new road, or timber? And, we would add, who will get the flooding, or the pulp and paper effluent and pollution? A fourth and final site of power involves *everyday lives*: in Ahlborg and Nightingale's case, how does electrification change the ways that people live, work, or go to school?

If we were to apply their approach to our introductory case of contaminated drinking water in Indigenous communities, we might ask: how does a lack of potable water affect household budgets or livelihoods? Who is spending time boiling or treating water, or travelling to refill bottles? Or, what are the implications of differential and unsafe water aspects for senses of self, belonging, political subjectivities, or senses of trust or governmental legitimacy (cf. Harris 2021; Wilson et al. 2022).

Centering on Science, Knowledge, and Ontologies

Recent work informed by critical Indigenous studies, critical socio-natures, and science and technology studies has asked questions such as "what is nature?" or "what is water?" and what are the scientific practices and knowledge formations that enable such understandings? These foundational ontological and epistemological questions are important to foreground. Writing about water, Jamie Linton's (2010) contribution calls to question the "hydrologic cycle" that is assumed to be similar the world over—instead, it is a model from temperate zones that doesn't necessarily apply to other geographic areas. What water is, and how it is understood (e.g., as molecular compound and substitutable H_2O or a complex set of relationships that vary geographically) can hold tremendous importance for water's condition, use, and management (see also Yates et al. 2017; Wilson and Inkster 2018; Venot and Jensen 2021). Addressing similar and fundamental ontological issues in forestry as well, Robbins (2001) famously asked scientists and villagers in India "what is a forest?"—highlighting key disjunctures in terms of what species or types of trees are viewed as central to valuing and using forests.

Learning from Indigenous governance and legal frameworks, other fundamental contributions and lines of inquiry have emerged. Among them, Caleb Behn, an Eh-Cho Dene and Dunne Za/Cree lawyer and thinker (Behn and Bakker 2019), has suggested that we might usefully reframe watersheds in terms of kin relationships: glaciers are grandparents, large rivers as parents, and the smaller tributaries as children. How might such a perspective alter the legal and scientific frameworks used to govern water? This is not a rhetorical question. Consider the example introduced at the beginning of this chapter. Canada's First Nations drinking water crisis engenders several assumptions. The first is about the country of Canada, which, before 1867, was a patchwork of European colonies superimposed on lands that had been violently emptied of some of its occupants. In this context it is unsurprising that the spaces to which displaced Indigenous people were forcibly relocated, were also at times among the least hospitable and difficult to access. A second assumption here surrounds "clean drinking water." Water is constantly circulating. Separating drinking water from other waters (source waters, aquifers, etc...) is an artificial separation that—intentional or not—privileges "end of pipe" treatment solutions over pollution reduction and broader ecosystem health (Yates et al. 2017). As well, water access and quality for many Indigenous communities involves a complex set of practices fundamentally tied to the territory (Wilson et al. 2019). These insights are particularly fruitful to reframe the issues at stake. Rather than water as a resource, how does water as a relative or living entity, or as lifeblood, to which we have responsibilities and should proceed in a relation of obligation and respect, shift ways of being with, and "managing" water? Is the idea of resource and management itself fundamentally flawed, contributing to the current crisis of nature? (See also Todd 2017 for rethinking fossil fuels as "kin," Whyte 2018 on rethinking resilience through focus on responsibility and justice from an Indigenous studies perspective, or Wall Kimmerer 2013 on the deep territorial connections between people, flora, fauna, and territory.)

Box 3.2 Diverse and mixed methods in political ecology

Methodologically we can also understand that the imperatives of political ecology invite diverse and multi-method approaches, from policy and document analysis, to multi-sited comparison, as well as literary criticism, storytelling, or analysis of popular culture.

For instance, work by Paul Robbins (2007) on the culture of lawns and landscaping in North America includes consideration of political economic considerations, cultural ideals, and socio-cultural notions of gender and family. Robbins' analysis involves everything from political economic consideration of profit imperatives of pesticide companies in the post-World War II era, to Victorian landscape ideals and gendered notions of domesticity that feed into idealized families and lawnscapes.

Another approach, taken by Anna Tsing in her book *The Mushroom at the End of the World* (2015) is to follow a particular "thing"—in her case, the matsutake mushroom—from ecological growth to commercialization to international trade. By braiding together ecologies and capitalist landscapes, Tsing shows how ecologies are extracted, abstracted, circulated in particular ways in relation to monetization.

In another recent and provocative example, *Underflows* by Cleo Wölfle Hazard (2022) builds from learnings when members of the Scott River Watershed council, Karuk tribe, Quartz Valley Indian community, university students, and others collaborate on a field course that surveys fish, water quality, and habitat. From those partnerships, this work blends "Indigenous storytelling and land-care methods, ecopoetics field writing, and spatial and hydrological analysis … (following) Karuk protocol and a queer praxis of articulating a 'river model of justice.'" Such innovative mixed-methods approaches, blending the arts, Indigenous protocols, and ecological methods help to show the excitement and possibilities with political ecology and allied approaches.

Moving forward, continuing innovation, collaboration, and mixed method research design are likely to remain important to the development of new understandings regarding the political ecology of resource governance, from integrated qualitative and quantitative approaches, to community engaged and arts-based research.

Box 3.3 Metrology and science and technology studies

Metrology, or the science and study of measurement and its application, is another fruitful lens important for political ecology and science studies approaches to governing nature. Robertson (2004), Cooper and Rosin (2014), and others highlight science of observation, monitoring, and quantification as key to resource governance (for wetlands, or agricultural greenhouse gas emissions, respectively). Here, political ecological studies query the ways that resources are monitored and counted, looking carefully at the science and practices associated with monitoring and measurement to consider the effects that this has, whether for sustainability, or for shifting power dynamics. Here we again understand that the science of ecology (or atmospheric chemistry, or hydrology) is not apolitical, but is deeply influenced by social, cultural, and economic norms and dynamics. Such detailed analysis of how science is invoked, practiced, and operationalized is again an important theme of ongoing political ecological work.

Conclusion: Emergent Trends and New Insights for Governing Natures

While we have aimed to highlight some key political ecological themes related to governing nature, there are many more issues of interest than we were able to cover here. For instance, exciting work from feminist political ecology has increasingly highlighted the political, embodied, and lived dimensions of governing natures (Sultana 2020), new and emergent trends such as those associated with climate grief, or other ways that shifting scales of resource governance can shift political subjectivities (e.g., as with the classic work on environmentalities by Agarwal 2001 or more recent contributions such as Truelove and Cornea 2021). Conceptually and methodologically, recent work has sought to learn from abolition theory and black radical thought (Pulido and de Lara 2018), or to consider novel ways to engage communities in arts-based practices such as storytelling to foster new insights and modes of engagement (Harris 2021; see Wölfle Hazard example, above). All told, there are many examples that show the value, indeed necessity, of analyzing governing natures in ways that extend far beyond technical approaches that assume the resources, or the science of those resources, is straightforward or given. Instead, we see politics, social and economic assumptions, and socio-cultural values and practices, all as important pathways to critically interrogate, and also reinvigorate possibilities for governing natures.

With an overview of some of the analytical tools from political ecology, we can return to our introductory example of First Nations water challenges, to ask: What are the historical and governance roots of this failure, and how do drinking water concerns fit with other key inequalities in contemporary Canada? What can we learn from this case and others about the complex challenge of governing water, or forests, fisheries, and other resources? What political, economic, historic, ecologic, or scalar issues play a role in the ongoing debate regarding how to remedy this situation? Responses to such questions require consideration of historic legacies and inequities (for example, the establishment of the reserve system), complex management imperatives involving federal institutions and First Nations governance, as well as other policies and regulatory frameworks, such as the regulation and science of risk assessment, property rights, or drinking water quality. The situation also emerges from much more diffuse social, cultural, and economic processes such as those associated with settler colonialism, racism, or Indigenous knowledge and legal frameworks, including notions of respect or sense of territory (e.g., Mascarenhas 2007; Wilson and Inkster 2018; Arsenault et al. 2018). If we take this broader perspective as invited by political ecology and critical Indigenous studies, we can understand that issues such as trust, sense of community, social relationships, and historic connections to land and territory are all important to ongoing water issues, and potential responses (or "solutions"). As well, water quality and quantity challenges on reserves are not only a function of technical infrastructures, but are also directly linked to land uses and economic processes more broadly (chemical use in industry, mining, or forestry, and associated pollution), colonial land relations (Liboiron 2021) as well as broader cultural and ideological imperatives (e.g., Indigenous notions of respect for water, or colonial notions of water as a "resource," or economic growth imperatives linked to ongoing mining and fossil fuel extraction; Liboiron 2021). Because Indigenous peoples experience disproportionate rates and frequencies of water insecurity (Wilson et al. 2021), the politics of reconciliation and resurgence also highlight the imperative for new approaches that work to establish and enliven Indigenous approaches and self-determination around water governance (Daigle 2018; Sarna-Wojcicki et al. 2019; Wilson et al. 2022).

It is abundantly clear that questions about environmental governance are not simply questions that require attention to policy, ecology, or engineering—political ecology opens up one pathway that allows us to understand that such issues are also deeply felt, experienced, and lived. As such, assessment of international policies and frameworks, and also attention to lived and embodied experiences are all necessary. Themes of interest range from broad ontological and philosophical issues (i.e., what is water, and how do we *know* or *value* water), to biophysical and ecological dimensions (how aquifers are linked or affected by pollution), as well as nuts-and-bolts concerns related to consultation processes, institutional efforts to foster inclusion, or use of policy instruments to design more equitable and sustainable governance. All such modes of inquiry have been, and will continue to be, important to political ecologies of governing nature.

References

Agarwal, B. (2001). Participatory Exclusions, Community Forestry, and Gender: An Analysis for South Asia and a Conceptual Framework. *World Development* 29(10): 1623–1648. https://doi.org/10.1016/S0305-750X(01)00066-3

Ahlborg, H., and Nightingale, A. (2018). Theorizing Power in Political Ecology: The "Where" of Power in Resource Governance Projects. *Journal of Political Ecology* 25(1). https://doi.org/10.2458/v25i1.22804

Alatout, S. (2008). States of Scarcity: Water, Space, and Identity Politics in Israel, 1948–1959. *Environment and Planning D* 26(6): 959–982.

Arsenault, R., Diver, S., McGregor, D., Witham, A., and Bourassa, C. (2018). Shifting the Framework of Canadian Water Governance through Indigenous Research Methods: Acknowledging the Past with an Eye on the Future. *Water* 10(1): 49. https://doi.org/10.3390/w10010049

Bakker, K. (2013). Neoliberal Versus Postneoliberal Water: Geographies of Privatization and Resistance. *Annals of the Association of American Geographers* 103(2): 253–260.

Bakker, K., and Bridge, G. (2006). Material Worlds? Resource Geographies and the "Matter of Nature". *Progress in Human Geography* 30(1): 5–27. https://doi.org/10.1191/0309132506ph588oa

Barnes, J. (2014). *Cultivating the Nile: The Everyday Politics of Water in Egypt*. Duke University Press.

Batterbury, S. P., and Fernando, J. L. (2006). Rescaling Governance and the Impacts of Political and Environmental Decentralization: An Introduction. *World Development* 34(11): 1851–1863.

Bauer, C. J. (1997). Bringing Water Markets Down to Earth: The Political Economy of Water Rights in Chile, 1976–1995. *World Development* 25(5): 639–656.

Bee, B. (2017). Gendered Spaces of Payment for Environmental Services: A Critical Look. *Geographical Review* 109(1): 87–107.

Behn, C., and Bakker, K. (2019). Rendering Technical, Rendering Sacred: The Politics of Hydroelectric Development on British Columbia's Saaghii Naachii/Peace River. *Global Environmental Politics* 19(3): 98–119. https://doi.org/10.1162/glep_a_00518

Belfer, E., Ford, J., Maillet, M., Araos, M., and Flynn, M. (2019). Pursuing an Indigenous Platform: Exploring Opportunities and Constraints for Indigenous Participation in the UNFCCC. *Global Enviornmental Politics* 19(1): 12–33.

Biza, A., Kooy, M., Manuel, S., and Zwarteveen, M. (2021). Sanitary Governmentalities: Producing and Naturalizing Social Differentiation in Maputo City, Mozambique (1887–2017). *Environment and Planning E: Nature and Space* 5(2): 605–624.

Blaikie, P. (1985). *The Political Economy of Soil Erosion in Developing Countries*. Abingdon: Routledge. https://doi.org/10.4324/9781315637556

Bridge, G. (2014). Resource Geographies II: The Resource-State Nexus. *Progress in Human Geography* 38(1): 118–130.

Brown, C. J., and Purcell, M. 2005. There's Nothing Inherent about Scale: Political Ecology, the Local Trap, and the Politics of Development in the Brazilian Amazon. *Geoforum* 36(5): 607–624.

Bryant, R. L. (1992). Political Ecology: An Emerging Research Agenda in Third World Studies. *Political Geography* 11(1): 12–36. https://doi.org/10.1016/0962-6298(92)90017-N
Budds, J. (2016). Whose Scarcity? The Hydrosocial Cycle and Changing Waterscape of La Ligua River Basin, Chile. In M. K. Goodman, M. T. Boykoff, and K. Evered (eds.), *Contentious Geographies*. Abingdon: Routledge.
Bulkeley, H. (2010). Cities and the Governing of Climate Change. *Annual Reviews of Environment and Resources* 35: 229–253.
Bumpus, A. G. (2011). The Matter of Carbon: Understanding the Materiality of tCO(2)e in Carbon Offsets. *Antipode* 43(3): 612–638.
Canadian Broadcasting Corporation. (2015). Justin Trudeau Vows to End First Nations Reserve Boil Water Advisories within 5 Years. www.cbc.ca/news/politics/canada-election-2015-justin-trudeau-first-nations-boil-water-advisories-1.3258058
Castree, N. (2008). Neoliberalising Nature: The Logics of Deregulation and Reregulation. *Environment and Planning A* 40(1): 131.
Chakraborty, R., and Sherpa, P. Y. (2021). From Climate Adaptation to Climate Justice: Critical Reflections on the IPCC and Himilayan Climate Knowledges. *Climatic Change* 167: 49.
Chalifour, N. (2010). A Feminist Perspective on Carbon Taxes. *Canadian Journal of Women and the Law* 22(1): 169–212.
Clarke-Sather, A. (2017). State Power and Domestic Water Provision in Semi-arid Northwest China: Towards an Aleatory Political Ecology. *Political Geography* 58: 93–103.
Cleaver, F., Franks, T., Maganga, F., and Hall, K. (2013). Institutions, Security, and Pastoralism: Exploring the Limits of Hybridity. *African Studies Review* 56(3): 165–189. https://doi.org/10.1017/asr.2013.84
Collard, R.-C., Dempsey, J., and Holmberg, M. (2020). Extirpation Despite Regulation? Environmental Assessment and Caribou. *Conservation Science and Practice* 2(4): e166. https://doi.org/10.1111/csp2.166
Cooper, M. H., and Rosin, C. (2014). Absolving the Sins of Emission: The Politics of Regulating Agricultural Greenhouse Gas Emissions in New Zealand. *Journal of Rural Studies* 36: 391–400. https://doi.org/10.1016/j.jrurstud.2014.06.008
Cumming, G. S., et al. (2006). Scale Mismatches in Social-Ecological Systems: Causes, Consequences, and Solutions. *Ecology and Society* 11(1): 14.
Daigle, M. (2018). Resurging through Kishiichiwan: The Spatial Politics of Indigenous Water Relations. *Decolonization: Indigeneity, Education and Society* 7(1): 159–172.
Dauvergne, P. (1997). *Shadows in the Forest: Japan and the Politics of Timber in Southeast Asia*. MIT Press.
Deere, C. D., and Léon, M. (2002). Individual versus Collective Land Rights: Tensions between Women's and Indigenous Rights under Neoliberalism. In J. Chase (ed.), *The Spaces of Neoliberalism: Land, Place and Family in Latin America*. Kumarian Press.
Dempsey, J., and Robertson, M. (2010). Ecosystem Services: Tensions, Impurities, and Points of Engagement within Neoliberalism. *Progress in Human Geography* 36(6): 758–779.
Dietz, T., Ostrom, E., and Stern, P. C. (2003). The Struggle to Govern the Commons. *Science* 302(5652): 1907–1912.
Fletcher, R. (2010). Neoliberal Environmentality: Towards a Poststructuralist Political Ecology of the Conservation Debate. *Conservation and Society* 8(3): 171.
Fletcher, R., and Cortes-Vazquez, J. A. (2020). Beyond the Green Panopticon: New Directions in Research Exploring Environmental Governmentality. *Environment and Planning E: Nature and Space* 3(2): 1–11.
Goldin, J. A. (2010). Water Policy in South Africa: Trust and Knowledge as Obstacles to Reform. *Review of Radical Political Economics* 42(2): 195–212. https://doi.org/10.1177/0486613410368496
Goldman, M. J., et al. (2018). A Critical Political Ecology of Human Dimensions of Climate Change: Epistemology, Ontology, and Ethics. *Wiley Interdisciplinary Reviews-Climate Change* 9(4): 526.
Government of Canada. (2021). Ending Long Term Drinking Water Advisories. www.sac-isc.gc.ca/eng/1506514143353/1533317130660
Harris, L. M. (2021). Towards Narrative Political Ecologies. *Environment and Planning E: Nature and Space*, 25148486211010676. https://doi.org/10.1177/25148486211010677

Heynen, N., et al., eds. (2007). *Neoliberal Environments: False Promises and Unnatural Consequences.* New York, Routledge.

Jacobs, H. M. (2003). The Politics of Property Rights at the National Level. *Signals and Trends Journal of the American Planning Association* 69(2): 181–189. https://doi.org/10.1080/01944360308976305

Jasanoff, S., and M. L. Martello, eds. (2004). *Earthly Politics: Local and Global Environmental Governance.* Cambridge, MA, MIT Press.

Kimmerer, R. W. (2013). *Braiding Sweetgrass* (First edition). Milkweed Editions.

King, B., et al. (2018). Political Ecology of Dynamic Wetlands: Hydrosocial Waterscapes in the Okavango. *The Professional Geographer* 71(1): 29–38.

Kull, C. A., Arnauld de Sartre, X., and Castro-Larrañaga, M. (2015). The Political Ecology of Ecosystem Services. *Geoforum* 61(C): 122–134. https://doi.org/10.1016/j.geoforum.2015.03.004

Larson, A. M., and Soto, F. (2008). Decentralization of Natural Resource Governance Regimes. *Annual Review of Environment and Resources* 33(1): 213–223.

Lemos, M. C., and Agrawal, A. (2006). Environmental Governance. *Annual Review of Environment and Resources* 31(1): 297–325. https://doi.org/10.1146/annurev.energy.31.042605.135621

Liboiron, M. (2021). *Pollution is Colonialism.* Duke University Press.

Linton, J. (2010). *What Is Water? The History of a Modern Abstraction.* UBC Press.

Loftus, A. (2018). Political Ecology II: Whither the State? *Progress in Human Geography* 44(1): 139–149. doi:10.1177/0309132518803421

Mascarenhas, M. (2007). Where the Waters Divide: First Nations, Tainted Water and Environmental Justice in Canada. *Local Environment* 12(6): 565–577. https://doi.org/10.1080/13549830701657265

McCarthy, J. (2002). First World Political Ecology: Lessons from the Wise Use Movement. *Environment and Planning A: Economy and Space* 34(7): 1281–1302. https://doi.org/10.1068/a3526

McElwee, P. D. (2012). Payments for Environmental Services as Neoliberal Market-Based Forest Conservation in Vietnam: Panacea or Problem? *Geoforum* 43(3): 412–426. https://doi.org/10.1016/j.geoforum.2011.04.010

Menga, F., et al. (2019). Space, State-Building and the Hydraulic Mission: Crafting the Mozambican State. *Environment and Planning C* 37(5): 868–888.

Moss, T. (2004). The Governance of Land Use in River Basins: Prospects for Overcoming Problems of Institutional Interplay with the EU Water Framework Directive. *Land Use Policy* 21(1): 85–94.

Nightingale, A. J. (2011). Bounding Difference: Intersectionality and the Material Production of Gender, Caste, Class and Environment in Nepal. *Geoforum* 42(2): 153–162. https://doi.org/10.1016/j.geoforum.2010.03.004

O'Leary, H. (2019). Conspicuous Reserves: Ideologies of Water Consumption and the Performance of Class. *Economic Anthropology*, sea2.12150. https://doi.org/10.1002/sea2.12150

Ostrom, E. (1990). *Governing the Commons: The Evolution of Institutions for Collective Action.* Cambridge University Press. https://public.ebookcentral.proquest.com/choice/publicfullrecord.aspx?p=1103736

Pahl-Wostl, C., Lebel, L., Knieper, C., and Nikitina, E. (2012). From Applying Panaceas to Mastering Complexity: Toward Adaptive Water Governance in River Basins. *Environmental Science and Policy* 23: 24–34. https://doi.org/10.1016/j.envsci.2012.07.014

Penning-Rowsell, E. C., and Johnson, C. (2015). The Ebb and Flow of Power: British Flood Risk Management and the Politics of Scale. *Geoforum* 62: 131–142.

Perreault, T. (2005). State Restructuring and the Scale Politics of Rural Water Governance in Bolivia. *Environment and Planning A* 37(2): 263–284.

Pulido, L. (2000). Rethinking Environmental Racism: White Privilege and Urban Development in Southern California. *Annals of the AAG* 90(1): 12–40.

Pulido, L., and De Lara, J. (2018). Reimagining "Justice" in Environmental Justice: Radical Ecologies, Decolonial Thought, and the Black Radical Tradition. *Environment and Planning E: Nature and Space* 1(1–2): 76–98. https://doi.org/10.1177/2514848618770363

Ranganathan, M. (2022). Caste, Racialization, and the Making of Environmental Unfreedoms in Urban India. *Ethnic and Racial Studies* 45(2): 257–277. https://doi.org/10.1080/01419870.2021.1933121

Ranganathan, M., and Bonds, A. (2022). Racial Regimes of Property: Introduction to the Special Issue. *Environment and Planning D* 40(2): 197–207.

Rice, J. (2014). An Urban Political Ecology of Climate Change Governance. *Geography Compass* 8: 381–394.

Robbins, P. (2001). Fixed Categories in a Portable Landscape: The Causes and Consequences of Land-cover Categorization. *Environment and Planning A* 33: 161–179.

Robbins, P. (2007). *Lawn People: How Grasses, Weeds, and Chemicals Make Us Who We Are.* Temple University Press.

Robbins, P. (2012). *Political Ecology: A Critical Introduction* (2nd ed). Chichester: John Wiley & Sons.

Robertson, M. M. (2000). No Net Loss: Wetland Restoration and the Incomplete Capitalization of Nature. *Antipode* 32(4): 463–493. https://doi.org/10.1111/1467-8330.00146

Robertson, M. M. (2004). The Neoliberalization of Ecosystem Services: Wetland Mitigation Banking and Problems in Environmental Governance. *Geoforum* 35(3): 361–373. https://doi.org/10.1016/j.geoforum.2003.06.002

Sarna-Wojcicki, D., Sowerwine, J., Hillman, L., Hillman, L., and Tripp, B. (2019). Decentring Watersheds and Decolonising Watershed Governance: Towards an Ecocultural Politics of Scale in the Klamath Basin. *Water Alternatives* 12: 241–266.

Sneddon, C., Magilligan, F. J., and Fox, C. A. (2022). Peopling the Environmental State: River Restoration and State Power. *Annals of the American Association of Geographers* 112(1): 1–18.

Steinberg, P. (2001). *The Social Construction of the Ocean.* Cambridge: Cambridge University Press.

Sultana, F. (2015). Emotional Political Ecologies. In R. Bryant (Ed.), *The International Handbook of Political Ecology*, 633–645. Edward Elgar.

Sultana, F. (2020). Embodied Intersectionalities of Urban Citizenship: Water, Infrastructure, and Gender in the Global South. *Annals of the American Association of Geographers 110*(5): 1407–1424. https://doi.org/10.1080/24694452.2020.1715193

Sultana, F. (2021). Critical Climate Justice. *The Geographical Journal* 188(1): 118–124.

Sultana, F., and Loftus, A. (2013). *The Right to Water: Politics, Governance and Social Struggles.* Abingdon: Routledge.

Swyngedouw, E. (2004). Globalisation or "Glocalisation"? Networks, Territories and Rescaling, *Cambridge Review of International Affairs* 17(1): 25–48. DOI: 10.1080/0955757042000203632

Taylor, P., and Buttel, F. (1992). How Do We Know We Have Global Environmental Problems? Science and the Globalization of Environmental Discourse. *Geoforum* 23: 405–416.

Todd, Z. (2017). Fish, Kin and Hope: Tending to Water Violations in Amiskwaciwâskahikan and Treaty Six Territory. *Afterall: A Journal of Art Context and Enquiry* 43: 102–107. https://doi.org/10.1086/692559

Truelove, Y., and Cornea, N. (2021). Rethinking Urban Environmental and Infrastructural Governance in the Everyday: Perspectives from and of the Global South. *Environment and Planning C: Politics and Space* 39(2), 231–246. https://doi.org/10.1177/2399654420972117

Tsing, A. L. (2015). *The Mushroom at the End of the World.* Princeton University Press.

Venot, J.-P., and Jensen, C. B. (2021). A Multiplicity of *Prek(s)*: Enacting a Socionatural Mosaic in the Cambodian Upper Mekong Delta. *Environment and Planning E: Nature and Space*, 251484862110268. https://doi.org/10.1177/25148486211026835

Whyte, K. (2018). Critical Investigations of Resilience: A Brief Introduction to Indigenous Environmental Studies and Sciences. *Daedalus* 147(2): 136–147.

Wilson, N. (2019). "Seeing Water Like a State?": Indigenous Water Governance through Yukon First Nation Self-Government Agreements. *Geoforum* 104: 101–113.

Wilson, N. J., and Inkster, J. (2018). Respecting Water: Indigenous Water Governance, Ontologies, and the Politics of Kinship on the Ground. *Environment and Planning E: Nature and Space* 1(4): 516–538. https://doi.org/10.1177/2514848618789378

Wilson, N. J., Harris, L. M., Nelson, J., and Shah, S. H. (2019). Re-theorizing Politics in Water Governance. *Water* 11(7): 1470. https://doi.org/10.3390/w11071470

Wilson, N. J., Montoya, R. A. T., and Curley, A. (2021). Governing Water Insecurity: Navigating Indigenous Water Rights and Regulatory Politics in Settler Colonial States. *Water International* 46(6): 783–801.

Wilson, N. J., Montoya, T., Lambrinidou, Y., Harris, L. M., Pauli, B. J., McGregor, D., Patrick, R. J., Gonzalez, S., Pierce, G., Wutich, A. (2022). From "Trust" to "Trustworthiness": Retheorizing Dynamics of Trust, Distrust, and Water Security in North America. *Environment and Planning E: Nature and Space* 25148486221101460.

Wölfle Hazard, C. (2022). *Underflows: Queer and Trans Ecologies of River Justice*. Seattle, WA, University of Washington Press.

Yates, J. S., Harris, L., and Wilson, N. (2017). Multiple Ontologies of Water: Politics, Conflict and Implications for Governance. *Environment and Planning D: Society and Space* 35(5): 797–815.

4 Contesting Nature

Nature as a Field of Power, Difference, and Resistance

Kate Derickson, Gabe Schwartzman, and Rebecca Walker

Understanding "Nature" as a Field of Power

On an island just off the coast of South Carolina, residents diligently turn off their lights at night during sea turtle nesting season. Newly hatched sea turtles orient themselves by the light of the moon reflecting off the ocean, and the lights of the new homes along the coastline can disorient the threatened species, causing them to wander off in the wrong direction and die. "Turtle crossing" signs line the roads, along with signs reminding people to protect the turtles and their habitat. Residents of Edisto Beach report a sense of duty and obligation to the well-being of the turtles in interviews and focus groups, and describe efforts they take to keep their homes dark at night and protect turtle nests along the shore. They express disgust at anyone who would knowingly damage a turtle nest, especially anyone who would "poach" turtle eggs for consumption.

At first blush, this story appears rather straightforward; a group of people with a sense of responsibility for environmental protection and care for nonhuman creatures are engaged in activities to protect a threatened species. The field of political ecology has shown, however, that the making, manipulating, and governing of nature is always an ideological project, and that nature is best understood as a field of power through which social relations are mediated. Political ecologists regard nature as a set of human and nonhuman power-laden relations, in sharp contrast to the colloquial meaning of the word "nature" to refer to things, spaces, and processes that are separate from or uninfluenced by people.

This framing offers a different vantage point from which to consider the conception of nature mobilized by the turtle protectors on Edisto Island. Edisto is a sea island south of Sullivan's Island where about 40 percent of enslaved Africans first entered the US (Janiskee 2009), many of whom would work on Sea Island cotton plantations on Edisto. These enslaved people established a language and culture that endures today, and Edisto is in the heart of what some people call Gullah/Geechee Nation. Many Gullah/Geechee people still live on Edisto, though they tend to live on family compounds set back from the waterfront. From Reconstruction, when many formerly enslaved people established their family land on the island, to the present, Gullah/Geechee people have kept their traditions alive, farming, fishing, sewing sweet grass baskets, and speaking the Gullah language. Some recall a time when turtle eggs were part of their diets (Atwell 2019).

The beachfront homes are newer, built in the second half of the twentieth century as highways and air conditioning made Edisto easier to get to and more pleasant for white people looking for a vacation spot. The residents of these homes—the ones who turn off their lights for the turtles—tend to be white and wealthier. When they refer to people who

DOI: 10.4324/9781003165477-6

might "poach" turtle eggs, or "disturb" turtle habitat, they are subtly referring to Gullah/Geechee people. There's little available evidence that turtle eggs remain a meaningful part of Gullah/Geechee diets; and whether they do or not is beside the point. The Gullah/Geechee Sea Island Coalition partners with agencies and organizations to participate in protecting turtle habitat, and there are scant arrest records showing natural law enforcement has found any evidence of Gullah/Geechee poaching, despite over policing of Gullah/Geechee people by the game warden.

In this chapter, we illustrate how political ecology understands nature as a field of power through which various actors and institutions attempt to navigate. In this sense, political ecology "denaturalizes" nature, attending to the always already social dimension of nature. Political ecology was first developed to situate nature in capitalist social relations and its commitment to centering the complexity of the relationship between nature and capital distinguishes it from other fields of study that consider the politics of struggles over nature. As understandings of the way that capitalism articulates with other structures of oppression deepened, political ecology also engaged with race and gender (see for example Sundberg 2016; Massé et al. 2021). This expansive understanding of the political dimensions of political ecology invites attention to the way nature articulates with forms of difference to become a site of struggle, contestation, oppression, resistance, and solidarity.

Political ecologists are interested in how nature both becomes a site of contestation and is a site *through which* power is contested. In the case of Edisto Island, a campaign to protect sea turtle habitat can be read as an ideological project that advances some ideas about nature, development, race, and power while subsuming others. For example, in the public discourse about sea turtle protection, the current socio-natural paradigm is taken as given, and the proposed solutions or mitigating approaches require no real disruption to the organization of society. Waterfront development, road construction, and automobile and boat traffic all introduce the profound habitat threats sea turtles face, and yet those dynamics remain unchallenged in efforts to "protect" sea turtles. Instead of being encouraged to not build or live near the beach, residents are encouraged to turn off lights. Instead of banning automobile traffic, residents are encouraged to drive slowly. Yet in focus groups, white residents express the most scorn for those who directly tamper with turtle nests, naming this form of disruption as perverse while normalizing other habitat threats.

A political ecology lens highlights the way that in this context, campaigns to protect nature have the effect of normalizing processes and relations that actually threaten habitats while framing themselves as protecting nature, and shifting blame for turtle population decline to racialized "others." Massé et al. (2021) extend this analysis of the intersections of race, capitalism, and nature to develop a feminist political ecology of wildlife poaching, exploring how gendered systems articulate with the politics, practices, and implications of wildlife poaching and the militarization of its enforcement (see also Mollett and Faria 2013).

Political ecologists studying diverse geographies have pointed to the ways that contestation over the control and access to various kinds of natures has regularly relied on claims about differently racialized "others." For instance, political ecology research has explored the centrality of struggles over nature in the perpetuation of white settler society and the dispossession of Indigenous people (see for example Todd 2018). Political ecology research in Southeast Asia has highlighted the ways that states rely on racialized claims about Indigenous and immigrant populations to extend control over forest resources (Vandergeest 2003; Li 2014; Peluso 2017). In another example, Ranganathan (2022) has shown how narratives about nature and environmental health enable the Indian state to clear urban slums where racialized Dalit caste laborers reside. Other political ecologists examine how colonial dispossession and ongoing colonialism operate through claims about racialized colonized peoples' inability to sustainably manage environments (e.g., Moore 2005; Kosek 2006; Wainwright 2011).

"There's No Such Thing as a Natural Disaster": Resilience and Resourcefulness in a Changing Environment

Political ecologists have long pointed to the ways that responses to and preparations for so-called natural disasters are always power-laden. These struggles are evident in the politics of resilience planning. In 2013, for instance, the Rockefeller Foundation launched the "100 Resilient Cities" program to help "cities around the world become more resilient to physical, social, and economic shocks and stresses" (Rockefeller Foundation 2022). The initiative was responding to an evolving understanding of the threat posed by increasingly common natural disasters resulting from rising global temperatures and associated ecosystem changes. The framework of "resilience" aimed to apply current thinking on the intertwining of social and ecological systems to developing social systems better equipped to respond and adapt to the effects of flooding, fires, storms, heat, and associated disruptions to labor markets, infrastructure, and the social safety net. Drawn to the "problem space" that resilience thinking identified, many communities welcomed resources and capacity to enhance their community connections, improve their infrastructure, and promote creative alternatives to fragile global networks by, for example, expanding local food systems.

At the GalGael Trust, a nonprofit in Glasgow, Scotland, a group of community members, activists and academics convened to consider what resilience might look like in the Govan neighborhood, where residents experienced higher than average rates of joblessness and poor health outcomes. What the group ultimately concluded, however, was that a resilience framework had the effect of naturalizing and amplifying the status quo of the social and environmental systems and distracted from pursuing meaningful system change. Like the example of sea turtle conservation on Edisto Island, resilience thinking and policy making ultimately mobilized concerns about environmental change in ways that failed to engage with the need for fundamental systemic change in social and natural relations. The inquiry group at GalGael opted to pursue "resourcefulness" as an alternative to resilience. Instead of promoting a rapid return to status quo in the face of taking "knock after knock" (as one participant put it), resourcefulness centers the uneven distribution of resources in order to disrupt the unequal ways that environmental change is experienced (Derickson and MacKinnon 2015).

The concept of resourcefulness builds on the political ecology maxim that there is "no such thing as a natural disaster" (Smith 2006). When Hurricane Katrina hit the Gulf Coast of the United States in 2005, images of mostly Black people stranded on rooftops amid rising flood waters and crowded into a sweltering stadium full of garbage highlighted the profoundly uneven exposure different groups have to natural disasters (Derickson 2014). When Neil Smith wrote after the storm that there is "no such thing as a natural disaster" he highlighted the way that the impact of storms and environmental events on communities is always mediated by uneven power relations that create disproportionate vulnerabilities and capacities for survival (Smith 2006). For thinkers like

Smith, this unevenness is a result of capitalism, which concentrates economic and political power in the hands of the few.

As communities face multifaceted, ongoing environmental crises wrought by a changing climate, political ecologists continue to explore how the impacts of rising seas and temperatures are borne unevenly by low-income communities and communities of color. Many of these communities are located in low-lying neighborhoods, which were designated or became less desirable real estate because they are vulnerable to flooding. These communities often live in low-quality housing like mobile homes that are less resilient to bad weather (Rumbach et al. 2020). And finally, as our research has shown (Walker et al. 2023), investments in green infrastructure in cities like Minneapolis that has the capacity to mitigate the impact of natural hazards and lower temperatures has often been concentrated in wealthy, white communities, while low-income communities of color experience structural disinvestment that has led to low levels of green infrastructure, further exacerbating their vulnerability.

Natural disasters can be consciousness raising events for communities, however, and can spark new waves of activism. That activism can provide mutual aid in the recovery process, advocate for new paradigms of collectivity and governance in the rebuilding process, and promote engagement in climate and environmental justice movements. For instance, after Hurricane Katrina, disaster response efforts sparked widespread community organizing efforts whose scope stretched far beyond immediate recovery needs. These included organizing efforts for racial equality in access to affordable housing as part of the People's Hurricane Relief Fund, organized by Black liberation movement leaders in New Orleans (Luft 2012). In the wake of "Superstorm Sandy," in New York City, dormant Occupy Wall Street networks regrouped as Occupy Sandy to create a disaster relief hub that not only provided mutual aid, but mobilized a grassroots movement for low-income groups and communities of color to take more political power (Bondesson 2022). In the Global South, disasters such as floods and dam failures have triggered regional and national environmental justice movements, as seen in the aftermath of the mining-dam failure in Mariana, Brazil, in 2015 (Losekann and Milanez 2021). Natural disasters can be moments that, as Rebecca Solnit writes, "shake loose the old order" (2020), and provide new sites of political possibility.

Disrupting Displacements

Capital, difference, and power shape—and are shaped by—struggles over land tenure, access to land, and its various uses. At the same time, land is often at the heart of debates over environmental protection, conservation, and sustainability. The notion of conserving nature and wilderness from destructive human use originated in Western ideas about the division of humans and nature and concretized in a set of violent dispossessions of Indigenous peoples in the North American West (Cronon 1996). Around the world, Indigenous, minoritized, racialized, and marginalized peoples have faced dispossession, displacement, and disciplining in the creation of protected areas for conservation purposes (West et al. 2006). Many of these communities, however, have leveraged conservation regimes to contest uneven power relations, and the Western notion that conservation must be devoid of people in their fights for sovereignty, territory, and autonomy (Escobar 1998; Li 2007). Conceptualizations of nature are contested terrains, and competing visions are often leveraged in support of competing claims to land, access, and resources. This is as true in the forests of the Brazilian Amazon as it is in debates over urban greening and sustainability projects, where debates over urban nature are at the heart of contested claims to urban space. In this section, we highlight the ways that debates over conservation, environmental protection, and sustainability have been contentious sites of struggle over dispossession, displacement, and access to land and resources.

In the Brazilian Amazon, for example, Indigenous peoples are engaged in ongoing contestation over regimes of conservation and the onslaught of a settler development frontier. In this contentious landscape, Indigenous peoples are struggling to secure and expand territorial sovereignty over some of the largest tracts of Indigenous-controlled lands in the world. When the Brazilian military dictatorship began building roads and incentivizing mining and agricultural development across the Amazon in the 1970s, there was a consensus in Brazilian society that Indigenous peoples would largely become peasants and that the state could administer parks for forest conservation (Moran 1976). In the ensuing 50 years, however, the inhabitants of the forest contested the assumed inevitability of both Indigenous removal and the expansion of a capitalist agricultural frontier (Cleary 1993; Hecht and Cockburn 2011; Brondízio et al. 2013). Through a concerted social movement of protest, direct action, and a war for public opinion, along with the support of international allies, Indigenous peoples today control 45 percent of the land in the Amazon Basin (FAO and FILAC 2021). Indigenous communities leveraged the idea that their chosen ways of life were tied to accessing extensive areas of standing tropical forest and that they were the best agents to ensure the conservation of Amazonian forests, advocating for the creation of "parks with people in them" (Nepstad et al. 2006), a conservation practice that has traveled far beyond the Amazon. Indigenous land struggles contest the Western principles that underlie mainstream conservation and the assumed inevitability of an unstoppable settler capitalist frontier.

Like the forest politics in the Brazilian Amazon, struggles for access to land, resources, and nature are central to the production of spaces of "nature" in cities. Particularly in light of environmental challenges posed by climate change, cities are increasingly turning to "green infrastructure"—such as parks, street trees, rain gardens, and living shorelines—to mitigate heat, reduce flood risk, and manage stormwater (Matthews et al. 2015). Urban greening has become a nearly ubiquitous mode of city planning, underpinned by an often unquestioned planning orthodoxy to make cities green, sustainable, and resilient (Kaika 2017; Wachsmuth and Angelo 2018; Anguelovski et al. 2020). Yet numerous scholars have noted that urban greening agendas have served to reproduce and deepen urban inequalities, in particular through centering the priorities of privileged actors and contributing to gentrification, dispossession, and displacement (Angelo et al. 2022; Anguelovski et al. 2016). This vision of urban nature, researchers have argued, serves to reproduce and reinforce political and economic agendas geared toward growth and development, mobilizing sustainability agendas as a new form of capital (McClintock 2020). The urban green growth agenda espouses politically neutral rhetoric in which urban greening is framed as a universal good (Angelo 2019; Checker 2011; Rigolon and Németh 2018).

Investments in green infrastructure, like other forms of infrastructure, are embedded in the power relations that shape cities (Finewood et al. 2019; Gandy 2003). While the value ascribed to urban nature is seen as a boon to city governments and developers hoping to raise property values (Angelo 2019), low-income, often racially minoritized residents of disinvested neighborhoods must navigate the tension between advocating for long-needed access to urban greenspaces and fears over rising rents and displacement

(Dooling 2009; Safransky 2014). Known as "green gentrification," investments in urban nature can drive increases in perceived local desirability and, as a result, higher property values and rents (Anguelovski et al. 2018). A political ecology lens challenges a-political urban greening rhetoric and points to the ways that urban greening is embedded in claims over access to urban space.

Activists and scholars have challenged the idea that greening is a-political, instead framing greening and sustainability efforts as a contested field of power, and pointing to the ways in which the "green growth" agenda reflects a vision of nature situated in the logic of capital accumulation (Quastel 2009). In some cases, activists articulating an alternative vision of urban nature that challenges the urban green growth agenda by forming coalitions among housing and environmental justice advocates to fight for both their right to urban nature and their right to affordable housing and protections from displacement (Lubitow and Miller 2013; Pearsall 2013). One alternative vision of urban nature includes the "just green enough" strategy, which draws on the strength of the environmental justice movement to articulate a vision of urban space in which all residents have access to urban nature. The just green enough strategy centers the needs and priorities of working-class residents by, for example, preserving industrial activity that provides working-class jobs while cleaning up industrial pollution and restoring habitat (Curran and Hamilton 2012). Still, others have criticized the "just green enough" framework, arguing that it ultimately preserves, rather than challenges, existing political and economic inequalities and framing it as an interim strategy as we advance new, transformative frameworks that fundamentally unsettle the political and economic systems at the root of persistent environmental injustices (Dooling 2017; Pulido 2016; Sze and Yeampierre 2017). Whether articulating a vision of urban green growth, "just green enough," or more radical and transformative visions of urban nature, urban greening is a contested field in which debates over claims to land, access, and resources in the urban environment play out.

Contestation Across Scales

Political ecology emphasizes the way that particular sites of contestation are made, experienced, and often contested in relation to broader processes that play out at a range of spatial and temporal scales. In the case of Edisto Island, where this chapter began, the specificity of Edisto matters, but it matters in *relation* to broader processes. The campaign to save the sea turtles is situated in relation to the history of transatlantic slavery, which distributed the African diaspora, created racial heirarchies and shaped the contemporary geographies, demographics, landscapes, and natures of Edisto Island. It is also situated in relation to the histories of global capitalism, which likewise shaped the demographics, landscapes, and natures of the island. From a political ecology perspective, then, *contesting nature* can happen at a range of scales.

Community-based activism, whether it is to contest the siting of a polluting facility or advocate for more green investments in a neighborhood, can sometimes be "scaled up" from what David Harvey called a "militant particularism"—or a concern related to a specific manifestation of structural dynamics—to engage in multi-sited coalitional politics. The environmental justice movement in the United States is an example of a movement that scaled up its activism from place-based matters of concern to a broader critique of the socio-natural relations that create uneven exposure to environmental toxins across race. When the United Church of Christ commissioned a study in 1987 that identified common trends in the co-location of neighborhoods of color and hazardous polluting facilities, environmental justice activists and advocates were able to create a shared, national-scale platform that became the "environmental justice movement." The movement has evolved to entail networks of researchers, activists and nonprofit professionals to share knowledge, craft legislation, and build solidarity to contest these outcomes at a national scale. Their efforts have led to the incorporation of environmental justice principles in state and federal government agencies (with mixed results; see Harrison 2019).

The movement was launched in 1982 when activists in Warren County, North Carolina protested the dumping of soil contaminated with hazardous chemicals in a site adjacent to a low-income and majority-Black community (McGurty 2000). Residents feared the new dump would leak the toxic chemicals—linked to birth defects, skin and liver problems, and cancer—into their groundwater, threatening their health and rendering their homes unlivable. Activists argued that the state chose to place the landfill in their community because its residents were "few, black, and poor" (Reimann 2017). Protesters called the siting decision a form of "environmental racism" and strategically connected their fight to a broader history of racial discrimination in the American South (Bullard 2004). Through their rhetoric, forms of protest, and alliances with church leaders and the National Association for the Advancement of Colored People, Warren County activists linked their protests to the broader Civil Rights movement, arguing that environmental risk was yet another arena in which systemic racism in America played out (McGurty 2000). In so doing, they named a link between the governance of nature and race, by identifying and contesting the tight coupling between environmental degradation and impacts on communities of color.

While the movement is varied and diffuse, many EJ activists aim to challenge the social relations that not only produced the *uneven* outcomes, but the broader processes that create toxic and polluting facilities. Many EJ activists embrace the "precautionary principle" which challenges the social relations that put the onus on communities to demonstrate that industries are harmful, and instead argues that industries should prove that they will not harm human health or the environment before being allowed to operate. The burden of proof should be on the entity that may produce the harm, rather than those that might be harmed (Morello-Frosch et al. 2002). Many have continued to push the EJ movement from within and from outside to continue to center and contest the social relations that produce uneven outcomes by being more attentive to the intersection between environmental outcome and other sectors like housing, infrastructure, and mass incarceration (Heynen 2016; Pellow 2016; Ranganathan 2016). By thinking more holistically about the production of uneven outcomes, activists and scholars hope to short-circuit the processes that produce harms in the first place.

The climate justice movement is another example of scaling up the contestation of nature. While individual communities of color have experienced the impacts of climate change, whether through the impact of sea level rise, intensified and more frequent natural disasters, or rising temperatures, the climate justice movement is aimed at creating international solidarity among communities on the "frontlines" of climate change to influence climate policy at the national and international scale. The climate justice movement, perhaps more so than the environmental justice movement in the US, has also developed a critique of government-led action, and in some cases contesting the notion that government bureaucracies should mediate collective social relations, arguing instead for alternatives. Unlike the example on Edisto, which takes as given that society will remain organized around private property and automobiles, some elements of the climate justice movement have promoted a different vision, in which communities work "in, against, and beyond" governments (Holloway and Sergi 2010). Routledge et al. (2018) show how climate justice activists both make claims "in and against" governments, by demanding action and legislation *from* governments, while also organizing to rework socio-natural relations in ways that are not mediated by, or are *beyond* governments. The GalGael Trust, in Glasgow, Scotland, for example, led the organization of the Govan Free State during the 2021 Council of Parties meeting. The "Govan Free State" was "a playful exploration of power and legitimacy in the face of climate emergency. A declaration of independence, interdependence and radical dependence." Visitors to the site were encouraged to "declare yourself welcome," a direct subversion of the role of state power in deciding who belongs, and who can enter and leave. "No one is coming to save us," they claim, inviting people to reimagine their own and their collective relationship to nature that is not mediated through governments.

Public Political Ecologies

The field of political ecology attracts scholar-activists who aim to use the conceptual resources of the field to inform and contribute to on-the-ground struggles and social movements. In this final section, we illustrate how political ecologists learn from and support marginalized, minoritized, and racialized peoples in their struggles for self-determination and a more even distribution of wealth and resources. Osborne proposes the phrase "public political ecology" as a way to practice "engaged scholarship in this moment of ecological crisis" (2017). Osborne draws on the work of Antonio Gramsci to advocate for a "praxis" (theory informed practice) of political ecology that entails engaged, problem-oriented research that produces scholarship that can resource social movements (see also Derickson and Routledge 2015). Scholars like Carolyn Finney (2014) and Jill Harrison (Ottinger 2022) use their expertise and critical perspectives to inform and shape government agencies, while others draw from experience in and relationships with social movements to inform their scholarship (see Routledge and Derickson 2015). Scholars like Laura Pulido have highlighted the way environmental social movements and the fields of environmental studies are often dominated not only by white people, but the knowledges, cultural norms, and utopian visions of Western forms of thought. These scholars have diversified the field of political ecology by thinking with Black (Roane 2018), Latinx (Pulido and Alaimo 2019), and Indigenous environmentalisms (Todd 2018) and political ecologies.

Nature as a Field of Power, Difference, and Resistance

Sitting in a Gullah/Geechee restaurant in Savannah, Georgia, Anita Collins recalls the ingredients of the "crab boil" of her youth: crab, potatoes, sausage, and turtle eggs. The eggs, she recalls, were in their shell in the stew, and she and her family would use a narrow pin to pierce them. She recalls the inside being sweet liquid, but can't quite remember because it has been decades since she has eaten a turtle egg. "We are not the reason they had to ban" the practice of eating eggs, she says. "It was subsistence for us, why would we have depleted it?" Earlier in the day we had joined Queen Quet, Chieftess and Head of State of the Gullah/Geechee Nation at the Chatham County Commissioners meeting, where a declaration recognizing Gullah/Geechee Nation Appreciation Week had been read. After the meeting, a man who had attended the meeting for an unrelated reason pulled Queen Quet aside and told her that when she spoke in Gullah, he immediately recalled his uncle, who had used the same words when the man was a child. He was emotional as he told her that in that moment he realized that his uncle's way of speaking was actually evidence of his family's connection to Gullah/Geechee culture, a connection that he had never before known.

In this chapter, we have sought to illustrate how political ecology interrogates the tight coupling between power and nature, to illustrate how the economy works alongside cultural dominance, and how claims about nature and the environment have consistently worked to reinforce and create relationships of racialization, marginalization, minoritization. To do so, political ecologists interrogate the relationship between culture, power, and nature, to illustrate how cultures and practices, like speaking Gullah and eating turtle eggs, are entwined with the overdevelopment of Edisto Island. The narratives that emerge to explain resource scarcity often obscure the way that our social and economic systems produce these outcomes. Political ecology aims to illustrate these relations, often in ways that learn from and resource community-based struggles.

This emphasis means that as social struggles and movements rework in response to changing political and ecological landscapes, so too does the field of political ecology. As a result, future directions in political ecology are increasingly focused on critically interrogating the value and role of scholarly knowledge for social justice and how forms of scholarly knowledge can support existing forms of experimentation. Increasingly, political ecologists are considering the role that scholarly institutions play in relation to struggles over land, nature, power, and capital, or what Derickson (2022) has called the "social relations engendered in the act of knowing." Work in this vein foregrounds actually existing social movements and struggles to rework the social relations that articulate through discourses around—and the governance of—nature while carefully considering whether and how scholarly forms of knowing meaningfully contribute, or "resource" those struggles (see Box 4.1 on the CREATE Initiative as well as this volume for work on public political ecologies).

Box 4.1 Knowledge for social change: the CREATE Initiative

The CREATE Initiative, housed at the University of Minnesota, aims to marshal the resources of the research university to ask and answer questions that are a priority to communities that have not historically shaped disciplinary research agendas. At the outset, the project assembled a Policy Think Tank, composed of environmental justice advocates in communities in Florida, South Carolina, Atlanta, Minneapolis and the greater Twin Cities, places where project co-Directors Kate Derickson and Bonnie Keeler had existing partnerships. Once assembled, the Policy Think Tank identified a set of research questions that would address the needs and priorities of their communities. The Think Tank identified "investment without displacement" and "affordable housing in quality environments" as collective goals that could benefit from scholarly research. CREATE then recruited and trained two cohorts of graduate students in community-engaged research practices, produced a public facing toolkit to address "green gentrification" and developed an overarching research agenda to explore the coupling of investments in the natural environment and the racial wealth gap (see Keeler et al. 2020; Ehrman-Solberg et al. 2022; Derickson, Klein, and Keeler 2021). By working with community partners to identify research priorities and carefully considering the experiences that communities have working with academics (see Derickson 2022), CREATE aims to center research and knowledge production as one site of potential transformative social change.

Work in this vein tends to focus on and express optimism about the proliferation of various forms of resistance in a range of places. Stephanie Wakefield's (2020) investigation of what she calls "back loop" formations, or a moment of system change where experimentation might lead to new system outcomes, directs attention to the myriad ways that reworking social and material relations in the context of the Anthropocene is already happening. Likewise, the Seeds of a Good Anthropocene project (https://goodanthropocenes.net), led by the Stockholm Resilience Centre and others, highlights existing "bright spots" of ways of being in relation to a changing environment that can provide insight into how we might imagine radically different futures that are also just.

Finally, influenced by an ascendent social movement to abolish the prison industrial complex and rethink policing and public safety, political ecologists (see Heynen and Ybarra 2021) have turned to the body of work in abolition geography (Gilmore 2022) to better understand how ecological and carceral landscapes are imbricated (Hall 2020). As climate change creates new forms of scarcity that are policed in new ways, the field of abolition geography and abolition ecology charts an urgent research agenda. In the Gullah/Geechee Nation, for instance, the state of South Carolina's Department of Natural Resources is leveraging threats from environmental change to advocate for more policing resources. Work on abolition geographies helps situate calls for more policing resources in the long history of Atlantic racial capitalism to better understand how nature is at the site of racialized dispossession, as well as the urgency of abolition for environmental futures.

Environmental change promises to shape the political future, both as a site of contestation itself, and a process in relation to the other sites identified. Struggles over gentrification, infrastructure, extraction and conservation are all profoundly shifting in the face of a changing climate. As communities attempt to navigate the unevenly borne consequences of changing climate, nature as a site of politics will come increasingly to the fore. Political ecology, with its focus on the relationship between power and nature, offers tools for understanding—and contesting—these dynamics.

References

Angelo, H. (2019). Added value? Denaturalizing the "good" of urban greening. *Geography Compass*, 13(8), e12459.

Angelo, H., MacFarlane, K., Sirigotis, J., and Millard-Ball, A. (2022). Missing the housing for the trees: Equity in urban climate planning. *Journal of Planning Education and Research*, 0739456X211072527.

Anguelovski, I., Connolly, J. J., Garcia-Lamarca, M., Cole, H., and Pearsall, H. (2018). New scholarly pathways on green gentrification: What does the urban "green turn" mean and where is it going? *Progress in Human Geography*, 43(6), 1064–1086. 0309132518803799.

Anguelovski, I., Shi, L., Chu, E., Gallagher, D., Goh, K., Lamb, Z., Reeve, K., and Teicher, H. (2016). Equity impacts of urban land use planning for climate adaptation: Critical perspectives from the global north and south. *Journal of Planning Education and Research*, 36(3), 333–348.

Anguelovski, I., Brand, A.L., Connolly, J.J., Corbera, E., Kotsila, P., Steil, J., Garcia-Lamarca, M., Triguero-Mas, M., Cole, H., Baró, F. and Langemeyer, J. (2020). Expanding the boundaries of justice in urban greening scholarship: Toward an emancipatory, antisubordination, intersectional, and relational approach. *Annals of the American Association of Geographers*, 110(6), 1743–1769.

Atwell, M. (2019). "Searching out the hidden stories of South Carolina's Gullah Country" *New York Times* www.nytimes.com/2019/04/15/travel/south-carolina-gullah-geechee-low-country.html (accessed October 1, 2022).

Bondesson, S. (2022). *Empowerment and Social Justice in the Wake of Disasters: Occupy Sandy in Rockaway After Hurricane Sandy*, USA. Routledge.

Brondízio, E. S., Cak, A., Caldas, M. M., Mena, C., Bilsborrow, R., Futemma, C. T., Ludewigs, T., Moran, E. F., and Batistella, M. (2013). Small farmers and deforestation in Amazonia. In *Amazonia and Global Change* (pp. 117–143). American Geophysical Union (AGU). https://doi.org/10.1029/2008GM000737

Bullard, R. (2004). *Environmental Justice and Communities of Color*. San Francisco: Sierra Club Books.

Checker, M. (2011). Wiped out by the "greenwave": Environmental gentrification and the paradoxical politics of urban sustainability. *City and Society*, 23(2), 210–229.

Cleary, D. (1993). After the frontier: Problems with political economy in the modern Brazilian Amazon. *Journal of Latin American Studies*, 25(2), 331–349.

Cronon, W. (1996). The trouble with wilderness: Or, getting back to the wrong nature. *Environmental History*, 1(1), 7–28.

Curran, W., and Hamilton, T. (2012). Just green enough: Contesting environmental gentrification in Greenpoint, Brooklyn. *Local Environment*, 17(9), 1027–1042.

Derickson, K. D. (2014). The racial politics of neoliberal regulation in post-Katrina Mississippi. *Annals of the Association of American Geographers*, 104(4), 889–902.

Derickson, K. D., and MacKinnon, D. (2015). Toward an interim politics of resourcefulness for the Anthropocene. *Annals of the Association of American Geographers*, 105(2), 304–312.

Derickson, K., Klein, M., and Keeler, B. L. (2021). Reflections on crafting a policy toolkit for equitable green infrastructure. *NPJ Urban Sustainability*, 1(1), 1–4.

Derickson, K. D. (2022). Disrupting displacements: Making knowledges for futures otherwise in Gullah/Geechee Nation. *Annals of the American Association of Geographers*, 112(3), 838–846.

Derickson, K.D. and Routledge, P. (2015). Resourcing scholar-activism: Collaboration, transformation, and the production of knowledge. *The Professional Geographer*, 67(1), 1–7.

Dooling, S. (2009). Ecological gentrification: A research agenda exploring justice in the city. *International Journal of Urban and Regional Research*, 33(3), 621–639.

Dooling, S. (2017). Making just green enough advocacy resilient: Diverse economies, ecosystem engineers and livelihood strategies for low-carbon futures. In *Just Green Enough* (pp. 47–60). New York: Routledge.

Ehrman-Solberg, K., Keeler, B., Derickson, K., and Delegard, K. (2022). Mapping a path towards equity: Reflections on a co-creative community praxis. *GeoJournal*, 87(Suppl 2), 185–194.

Escobar, A. (1998). Whose knowledge, whose nature? Biodiversity, conservation, and the political ecology of social movements. *Journal of Political Ecology* 5(1), 53–82.

FAO and FILAC. (2021). *Forest Governance by Indigenous and Tribal Peoples. An Opportunity for Climate Action in Latin America and the Caribbean*. United Nations: Food and Agriculture Organization (FAO). https://doi.org/10.4060/cb2953en

Finewood, M. H., Matsler, A. M., and Zivkovich, J. (2019). Green infrastructure and the hidden politics of urban stormwater governance in a postindustrial city. *Annals of the American Association of Geographers*, 109(3), 909–925.

Finney, C. (2014). *Black Faces, White Spaces: Reimagining the Relationship of African Americans to the Great Outdoors*. UNC Press Books.

Gandy, M. (2003). *Concrete and Clay: Reworking Nature in New York City*. MIT Press.

Gilmore, R. W. (2022). *Abolition Geography: Essays Towards Liberation*. London: Verso Books.

Hall, C (2020). *A Prison in the Woods*, Amherst, MA: University of Massachusetts Press.

Harrison, J. L. (2019). *From the Inside Out: The Fight for Environmental Justice within Government Agencies*. Cambridge: MIT Press.

Hecht, S., and Cockburn, A. (2011). *The Fate of the Forest: Developers, Destroyers, and Defenders of the Amazon, Updated Edition* (Updated ed. edition). Chicago: The University of Chicago Press.

Heynen, N. (2016). Urban political ecology II: The abolitionist century. *Progress in Human Geography*, 40(6), 839–845.

Heynen, N., and Ybarra, M. (2021). On abolition ecologies and making "freedom as a place". *Antipode*, 53(1), 21–35.

Holloway, J., and Sergi, V. (2010). *Crack Capitalism* (Vol. 40). London: Pluto Press.

Janiskee, Bob (2009). Sullivan's Island was the African American Ellis Island. www.nationalparkstraveler.org/2009/03/sullivan-s-island-african-american-ellis-island (accessed January 10, 2022).

Kaika, M. (2017). 'Don't call me resilient again!': The New Urban Agenda as immunology ... or ... what happens when communities refuse to be vaccinated with 'smart cities' and indicators. *Environment and Urbanization*, 29(1), 89–102.

Keeler, B. L., Derickson, K. D., Waters, H., and Walker, R. (2020). Advancing water equity demands new approaches to sustainability science. *One Earth*, 2(3), 211–213.

Kosek, J. (2006). *Understories: The Political Life of Forests in New Mexico*. Durham, NC: Duke University Press.

Li, T. M. (2007). *The Will to Improve*. Durham, NC: Duke University Press.

Li, T.M. (2014). *Land's End: Capitalist Relations on an Indigenous Frontier*. Durham, NC: Duke University Press.

Losekann, C., and Milanez, B. (2021). Mining disaster in the Doce River: Dilemma between governance and participation. *Current Sociology*, 71(7), 1255–1273.

Lubitow, A., and Miller, T. R. (2013). Contesting sustainability: Bikes, race, and politics in Portlandia. *Environmental Justice*, 6(4), 121–126.

Luft, R. (2012). 14. Community organizing in the Katrina diaspora: Race, gender, and the case of the People's Hurricane Relief Fund. In: Weber, L. and Peek, L. (ed.). *Displaced: Life in the Katrina Diaspora*. New York, USA: University of Texas Press, pp. 233–256. https://doi.org/10.7560/735774-019

Massé, F., Givá, N., and Lunstrum, E. (2021). A feminist political ecology of wildlife crime: The gendered dimensions of a poaching economy and its impacts in Southern Africa. *Geoforum*, 126, 205–214.

Matthews, T., Lo, A. Y., and Byrne, J. A. (2015). Reconceptualizing green infrastructure for climate change adaptation: Barriers to adoption and drivers for uptake by spatial planners. *Landscape and Urban Planning*, 138, 155–163.

McClintock, N. (2020). Cultivating (a) sustainability capital: Urban agriculture, ecogentrification, and the uneven valorization of social reproduction. In *Social Justice and the City* (pp. 279–290). Routledge.

McGurty, E. M. (2000). Warren County, NC, and the emergence of the environmental justice movement: Unlikely coalitions and shared meanings in local collective action. *Society and Natural Resources*, 13(4), 373–387.

Mollett, S. and Faria, C. (2013). Messing with gender in feminist political ecology. *Geoforum*, 45, 116–125.

Moore, D. (2005). *Suffering For Territory: Race, Place, and Power*. Durham: Duke University Press.

Moran, E. F. (1976). *Agricultural Development in the Transamazon Highway*. Bloomington, IN: Indiana University Press.

Morello-Frosch, R., Pastor, M., and Sadd, J. (2002). Integrating environmental justice and the precautionary principle in research and policy making: The case of ambient air toxics exposures and health risks among schoolchildren in Los Angeles. *The ANNALS of the American Academy of Political and Social Science*, 584(1), 47–68.

Nepstad, D., Schwartzman, S., Bamberger, B., Santilli, M., Ray, D., Schlesinger, P., Lefebvre, P., Alencar, A., Prinz, E., Fiske, G., and Rolla, A. (2006). Inhibition of Amazon deforestation and fire by parks and indigenous lands. *Conservation Biology: The Journal of the Society for Conservation Biology*, 20(1), 65–73.

Osborne, T. (2017). Public political ecology: A community of praxis for earth stewardship. *Journal of Political Ecology*, 24(1), 843–860.

Ottinger, G (2022). Why does the state allow environmental inequalities to persist? Talking with Jill Lindsey Harrison. www.publicbooks.org/environmental-justice-government-agencies-jill-lindsey-harrison/ (accessed June 2, 2022).

Pearsall, H. (2013). Superfund me: A study of resistance to gentrification in New York City. *Urban Studies*, 50(11), 2293–2310.

Pellow, D. N. (2016). Critical environmental justice studies. In *Resilience, Environmental Justice and the City* (pp. 25–44). New York: Routledge.

Peluso, N. L. (2017). Plantations and mines: Resource frontiers and the politics of the smallholder slot. *The Journal of Peasant Studies*, 44(4), 834–869.

Pulido, Laura. (2016). Flint, environmental racism, and racial capitalism. *Capitalism Nature Socialism* 27(3): 1–16.

Pulido, L., and Alaimo, S. (2019). *Latinx Environmentalisms: Place, Justice, and the Decolonial*. Philadelphia: Temple University Press.
Quastel, N. (2009). Political ecologies of gentrification. *Urban Geography*, 30(7), 694–725.
Ranganathan, M. (2016). Thinking with Flint: Racial liberalism and the roots of an American water tragedy. *Capitalism Nature Socialism*, 27(3), 17–33.
Ranganathan, M. (2022). Caste, racialization, and the making of environmental unfreedoms in urban India. *Ethnic and Racial Studies*, 45(2), 257–277.
Reimann, M. (2017). The EPA chose this county for a toxic dump because its residents were "few, black, and poor". https://timeline.com/warren-county-dumping-race-4d8fe8de06cb
Rigolon, A., and Németh, J. (2018). "We're not in the business of housing": Environmental gentrification and the nonprofitization of green infrastructure projects. *Cities*, 81, 71–80.
Roane, J. T. (2018). Plotting the Black commons. *Souls*, 20(3), 239–266.
Rockefeller Foundation (2022). 100 resilient cities. www.rockefellerfoundation.org/100-resilient-cities/ (accessed June 23, 2022).
Rumbach, A., Sullivan, E., and Makarewicz, C. (2020). Mobile home parks and disasters: Understanding risk to America's third housing type. *Natural Hazards Review*, 21(2).
Routledge, P., and Derickson, K. D. (2015). Situated solidarities and the practice of scholar-activism. *Environment and Planning D: Society and Space*, 33(3), 391–407.
Routledge, P., Cumbers, A., and Derickson, K. D. (2018). States of just transition: Realising climate justice through and against the state. *Geoforum*, 88, 78–86.
Safransky, S. (2014). Greening the urban frontier: Race, property, and resettlement in Detroit. *Geoforum*, 56, 237–248.
Smith, N. (2006). There's no such thing as a natural disaster. Understanding Katrina: Perspectives from the social sciences https://items.ssrc.org/understanding-katrina/theres-no-such-thing-as-a-natural-disaster/ (accessed July 28, 2022).
Solnit, R. (2020). Who will win the fight for a post-Coronavirus America? *The New York Times*, March 29.
Sundberg, J. (2016). Feminist political ecology. In *International Encyclopedia of Geography: People, the Earth, Environment and Technology*. Wiley.
Sze, J., and Yeampierre, E. (2017). Just transition and just green enough: Climate justice, economic development and community resilience. In *Just Green Enough* (pp. 61–73). New York: Routledge.
Todd, Z. (2018). Refracting the state through human–fish relations. *Decolonization: Indigeneity, Education and Society*, 7(1), 60–75.
Vandergeest, P. (2003). Racialization and citizenship in Thai forest politics. *Society and Natural Resources*, 16(1), 19–37.
Wainwright, J. (2011). *Decolonizing Development: Colonial Power and the Maya*. New York: Wiley-Blackwelll.
Wachsmuth, D. and Angelo, H. (2018). Green and gray: New ideologies of nature in urban sustainability policy. *Annals of the American Association of Geographers*, 108(4), 1038–1056.
Wakefield, S. (2020). *Anthropocene back loop: Experimentation in unsafe operating space* (p. 215). Open Humanities Press.
Walker, R. H., Ramer, H., Derickson, K. D., and Keeler, B. L. (2023). Making the City of Lakes: Whiteness, nature, and urban development in Minneapolis. *Annals of the American Association of Geographers*, 113(7), 1615–1629.
West, P., Igoe, J., and Brockington, D., (2006). Parks and peoples: The social impact of protected areas. *Annu. Rev. Anthropol.*, 35, 251–277.

Part II

Making Nature Knowable

5 Constructing Nature

Construal, Constitution, Composition, and the Politics of Ecology

Noel Castree

For those of us fascinated by the relationships between people and the biophysical world, the themes of destruction and loss seem ever present. Anxious since my childhood about the escalating scope, scale and magnitude of human impacts on the Earth, over 25 years ago—at that point a PhD student in Vancouver—I looked to the past for lessons. The virtual extermination, by 1890, of the once huge North Pacific seal herd at the hands of Americans, Canadians and Russians seeking valuable seal pelts caught my attention. Killed both at sea and on the tiny Pribilof Islands (where mating and birthing occurred each summer), a population of over 3.5 million seals was reduced to just a few thousand within 30 years. Why did competition for shares of a finite natural resource lead these three countries to almost destroy it before, in the nick of time, they agreed a plan to manage the herd sustainably? In one of my first published articles, I sought to provide an answer (Castree 1997)—this was a time when the world's governments were only just waking up to the seriousness of anthropogenic climate change, an altogether larger human disruption of a natural phenomenon. The seal case, I reasoned, might provide an instructive precedent. But, of course, destruction isn't a one-way street. Living, as I now do, in New South Wales, I know this all too well: in the last 36 months enormous wildfires and once-in-500-years floods have wreaked havoc on houses, farms, roads, electricity networks, railway lines and peoples' livelihoods. Climate change will intensify the harm in future. Just as we have the capacity to destroy the biophysical world that sustains us, so that biophysical world continues to tear the fabric of economy and society in many places around the globe—including places in very wealthy nations like Australia (Figures 5.1 and 5.2).

And yet in this chapter I will write about the *construction* of (what we by convention call) "nature"—how it occurs, why it matters and how political ecologists (broadly defined) have sought to investigate it. This may seem odd. Surely, nature can be variously utilized, accessed, degraded, exploited, polluted, damaged, protected, conserved, fought over, and protected by people, but rarely (if ever) "constructed"? For to construct is to fabricate, and human constructions are therefore by definition not "natural" (unless one classifies *all* human activities as natural!). What's more, an awful lot of the biophysical world is too complicated and large to be amenable to "construction," as the New South Wales fires and floods remind us. The ideal (or nightmare) of a fully programmed Earth where humans control nature and avoid its vagaries is the stuff of science fiction.

When I researched my article about the "war against the seals" in the 1990s, a "social constructionist" turn had recently occurred in the Anglophone social sciences and humanities. Publications with arresting, counter-intuitive titles appeared—such as "the social construction of the ocean" (Steinberg 2001). As I will explain in this chapter, the turn was

DOI: 10.4324/9781003165477-8

Figure 5.1 Timber harvesting in Germany.
Source: https://commons.wikimedia.org/wiki/File:Deforested_area,_Goldisthal,_2023-05-20.jpg

Figure 5.2 A 2020 post-flood clean-up in Hyderabad, India.
Source: https://commons.wikimedia.org/wiki/File:2020_Hyderabad_floods.jpg

already evident in the field of political ecology (by then around 25 years old), even in cases when the term "social constructionism" was not used by practitioners. In the last 20 years or so the social sciences and humanities have, in many respects, taken a "post-constructionist" turn (e.g., Whatmore 2002). Much more attention has been paid to the material properties of the non-human world and their role in enabling, or disrupting, social thought and action (see, for instance, the book *New Materialisms* edited by Coole and Frost 2010). Again, political ecologists have been part of this (as Rosemary Collard explains in Chapter 18). But, as I will show below, "the social construction of (what we call) nature" rightly remains an important theme in the research undertaken by political ecologists. It's not been—and should not be—eclipsed, though these days many rightly prefer less muscular words than "construction." This sentiment finds an echo in the chapters on "counting," "representing" and "producing" in this volume.

As demonstrated in the sections that follow, to understand things like environmental use, access, loss, damage, pollution, conservation, conflict and so on we (still) need to keep a close eye on the what, why, and how of "social construction." This means that the process and outcomes of social construction are integral to the politics surrounding peoples' interactions with the biophysical world at a range of spatial and temporal scales. It means too that there is complementarity, rather than a contradiction, between the proposition that "nature" is socially constructed and the proposition that it's *also* possessed of attributes and powers irreducible to that process. Terms such as "hybridity," "assemblages," "actor-networks," and "socio-nature" have been employed by political ecologists and others to capture this intimate entanglement of "social" phenomena and "environmental" phenomena. In light of this, the terms "construal," "constitution," and "composition" are, in my view, now apt replacements for the portmanteau term "construction" as political ecologists refine their terminology to make sense of a world of intersecting societal and non-human phenomena. Put differently, they refer to three related dimensions of construction as part of processes that are more-than-social. Construal is about depiction (or framing); constitution is about the material making of phenomena; and composition is about assembling materials that are made or found. This is the conclusion I drive towards as the chapter proceeds, some 25 years after the term "social construction" was first embraced by political ecologists (Escobar 1996).

Constructive Critique: Deconstruction, Denaturalization, Anti-essentialism, and the Politicization of Ecology

Before we consider various examples of constructionist analysis in political ecology, it's useful to define our terms and to explain why social constructionism became a hugely influential approach in multiple fields of university research and teaching from the mid-1980s onwards. This—bear with me—involves a bit of abstract thinking. "Construction" is a process but also a product, as when a builder combines concrete, timber, bricks, windows, wiring, nails, glue, tiles etc. that result in a new house. So construction is about *human agents* (possessed of intentions defined by their contexts), *materials* (real world phenomena of various kinds, be they physical or mental, that are made or used by those agents) and *outcomes* (physical things but also "immaterial" things, such as maps and scientifically authorized public policy targets). There are many arenas in which construction occurs, many reasons why it occurs and many particular forms it can take (compare a work of art to a motorbike to a tunnel to a book like this one). Philosophically, a "constructionist" approach is the opposite of a "naturalist" one. The latter commits researchers to (i) the ontological belief that any given part of material reality has one (and just one) set of characteristics and (ii) the epistemological belief that, with the "right" analytical methods, these characteristics can be understood truthfully and false knowledge about them eliminated over time.

In Anglophone social science and the humanities, the social constructionist approach was first codified in the late 1960s. Sociologists Peter Berger and Thomas Luckerman published *The Social Construction of Reality* (1966). They argued that specific social groups (e.g., religious fundamentalists), but also whole societies, come to perceive reality in various ways that become "sedimented" and institutionalized over time. This implied that even when inhabiting the same material environment, different collectives of people would not necessarily think and behave in the same way: their "reality" would vary, reflecting humans' capacity to create their own mental worlds and act accordingly. In the 1970s and 80s, the constructionist approach gathered momentum in several disciplines (e.g., philosophy and literary criticism, where translations of the late Jacques Derrida's "deconstructive" approach to language and sense-making had a huge impact in British and North American academia). Major works followed about the social construction of crime, illness, gender, sexuality, science, "race," risk, ignorance, intelligence, genes, and so on. Many books have been written that aim to map the varieties of constructionist thinking and to address critiques of the same (e.g., Hacking 2000; Searle 2010; Soper 1995). Such thinking proved very controversial from early on, being variously criticized for promoting relativistic, voluntarist, anti-realist, anti-scientific and anti-environmental beliefs—as if societies can think and act just as they please, regardless (see, for instance, Sokal and Bricmont 1999; Soule and Lease 1995). Critics insisted that there is a reality existing *beyond* people's contingent constructions, one that can be known and valued independently (e.g., using scientific methods).

This is not the place to revisit the heated debates of the past, which have helped to spur a "post-constructionist" (or so-called "new materialist") turn since around 2000 (e.g., see Jane Bennett's influential book *Vibrant Matter*—Bennett 2010). Instead, it's important to understand some of the key reasons why the constructionist approach was so widely advocated from the 1970s onwards, its legacy still being very evident half a century later. A path-breaking article authored by the Marxist geographer David Harvey way back (1974) usefully illustrates the rationale for "constructionism" (even though he did not use this word at that time and his contribution was limited to his own discipline, Geography).

The article was called "Population, resources and the ideology of science." It noted the dominance of neo-Malthusian thinking in the West about the supposed crisis caused by looming scarcities of oil, food and other key resources as world population numbers exploded after 1945. Such thinking was predicated on the ideas that (i) "natural resources" were finite and (ii) their "limits" could be specified scientifically so as to indicate the maximum human population that could be sustained at a reasonable standard of living. Against this, Harvey showed that what *counts* as a "resource," so too its "limits," was defined *relative to* the social categorizations, social relationships, social values, and particular technologies prevailing in any given society. By pretending to specify *absolute* resource constraints and *absolute* population maxima, Harvey argued that powerful actors in society threatened to reduce the range of choice that humans possess in how they think and act in the material world. Authoritative (scientific) references to nature were used to justify important, contentious policy choices (as such the male sterilization programs rolled-out in parts of India in the 1970s). The space for alternative thought and action was thereby foreclosed by hitching the "is" and the "ought" together too tightly.

In sum, the constructionist perspective allows its advocates to (i) critique that which seems natural, normal, objective or given, and more positively (ii) show that alternative ways of knowing and acting towards the biophysical world are both available and worthy of consideration. It opposes attempts to attribute events, actions, or political decisions to supposed facts about the intrinsic properties of a world beyond our discourses and modes of social organization (see Daston 2019). Properly specified, this approach retains considerable analytical and normative value for political ecologists and others. Indeed, a constructionist perspective lives on within a range of erstwhile post-constructionist approaches now central to political ecology (e.g., see Collard, Dempsey, and Sundberg 2015). The task is to understand what, in different contexts, is meant by "construction" and by "nature" (or its semiotic stand-ins like ecology, resources, land, environment, biology and so on). The early political ecologists did not use the term "social construction," and tended to focus on land use and natural hazards in the developing world (Box 5.1). But from the 1990s onwards that changed, as the field became influenced by post-structural, post-colonial and post-modern thinking (see, for instance, Escobar 1996) to complement its prior focus on the political economy of farming and natural hazard impacts. Let us now unpack what constructionism means and how it retains analytical and political value for political ecologists. In the next three sections I cite classic and more recent studies.

Box 5.1 Politicizing "ecology" using a constructionist approach in all but name: key early studies

Political ecology is today a very large, vibrant, and diverse field of research, but its founding studies were clearly constructionist in their ethos (even though "social constructionism" was not a common term of use in those early publications, whereas "political economy" was). These studies took issue with dominant understandings of "environmental realities," and thereby opened up new analytical and political possibilities. At that time (the 1970s), ecology was often called "the subversive science" because various facts about species loss and ecosystem disruption at human hands were said to "speak truth to power," mandating governments who chose to pay attention to strictly regulate how people used the non-human world. The early political ecologists, however, turned this on its head. They showed that social power and the political coercion of people *frequently worked through authoritative claims made about "ecological facts."* Mike Watts's (1983) *Silent Violence* and Piers Blaikie's (1985) *The Political Economy of Soil Erosion in Developing Countries* are classic examples. Watts studied the heightened incidence of famines in rural northern Nigeria subsequent to British colonization in the late nineteenth century. Typically understood as "natural disasters" triggered by droughts and reflecting poor adaptation practices by rural Nigerians, Watts showed that these Nigerians had, in fact, historically been very *well adapted* to the vagaries of climate in the region. Food shortages were, he showed, caused by the impacts of British land use and taxation policies, which altered colonized people's relationships to crop production, land availability, water conservation and food storage. The "ecological fact" of a drought, Watts demonstrated, only led to food shortages when the prevailing societal arrangements (the "social formation," as he called it using Marxist language) exposed subordinated groups to climatic vulnerability by impairing their livelihoods. Blaikie's book, meanwhile, focused on how people impacted the physical environment rather than vice versa. Since the United States' "dustbowl" disaster of the 1930s, there has been global concern about soil erosion, especially in the developing world. After 1945, erosion was often blamed by Western-led agencies (e.g., the World Food Programme of the United Nations) and colonial governments on poor land management by peasant farmers. Blaikie, however, showed that local erosion cases could typically be traced to global forces that pressured farmers into making sub-optimal decisions about arable and pastoral production. The farmers, he demonstrated, did not lack an understanding of their land but were locked into decisions conditioned by national and overseas markets their governments had incentivized them to sell their produce to. In short, Watts and Blaikie questioned expert claims made about farmers' apparent failure to understand drought risk and soil capacities; and they showed how sets of societal relations, power dynamics and institutions served to define the material engagements that front-line people experienced with nature. The impacts caused by (and inflicted on) the non-human world were, Watts and Blaikie demonstrated, only partly explained by the material actualities of drought and soil quality.

Representing the Non-human World: The Construction and Contestation of Meanings (Linguistically and Otherwise)

The non-human world cannot speak for itself. It must be made sense of or *construed*: meaning is extracted from it—or, we might say, imposed upon it—by using language and other modes of depiction (e.g., aerial photographs; soil quality surveys). Talk about "nature" is thus, in some measure, always talk about ourselves. Yet there is no universal mode of representation shared by all humans. This immediately raises the question: who is representing what, why and with what goals in mind? In turn, this question alerts us to the possibility that certain representations of the non-human world may, in any given situation, become normalized over time while others are marginalized or resisted. Is this balance of dominance and exclusion justified? As we will see below, political ecologists have long focused on "the politics of representing nature": that is, the values, norms and desires baked into the linguistic and visual cakes of various groups seeking to speak for land, water, sub-surface minerals, earthquakes, volcanoes and so on. Even scientists, supposedly neutral in their search for truth, encase value judgements in their research, while this research can be used by others for political purposes. A number of now classic works focused on certified and authorized "experts" in developing world contexts—they were shown to use their authority to (innocently or knowingly) perpetuate partial or false depictions of the "realities" of vegetation, soils and water. The depictions, in turn, undergirded political actions that negatively affected peasants, small holders and other relatively powerless land- and resource-users.

For instance, consider James Fairhead and Melissa Leach's influential book *Misreading The African Landscape* (1996). It was a study of land cover in West Africa, a region colonized by the British and the French from around 1880 onwards. Fairhead and Leach focused on the transition zone between tropical forest and savannah grasslands. They showed how colonial officials mistook isolated clusters of trees as remnants of formerly forested areas. The officials inferred that villagers had mismanaged the land, degrading it to a savannah state. Against this, Leach and Fairhead assembled aerial data, archival information and ethnographic data to show that, in fact, local people were responsible for the forest islands' very existence. The islands were evidence of deliberate *afforestation* not deforestation, in a semi-arid environment. Fairhead and Leach's study showed how "group think" can occur on the basis of limited evidence, to the point that prejudicial myths get confused for environmental realities. In the same vein, other political ecologists debunked myths about soil erosion, biodiversity loss and desertification in the developing world (e.g., Davis 2016; Forsyth 1996; Simon and Peterson 2019; Benjaminsen 2021). Still others have shown the arbitrary (non-natural) definitions of ecological baselines used to define a "natural ecosystem" (Neumann 1998; Sayre 2008) or natural resource (Demeritt 2001).

This kind of research has used alternative bodies of evidence to "debunk" authoritative (mis)understandings of ecology, and to show the harm these representations can do. Other work has sought to valorize the understandings of ordinary land and water users, whose "framing" of reality reflects different values, norms and priorities. This framing has meant that different "matters of fact" become "facts that matter" when compared to what experts and bureaucrats choose to see on the ground. Political ecologists have illuminated the categories, evidence and knowledge possessed by peasants, small holders, Indigenous peoples and others, including the "counter-maps" that get fashioned and used to push-back against powerful actors (e.g., Bryan and Wood 2015; Kelley 2018). Their research deliberately challenges the belief that "scientific" or "expert" representations of biophysical reality are somehow superior compared with other ways of knowing and depicting the world. An early example was Nancy Peluso's *Rich Forests, Poor People* (1992). A now classic study of government-led forest management in Indonesia, Peluso showed how rural peoples' knowledge was ignored, and their actions criminalized, because they were deemed threatening to the state's favored resource practices. Yet this knowledge reflected those peoples' legitimate desire to maintain their livelihoods and culture. It's being discounted was not a question of it being less scientific or truthful, and more that it didn't conform with government-imposed ways of defining "proper" land use. Peluso's and other people's studies have shown us that struggles over whose knowledge of ecology counts can be violent or more a battle for ideational "hegemony" to secure consent. In part, what's at stake in these struggles is the right to define *counts* as "destruction," "restoration" or "conservation" of particular resources and landscapes, and who is responsible for each. A recent, powerful example can be founded in Mara Goldman's book *Narrating Nature: Wildlife Conservation and Maasai Ways of Knowing* (2020).

Constitutive Fictions: Enacting Eco-governmentalities

The research highlighted above has accented the so-called "materiality" and "performativity" of linguistic and pictorial representations. That is, these representations are shown to be as real and potentially influential as the realities they ostensibly mirror. But some political ecologists have gone further. They've argued that representations of the biophysical world do not merely *construe*; they also *constitute* that which they depict, at least in part. Many have been inspired by Michel Foucault's (1926–84) hugely influential research into modalities of "power-knowledge" and "regimes of truth." Foucault's historical analyses of pre- and post-eighteenth-century Europe revealed that different frameworks of knowing seeped into the pores of society, thereby exerting power by constituting certain sorts of persons (or subjects) who were habituated to thinking and acting in particular ways. Power thereby did not present itself *as* power because it assumed quite mundane, quotidian forms. In other words, it was not only operative in "obvious" arenas like prisons. Several political ecologists have adapted Foucault to argue that ways of knowing the non-human world have been vehicles for controlling "the conduct of conduct" in different societies—or what's sometimes called "eco-governmentality," "green governmentality" or else "environmentality" (Agrawal 2005). These terms riff-off Foucault's notion of "governmentality." This describes processes of governing "diffusely" or in a "capillary" way beyond the formal apparatus of a leader, laws, the military or government agencies.

A key early study was anthropologist Tania Li's (2007) book *The Will to Improve*. It analyzed decades of government efforts in colonial and post-colonial Indonesia to improve livelihoods in rural areas. The ostensibly beneficial programs Li analyzed had two elements: the identification of "problems" (e.g., insufficient rice production) and the rendering technical of "solutions" such that ordinary land users were enrolled into schemes of "rational management." Li's book explored the "contact points" between governmental ways of seeing and the conduct of rural residents who were being invited to see themselves as the recipients of government assistance and vectors of national "development." In a different context, Canadian geographer Stephanie Rutherford (2011) explored the subject-forming impacts of different sites of eco-governmentality. In her book *Governing the Wild*, she looked at the American Museum of Natural History, Disney's "Animal Kingdom" theme park, the Yellowstone National Park "experience," and former US Vice President Al Gore's (2006) film *An Inconvenient Truth*. In each case, she investigated the way these sites/sights instructed US citizens how to "correctly" understand the non-human world, with each site/sight secreting its own ensemble of cognitive and normative taken-for-granteds.

In these and other works, it's shown that there's no guarantee that governmentality will succeed (see Meehan and Molden 2015). People and non-human entities may fail to conform to the "scripts" they are given and the roles they are expected to play (e.g., see Birkenholtz 2009 for a study of Rajasthan, India). For British geographer Jamie Lorimer (2015), this implies the need for a more experimental and less "control-minded" approach to people-environment relations (see also Braun 2015). Lorimer explores recent Dutch attempts to rewild part of the country, highlighting the surprises restoration ecologists encountered along the way. These and other studies have forced an evolution of how green governmentality is understood to operate (for a review see Fletcher and Cortes-Vazquez 2020).

Compositions of Matter: The Potency and Limitations of Social Construction

If attempts to fully constitute action and matter run-up against limitations, then perhaps "composition" is a better way to think about how thought and practice can combine with the stuff of the world. To compose is to assemble things for particular reasons and purposes, without necessarily having full control over them, their preconditions or their effects.

A classic early study in "global political ecology" (before the term was even coined) was *First the Seed*, written by American rural sociologist Jack Kloppenburg (1988). Kloppenburg explained how and why capitalist firms were able to penetrate the agricultural sector in the USA and beyond from around 1900. Using insights into plant reproduction generated by biologists, and driven by government agencies keen to increase crop yields, certain firms began to patent the seeds of high-yielding "hybrid varieties" of corn and other cash crops. They went on to manufacture special herbicides, pesticides, and machinery for them, over time cornering large, lucrative new markets in agricultural inputs. In turn, farmers became "locked in" to these new arrangements for growing commercial crops. Kloppenburg's book reveals the progressive composition of extended commodity networks linking individual farms, soils, crops, banks, agro-food firms and consumers at a range of spatial scales. These networks were not accidental; rather they were deliberate arrangements of knowledge, people and materials designed to generate profits, sustain farmer livelihoods and feed consumers. Relatedly, Paul Robbins's (2007) remarkable book *Lawn People* details the construction of a complex set of norms and products that rendered the front and back lawn thoroughly conventional in post-1945 American white society.

These and other studies are alive to the role non-human entities play in enabling and constraining the process of composition through time and over space. For instance, in their brilliant studies of how private firms make money from selling water and trees respectively, Karen Bakker (2004) and Scott Prudham (2005) show how the non-human world presents "barriers to accumulation" that are not readily circumvented.[1] These barriers stand as an incitement to further human ingenuity for some, most especially capitalist enterprises, while others prefer to retain the barriers for the good of people and nature. In some cases, "preference" is not the right word. For instance, in her brilliant book *The Mushroom at the End of the World* (2015), American anthropologist Anna Tsing takes one species to exemplify how "unplanned composition" is unavoidable. Her study of a prized mushroom (the matsutake) found in different countries, shows how it thrives in landscapes changed by human action yet resists deliberate cultivation.

The Construction of Political Ecology Itself: Reflexivity, Auto-critique, and Post-constructionism

Since the first seeds of political ecology were sown 50 years ago, the field has grown prodigiously. It has, quite self-consciously, been cultivated as an alternative to other ways of understanding people–environment relationships. This raises a key question: are the knowledge constructions of political ecologists somehow more objectively truthful than other peoples', or do these constructions contain their own biases and agendas? Increasingly, political ecologists are reflexive about their research practices: they realize that doing rigorous, evidence-based research does not circumvent the need to remember that *all* knowledge is partial, incomplete and predicated on particular (contestable) values. In this sense, all avowed constructionists cannot escape the logic of their own position without contradicting themselves. In other words, *they are themselves engaged, variously, in acts of construal, constitution and composition.* Given that political ecology has been largely academic for most of the field's five decades, construal is arguably the most prominent of the three auto-constructions in question, with constitution focused on changing the mind-sets of undergraduate and post-graduate students—many of whom end-up working in resource management professions, government environment agencies or ENGOs. For instance, political ecologists have done much to challenge conventional understandings of "resilience," "vulnerability," "risk" and "adaptation" in their analyses of so-called "natural hazards" (e.g., see Watts 2015). They are engaged in epistemic contests or a battle of ideas about whose definitions of key terms will count in the world at large.

But these "external" contests have "internal" equivalents as political ecologists seek to keep an open mind about their own epistemic axioms and practices. There's been debate about the hidden politics cemented within political ecologists' own knowledge constructions. This debate has focused on the (undue?) dominance of Anglophone researchers, on the anthropocentrism of political ecology, on the relative influence of radical perspectives (like Marxism and feminism), on the possible lack of engagement with policy makers, on the need for more "action research," and more besides (see Chapters 4 and 5, and part II of Perreault, Bridge, and McCarthy 2015 for especially rich explorations of these and other issues). Equally, there is recognition that in our "post-truth" times, political ecologists need to make robust knowledge-claims that can change the understanding of key actors in government, business, the third sector and civil society (see, for instance, Neimark et al. 2019; Osborne 2017). In short, political ecologists need to understand the contingency, relevance and potential potency of their own epistemic constructions.

The metaphor of house building allows us to understand the processes of construal, constitution and composition, each being of interest to political ecologists for analytical and political reasons.

Conclusion

To understand the social construction of nature—what it involves and why it matters—we need to understand the nature of construction: this much we've seen in the previous pages. There are different kinds and degrees of "social construction"; political ecologists can usefully research their causes and outcomes without venturing the implausible claim that the material world beyond our various constructions is reducible to them (see Box 5.2). We can never inhabit a wholly "second nature" that's made by people; presuming this can blind us to important biophysical forces and events that shape peoples' lives regardless of their thoughts and actions (Malm 2017; Neyrat 2019). Where, in the 1900s and early 2000s, "social construction" was an analytical buzzword among political ecologists, these days—as this chapter has shown—it's more accurate to talk of "construal," "constitution" and "composition" as the processes that should command our analytical and political attention (summarized in Figure 5.3). These terms help us get a handle on the social aspects of things like Pacific fur seal exploitation, wildfire management, peoples' exposure to flood impacts, deforestation, nature conservation and much more besides.

Box 5.2 Data and analysis in the social constructionist approach

Given the diversity of analyses that operationalize social constructionism in political ecology, there's no one methodological tool kit we can refer to. Instead, there are many ways and means to interrogate how the three aspects of "construction" distinguished in this chapter—namely, construal, constitution and composition—occur. The key constructionist questions pertain to who constructs what, why, how and with what impacts? They can be answered using data gleaned from interviews, ethnographic immersion, analysis of spoken and written text, analysis of imagery (e.g., land cover photographs; participatory maps), focus groups, questionnaires and historical archives. Combinations of quantitative and qualitative information can allow researchers to show the way that particular depictions and arrangements of the non-human world get instituted or marginalized, depending on the situation.

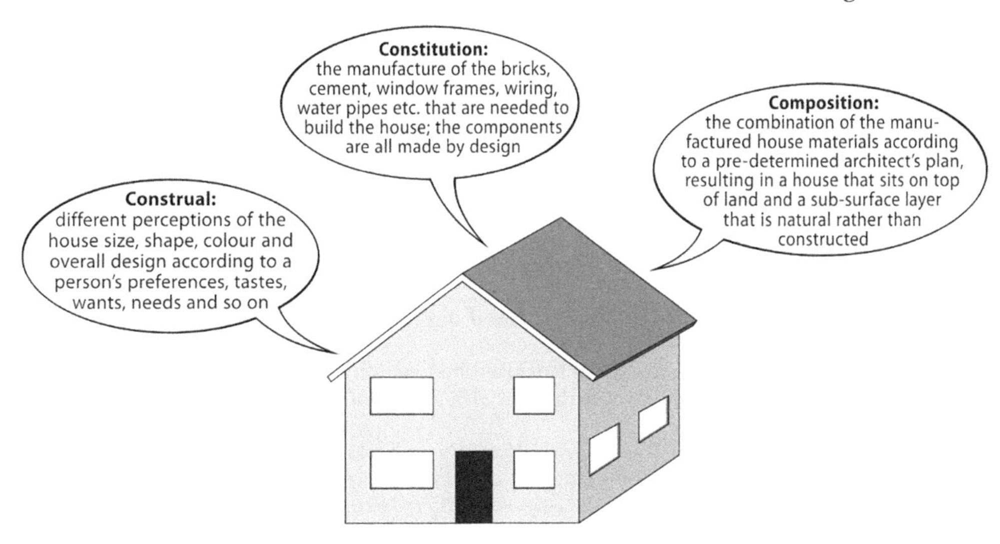

Figure 5.3 Deconstructing social constructionism.

The processes the terms describe are central to the questions of authority, identity, environmental change, resource availability/scarcity, exclusion and conflict that continue to preoccupy political ecologists in rural, urban, "Third World" and developed world settings. Illuminating the political dimensions, strategies and outcomes that course through these processes remains important work. In the end, what's political about ecology is that it gets routinely snarled-up in social power, social aspirations, social relations, social protest, socio-environmental in/justice and choices about different ways of living with the non-human world. The politics infuse *all* aspects of the people-environment relationship, though are often hidden or tacit. The research surveyed (albeit selectively) in this chapter helps us see how the de-politicization of certain aspects is a key way in which power, authority and potential injustices get enacted. *Politicization*, conversely, is about creating space for potential choice and change through alternative forms of construal, constitution and composition. By constructing their own particular forms of knowledge, political ecologists' agenda is precisely such politicization in the interests of greater equity, fairness, compassion, self-determination and democracy in our troubled world (aka what Peet and Watts 1996 called "liberation ecologies)." This is particularly important in a "post-truth" era of rampant disinformation, "fake news" and "alternative facts" where want counts as "reality" is up for grabs. Competing constructions of reality, as well as manufactured ignorance about various of them, are high stakes activities (Neimark et al. 2019; Simon 2022).

Note

1 Somewhat confusingly, geographers studying how capitalism "composes" the world have often talked not of the "construction" of nature but the *production of nature*. This reflects the influence of Marxist Neil Smith's thinking in Anglophone academic geography. See Castree (2015) and Chapter 2, this volume.

References

Agrawal, A. (2005) *Environmentality* (Stanford: Stanford University Press).
Bakker, K. (2004) *An Uncooperative Commodity* (Oxford: Oxford University Press).
Benjaminsen, T. A., (2021) "Depicting decline: Images and myths in environmental discourse analysis," *Landscape Research* 46, 2: 211–225.
Bennett, J. (2010) *Vibrant Matter: A political ecology of things*. Durham and London, UK: Duke University Press.
Berger, P. and Luckerman, T. (1966) *The Social Construction of Reality* (New York: Doubleday).
Birkenholtz, T. (2009) "Groundwater governmentality," *The Geographical Journal* 175, 2: 208–220.
Blaikie, P. (1985) *The Political Economy of Soil Erosion in Developing Countries* (London: Longman).
Braun, B. (2015) "From critique to experiment? Rethinking political ecology for the Anthropocene" in T. Perreault, G. Bridge and J. McCarthy (eds) *The Handbook of Political Ecology* (New York: Routledge), 102–113.
Bryan, K. and Wood, D. (2015) *Weaponizing Maps* (New York: Guilford Press).
Castree, N. (1997) "Nature, economy and the cultural politics of theory: The "war against the seals" in the Bering Sea, 1870–1911," *Geoforum* 28, 1: 1–20.
Castree, N. (2015) "Capitalism and the Marxist critique of political ecology," in T. Perreault, G. Bridge and J. McCarthy (eds) *The Handbook of Political Ecology* (New York: Routledge), 279–292.
Collard, R. C., Dempsey, J. and Sundberg, J. (2015) "A manifesto for abundant futures," *Annals of the Association of American Geographers*, 105, 2: 322–330.
Coole, D. and Frost, X. (eds) (2010) *New Materialisms* (Durham, NC: Duke University Press).
Daston, L. (2019) *Against Nature* (Cambridge, MA: MIT Press).
Davis, D. K. (2016) *The Arid Lands: History, Power, Knowledge* (Cambridge, MA: The MIT Press).
Demeritt, D. (2001) "Scientific forest conservation and the statistical picturing of nature's limits in the progressive-era United States," *Environment and Planning D: Society and Space* 19, 4: 431–459.
Escobar, A. (1996) "Constructing nature: A post-structural political ecology," in *Liberation Ecologies* (eds) R. Peet & M. Watts, (New York. NY: Routledge), 58–80.
Fairhead, J. and Leach, M. (1996) *Misreading The African Landscape* (Cambridge, MA: Cambridge University Press).
Fletcher, R. and Cortes-Vazquez, J. A. (2020) "Beyond the green panopticon: New directions in research exploring environmental governmentality," *Environment and Planning E: Nature and Space* 3, 2: 289–299.
Forsyth, T. (1996) "Science, myth and knowledge: Testing Himalayan environmental degradation in Thailand," *Geoforum* 27, 3: 375–392.
Goldman, M. (2020) *Narrating Nature: Wildlife Conservation and Maasai Ways of Knowing* (Tucson: University of Arizona Press).
Hacking, I. (2000) *The Social Construction of What?* (Cambridge, MA: Harvard University Press).
Harvey, D. (1974) "Population, resources and the ideology of science," *Economic Geography* 50, 3: 256–277.
Kelley, L. C. (2018) "The politics of uneven smallholder cacao expansion: A critical physical geography of agricultural transformation in Southeast Sulawesi, Indonesia," *Geoforum* 97, 1: 22–34.
Kloppenburg, J. (1988) *First the Seed* (Madison: University of Wisconsin Press).
Li, T. (2007) *The Will to Improve* (Durham, NC: Duke University Press).
Lorimer, J. (2015) *Wildlife in the Anthropocene* (Minneapolis: University of Minnesota Press).
Malm, A. (2017) *The Progress of This Storm* (London: Verso).
Meehan, K. and O. Molden (2015) "Political ecologies of the state," in *The Companion to Political Geography* (eds) J. Agnew, V. Mamadouh, J. Sharp and A. J. Secor (Hoboken, NJ: Wiley Blackwell), 438–450.

Neimark, B., Childs, J., Nightingale, A. J. et al. (2019) "Speaking power to 'post-truth': Critical political ecology and the new authoritarianism," *Annals of the Association of American Geographers* 109, 2: 613–623.

Neumann, R. P. (1998) *Imposing Wilderness: Struggles over Livelihood and Nature Preservation in Africa* (Berkeley: University of California Press).

Neyrat, F. (2019) *The Unconstructable Earth: An Ecology of Separation* (Fordham University Press, New York).

Osborne, T. (2017) "Public political ecology: A community of praxis for earth stewardship," *Journal of Political Ecology* 24, 1: 843–860.

Peet, R. and Watts, M. (eds) (1996) *Liberation Ecologies* (New York: Routledge).

Peluso, N. L. (1992) *Rich Forests, Poor People: Resource Control and Resistance in Java* (Berkeley: University of California Press).

Perreault, T., Bridge, G. and McCarthy, J. (eds) (2015) *The Handbook of Political Ecology* (New York: Routledge).

Prudham, S. (2005) *Knock on Wood* (New York: Routledge).

Robbins, P. (2007) *Lawn People: How Grasses, Weeds, and Chemicals Make Us Who We Are* (Philadelphia, PA: Temple University Press).

Rutherford, S. (2011) *Governing the Wild* (Minneapolis: University of Minnesota Press).

Sayre, N. (2008) "The genesis, history, and limits of carrying capacity," *Annals of the Association of American Geographers* 98, 1: 120–134.

Searle, J. (2010) *The Construction of Social Reality* (New York: Simon and Schuster).

Simon, G. L. and Peterson, C. (2019) "Disingenuous forests: A historical political ecology of fuel-wood collection in South India," *Journal of Historical Geography* 63: 34–47.

Simon, G. L. (2022). "Disingenuous natures and post-truth politics: Five knowledge modalities of concern in environmental governance," *Geoforum*, 132: 162–170.

Sokal, A. and Bricmont, J. (1999) *Fashionable Nonsense* (New York: Picador).

Soper, K. (1995) *What Is Nature?* (Oxford: Blackwell).

Soule, M. and Lease, G. (1995) *Reinventing Nature?* (Washington, DC: Island Press).

Steinberg, P. (2001) *The Social Construction of the Ocean* (Cambridge: Cambridge University Press).

Tsing, A. L. (2015) *The Mushroom at the End of the World: On the Possibility of Life in Capitalist Ruins* (Princeton: Princeton University Press).

Watts, M. (1983) *Silent Violence* (Berkeley: University of California Press).

Watts, M. (2015) "Now and then: The origins of political ecology and the rebirth of adaptation as a form of thought," in T. Perreault, G. Bridge and J. McCarthy (eds) *The Handbook of Political Ecology* (New York: Routledge), 19–49.

Whatmore, S. (2002) *Hybrid Geographies* (London: Sage).

6 Narrating Nature

Decolonizing Socio-ecological Assemblages

Diana K. Davis

As I write this in January 2023 while listening to the endless rain in Davis, California, my state has lost thousands (probably tens of thousands) of trees or more, a significant number of them very large and very old. These trees have come down during the relentless parade of atmospheric rivers onto buildings, rails and roads, peoples' homes and tents. The downed trees and flooding have caused widespread power outages, blocked transport, and resulted in huge losses of property and infrastructure. Twenty people have been killed so far, nearly half by falling trees, in what is being called the worst natural disaster in California history for many decades.

Just a couple of years ago California lost hundreds of millions of trees, in the "megafires" that raged over more than four million acres in the fall of 2020, the largest in California history. These fires were so massive and hot that they produced the newly named "giga-fire" of more than one million acres burned by a single fire. These gigantic, extremely hot fires also generated a new weather phenomenon: the pyrocumulonimbus, a fire-fueled thunderstorm-like cloud that produces high winds and lightning but little or no precipitation. Most of the forest trees damaged were naturally occurring whereas the majority of the trees in urban and peri-urban areas had been planted over the last century and a half by well-meaning humans. Both, however, were strongly influenced by human actions driven by particular arboreal environmental narratives and imaginaries.

Climate change is widely acknowledged to have exacerbated both the storms and the fires, but the roots of these problems are deeper and much more complex. In both of these cases, the long history of society's adoration of trees plays a critical role in both understanding and ameliorating the problems.

It is only by analyzing our stories about (socio-)nature over multiple temporal scales that we can start to adequately understand these problems and try to creatively construct better paths forward. Most political ecological research does this over time periods that vary, sometimes short and sometimes long, depending on the questions being asked. It does so utilizing multiple, mixed research methods including archival, documentary and textual, ethnographic, survey/interview, and biophysical methods while being informed by various bodies of critical social theory.

Stories about nature, often called environmental narratives, but also referred to as received wisdoms, discursive formations and storylines among others, are quite important for political ecologists.[1] This is because they frequently shape influential policies that govern millions of people and how, for example, they grow crops, fight disease and protect public health, provide water to towns and cities, graze livestock, tend forests, protect nature, manage wildfire, or regulate carbon emissions. Who wins and who loses access to

DOI: 10.4324/9781003165477-9

land and resources, as well as policy and legal debates (and thus implementation of new policy), frequently depends on how our stories about socio-nature are told as well as *who* is allowed to tell them.

Environmental narratives that become dominant, even when demonstrably false, often become normalized, widely adopted and cause various harms to both people and environments. Political ecologists have long considered work that analyses such stories over the *longue durée*, from their origins, to be "historical political ecology" (HPE) (Offen 2004). What is frequently overlooked, however, is that political ecology as a subdiscipline has been deeply historical since its inception and that political ecology research is rarely at its best without appropriate, critical historical context (Davis 2015).

Indeed, it is precisely the historical contextualization of political ecology research that helps to prevent apolitical analyses and conclusions, perhaps our most common and important primary goal. The historical development of landscapes, knowledge production, social relations, and economic and political systems need to be excavated and made clear if we are to reveal the current hidden relations of power that most political ecologists try to analyze with their research. In the words of Erik Swyngedouw, "political ecology is intrinsically historical in the sense that it is precisely the myriad power relations that shape the dynamics of socio-ecological transformation and the making of new socio-ecological conditions" (Gorostiza 2015).

Political ecology has been, since the emergence of the subdiscipline in the 1970s, strongly informed by historical research. This is in large part because it is a neo-Marxist subdiscipline and understanding international development and political economy, which was the focus of much early research, requires sophisticated historical context and background. We see in the most influential classics of political ecology, for example the work of Michael Watts in rural Africa, and Nancy Peluso in Java, Indonesia, great attention to detailed historical explanation. This is in order to illustrate the reordering of social relations and agriculture under colonialism and capitalism and the grave consequences for local peoples and the environment (Peluso 1992; Watts 2013). This early focus on rural agricultural political economy and conservation in the Global South was complemented by many other historical political ecological works on these topics in the following decades (Carney 2001; Hecht and Cockburn 1989; Jarosz 1993; Schroeder 1993; Sluyter 2012; Zimmerer 1996).

A subsequent generation of political ecology researchers deepened analyses of socio-environmental change in many parts of the globe by examining representations of the environment and the power of narratives over time. Often deconstructing "received wisdoms" about landscape change, these authors incorporated deep dives into the historical record to trace their revealing stories about socio-natures (Bassett and Zuéli 2003; Fairhead and Leach 1996; Goldman 2005; Hecht et al. 2014; Kull 2004; Showers 2005; Wainright 2008). Some like Rod Neumann showed how certain representations of landscapes and wildlife with complex colonial histories in effect "sold" certain landscapes for tourism, national parks, and "development" (Neumann 1998). His examination of the development of Arusha National Park and the attendant problems for the local peoples was only possible and convincing with deep historical analysis.

Declinsionist environmental narratives of environmental degradation have also received attention by political ecologists, particularly in formerly colonized regions. Historical research is crucial to these works which have examined, for instance, fire narratives and their use in controlling and marginalizing colonized peoples in Madagascar (Kull 2004) and stories of desertification deployed in colonial North Africa (Davis 2007) and sub-Saharan Africa to appropriate land and resources (Fairhead and Leach 1996).

Some of this influential political ecology research has drawn attention to the problematic nature of "baseline" data sets like deforestation, erosion, or desertification statistics—in particular from the colonial era—which quite often contain errors, exaggerations, biases, and other weaknesses that should prevent them from being invoked today (Davis 2016; Hecht et al. 2014; Kull 2004; Leach and Mearns 1996; Sayre 2008; Showers 2005; Walker 2004). The widespread policy-driving notion that about 30% of most parts of our landmass should be forested to be "ecologically healthy" is rooted in hundreds of years of problematic European and colonial history that needs to be rolled back (Davis and Robbins 2018).

As the subdiscipline matured and expanded, urban political ecology and political ecological work in the Global North has also been attentive to the historical evolution of discourses and narratives about landscapes and environmental change. Topics as varied as water provision (Swyngedouw 1997; Swyngedouw 2015), food and agriculture (Cooper 2017; Freidberg 2009; Galt 2014; Guthman 2004; Guthman 2019; Hollander 2008), landscaping (Robbins 2007), climate change (Bumpus and Liverman 2011), settler colonialism and resistance (Kirk in press), and public health (Biehler 2009; Senanayake 2022b; Werner et al. 2022) all demonstrate careful attention to narrating nature only possible with historical detail. Newer topics including range science (Sayre 2017) and environmental decolonization (Braun 2002; Kosek 2006; McCarthy 2001; Wainright 2008) have been treated by political ecologists with careful diachronic research in order to deconstruct problematic narrations of nature still doing work today. Some of the newest work is most interesting in political ecology in the ways that it is interfacing with science and technology studies (Goldman et al. 2011), with health and medical geography (Connolly et al. 2017; Senanayake 2022a), and with physical geography in a theoretically and historically informed fashion: critical physical geography (Lave and Doyle 2021).

As all this research demonstrates, humans have been telling stories about nature for thousands of years and for a great many reasons, from the fertility or barrenness of land, to who has been tending nature "well," who is blamed for "ruining" it, and how it can be "improved." Telling such stories about the environment and the human and non-human creatures who live there, narrating socio-nature, has been, and remains, intimately connected to how we value nature and, sometimes, to how we fear nature.

While desertification was the first major environmental crisis narrative to be officially recognized at a planetary level, the most prominent global environmental narrative today is the meta-crisis narrative of climate change in which the desertification and deforestation narratives, among others, are strong. The stakes are very high for many actors, including our planet, depending on how this story is told. Analyzing some key aspects of "narrating socio-nature" within political ecology, including representation, imaginaries, narratives, deconstruction, and related socio-political economy, is the goal of this chapter. Several examples drawn from or related to our dominant and complex climate change narrative are utilized. Understanding the foundational importance of narrating socio-nature can begin with one seemingly simple question: why is our world today dominated by arborocentric thinking and what are the consequences and alternatives?

Tree as Savior? A Political Ecological Lens

The trillion tree campaign is emblematic of the pervasive global ideology driving arborocentric thinking and policy today (Ellis et al. 2020). It is also a textbook example of the powers of narrating socio-nature in political ecology and why such complex narratives need to be deconstructed and critically analyzed. This campaign, and the related Bonn Challenge (Temperton et al. 2019), embodies centuries old "received wisdoms" about trees and their assumed positive roles in our planetary ecosystem that are so strong that scientists have recently been reported to say "if we found [that] forest loss cooled the planet, we wouldn't publish it" (Popkin 2019). Three years later, other brave scientists did publish research demonstrating that deforestation above 50 degrees North latitude does indeed lead to net global cooling (Lawrence et al. 2022).

The trillion tree campaign advocates and funds arboreal expansion in the name of sequestering carbon and improving the global environment with the goal of planting a trillion trees in many parts of the world. Most of the trees planted, though, are in plantations designed for commodity production, like wood pulp for paper and other products. Some of these plantations are expropriating marginalized people from their lands and resources (St. George 2022). Other tree plantations are wreaking havoc on ecosystems that are in fact ecologically non-arboreal, like grasslands and shrublands, by decreasing water tables, altering soil nutrient cycles, disrupting hydrological regimes, fueling fires, and lowering biodiversity (Fleischman et al. 2020; Veldman et al. 2019; Veldman et al. 2015; Vetter 2020). That is to say, the trillion tree campaign and many like it, fueled by an uncritical, capitalist arborocentrism, are causing a lot of problems and not significantly ameliorating the climate crisis.

How did we get here? Analyzing the development of arborocentrism allows us to explore many of the primary strengths of taking environmental narratives and imaginaries, and their usually deep histories, seriously in political ecological research. How did this environmental imaginary develop, who wins and who loses when it is deployed, and why does it matter today? As many political ecologists argue, since "socio-ecological relations and configurations are the historical—geographical result born out of a process of historical production, undertaking the archaeology of the production of these socio-ecological assemblages is of necessity a historical endeavor" (Swyngedouw quoted in Gorostiza 2015).

Arborocentrism is the "tree-centric" attitude that became dominant in Western thought in the seventeenth century, although strains of tree-adoration have much older histories. Sometimes termed arboreal chauvinism, this thinking often functions as an ideology that favors trees over other forms of vegetation. I have traced the complex history and development of arborocentrism, and its ideological sibling desertification, in detail elsewhere (Davis 2016, 2021; Davis and Robbins 2018). It became dominant in Anglo-European regions by about the 1600s in large part because timber was vital for the navies of many countries, bolstered by the elite obsession with hunting.

This period also marked the beginnings of Western environmental thinking which took place mostly in the humid and continental climates of Anglo-Europe. What began in Anglo-Europe with a concern over deforestation and perceived extremes in both droughts and floods which impacted agricultural and economic life, was reinforced and complexified during imperial/colonial escapades around the world. In Europe, the nineteenth-century enthusiasm for afforestation driven by arborocentrism, though, has resulted in the contemporary exacerbation of warming in some regions (Naudts et al. 2016).

Abroad, imperial explorations at first resulted in nature being described, especially in the tropics, as exotic, luxuriant, even sublime. Experience living in these territories, though, turned attitudes darker and the representations of nature became more negative, frequently either threatening or portraying widespread destruction. Anglo-Europeans essentially defined themselves and their home country environments as normal and desirable against the perceived difference and abnormalities/defects of their imperial territories, a process I have termed "environmental orientalism" (Davis 2011, 4; Gregory 2001).

Putting Narratives of Socio-Nature into Action

Being unfamiliar with arid and semi-arid zones to any great extent, the early trials of Anglo-European imperialism into the drylands of Africa, Asia, and the Middle East resulted in notions of desiccation and the widespread belief that arid environments had once been tree-covered, then deforested, overgrazed, and turned into desert. This went hand in hand with the common view that the local people had ruined the land and were somehow less than human and incapable of "proper" environmental management. Livestock grazing was identified by many imperial/colonial powers to be the most destructive force in these regions. Desiccation thinking also included the belief that planting trees in arid lands would "bring back the rains" and improve the climate and the environment to Anglo-European norms. Altogether, these narratives and framings created an environmental imaginary, a constellation of ideas, representations, and narratives of arid lands, that portrayed them as ruined at the hands of incompetent locals and therefore in need of "improvement" to be productive.

A parallel and related series of events in knowledge production took place in the more humid, tropical areas as imperial forces spread around the globe. In these regions the narratives were usually different and tended to center on the luxuriant vegetation at first seen as fertile. This abundant nature had to be "tamed" to be productive, though, and many Indigenous land uses came to be seen as destructive by the colonial agents. The use of fire, crucial for Indigenous forms of swidden agriculture, was particularly vilified. These imperial environmental narratives shaped many aspects of the colonial project in the humid tropical zones as they did in the arid and semi-arid lands. It was in these early colonial/imperial territories that the power of environmental narratives and imaginaries, and the central role of trees/forests, becomes readily apparent.

In short, environmental imaginaries, often informed by multiple environmental narratives, helped to pave the way for colonial and other forms of imperial domination in the form of agricultural, and sometimes timber, plantations, among others, and a reordering of society accordingly (Davis 2011). In many imperial territories, "repairing" the environment to Anglo-European norms also took on an urgency because an "appropriate" climate/environment was believed necessary to support "civilization" and prevent "degeneration" of the colonists. Thick tree cover came to be equated with civilization itself and 30% forestation rates traveled from colonial Algeria to become a global norm (Davis and Robbins 2018). Local, Indigenous people, their voices and knowledges, were, for the most part, deemed inferior and silenced. Political ecologists have been particularly prolific in research and writing about the role and importance of narrating socio-nature in the colonial world as indicated above.

How is this kind of research conducted?

Methods for Narrating Socio-Natures

Taking narratives and imaginaries of socio-nature seriously in political ecology requires a critical triangulation of several sources derived from multiple, mixed methods (Box 6.1). It is practically a cliché in political ecology that "who wins and who loses depends on how the story is told." That being given, all political ecology research should interrogate the dominant environmental discursive formations for hidden relations of power that may be privileging some while marginalizing others, and that may be harming environments.

Doing so involves meticulous research that triangulates the historical record of texts, usually found in archives and publications, with the available biophysical data in the form of fossil pollen core data, sedimentation studies, isotopes of oxygen and other atmospheric gases, charcoal analysis, among many others (Davis 2007; Fairhead and Leach 1995; Klinger 2017; Leach and Mearns 1996; Mathavet et al. 2015).[2] Maps and related data sets, whether colonial products or recent GIS with remote sensing data, of potential vegetation, deforestation, desertification, etc. are equally important (Robbins 2001a, 2001b; Turner 2003).

Because the archival records usually contain a bias towards the literate elite, oral histories among those still living (or recordings) are an invaluable addition as is ethnography where possible. Uncovering Indigenous voices, knowledges, and practices is important since they were usually silenced during imperial domination (Goldman 2020; Kosek 2006; Ogden 2011; Wainright 2008). Decolonial political ecology needs to actively decolonize many knowledge constructs about socio-natures that were formulated during the colonial period because they still inform our debates and policy today (Youdelis et al. 2021).

To unravel the power of a particular environmental narrative or imaginary necessitates understanding the history of the period(s) during which it was constructed and deployed. Relying on mainstream Anglo-Euro-triumphalist histories often confounds critical understanding and risks reinforcing existing unequal power structures. Not infrequently, this requires deconstruction or reconstruction of mainstream histories—and sometimes both—to result in critical historical analysis (Davis 2016; Sayre 2017). This kind of careful diachronic research has better potential to expose situated, political histories (and situated practices) that have been represented as apolitical histories (Peluso 2012).

Sources from the humanities such as stories, novels, and art (especially landscape painting and photographs) can be very useful in political ecological analyses of environmental imaginaries and narratives. This is because they often represent a widely accepted view or understanding of particular places and how they came to be in their current condition (Gandy 2003; Neumann 1998). Caution must be taken, though, to gather enough information to understand the context and positionality of the artist and their society in order to successfully interpret the work. Feature films and other media can also be of use for more contemporary topics.

In the case of arborocentrism, it takes consideration of contemporary cutting edge ecological science combined with careful analysis of a wide variety of historical texts including histories of various countries and their colonial territories, newspapers and journals, scientific texts and manuals of forestry and agriculture, maps and data sets, archival sources including notes, letters, and reports, analysis of landscape paintings and photos over time, and even philosophical treatises. All this had to be analyzed against the twentieth-century record of forestry and agricultural policy, ecological science, as well as the contemporary rhetoric of tree-planting as environmental panacea.

This kind of research is often time-consuming and difficult. It is worth emphasizing, however, that simply conducting a quick but shallow historical analysis risks maintaining or reinforcing the status quo. Careful, critical historical analysis in political ecology is crucial because it is both radical and emancipatory. It facilitates thorough understanding of, and potential escape from, hegemonic ideologies like arborocentrism.

Box 6.1 Narrating nature in political ecology: getting historical

For many political ecologists, the historical research necessary for "narrating nature" begins with an anomaly in the present. Being curious researchers, these political ecologists will then trace back and ferret out the origin and trajectory of the anomaly if they can. How they do this depends heavily on the location of the research, what the anomaly is, and what resources are available.

In my own work, which began as a political ecological analysis of resource use among nomads in Southern Morocco, the anomaly was that what the nomads told me about their environment (it was healthy, as much of the most recent scientific literature indicated) did not match at all what the Moroccan government agents and much of the development literature told me (it was an overgrazed, desertified disaster).

Trying to figure out what caused the government agents and many others to say that desertification was such a big problem led me to government reports and documents and also to those by international consultants for the likes of the World Bank and several different international NGOs. In these documents I found many references to French colonial-era sources going back to the early twentieth century. While reading these primary sources, I then found many references to much earlier sources going back to French colonial Algeria, some early in the nineteenth century. This, in turn, led to serious and lengthy research in the colonial archives in France. About a year of field work in Morocco was complemented with another year of archival research in France (with some in Morocco).

Triangulating what had been written about the environment since 1830, under a colonial regime, against what the nomads reported and the most current, cutting-edge arid lands ecology and range science allowed me to trace the situated context and the agendas of many who created and then utilized the discursive formation of desertification over the last two centuries (Davis 2007).

As Paul Robbins has pointed out, most political ecologists fall somewhere between social constructivism and realism (Robbins 2020). Narrating nature requires the delicate balance of tracing the ways the landscape has actually changed as well as how ideas about it and its use have changed over time and how and why those ideas have changed (Forsyth 2003). Leave no questions or assumptions unexamined and follow the money and power.

Traveling Narratives of Socio-Natures

Arborocentrism, which rose to dominance in the nineteenth century, became even more widely influential over the course of the twentieth century. That is, like many narratives of socio-nature, it traveled around the world. The ways this happened and the reasons driving the spread are complex, as is the case with most traveling narratives (Goldman et al. 2011; Lave et al. 2018). In nearly all climate zones, though, scientific forestry played a key role in placing trees in a dominant and privileged role. Scientific forestry, which became influential primarily in the early to mid-nineteenth century, was deeply imbued with arborocentrism and demonized fire since it was believed fires permanently harmed forests, and even created deserts. As we now know scientifically, suppression of fire over a century and more, has led to the overgrown, overcrowded, weakened and unhealthy forests of today that have become tinder boxes awaiting a spark.

Through scientific and forestry treatises and teaching at universities, as well as through scientific and colonial networks of "experts," fire suppression as a key component of good forestry technique spread around the world to become dominant in the twentieth century. The result has been ferocious and destructive wildland fires in many parts of the world. As this problematic part of the imaginary has become scientifically well-established and more widely known, though, there has been a new respect for long-standing Indigenous practices like targeted burning, often called prescribed burning in many places including California. This is a small step, hopefully, towards decolonizing knowledge and practice and bringing in what many Indigenous groups in California call "good fire" (Schelenz 2022).

Another key component of scientific forestry, however, is little recognized in the literature: grazing suppression. Just as fire was banned (and discipline meted out for infractions), grazing was banned or strictly regulated wherever scientific forestry spread around the world. Indeed the development of range management is closely related to grazing management in forested regions, not just on the "open range" (Sayre 2017). The ways that socio-nature has been narrated in the arborocentric imaginary, vilifying grazing as nearly always destructive, has resulted in grazing suppression as much as fire suppression. As this neglected part of the imaginary becomes more well-known, and grazing in many ecosystems exonerated, policies like targeted grazing for fuels management in overgrown forests are beginning to be utilized in areas where targeted burning is too risky or expensive (Huntsinger and Barry 2021).

Just as scientific foresters were planting and conserving forests, other aspects of arborocentrism were driving the afforestation of urban and peri-urban locations in many parts of the world. In arboreal regions, this often had beneficial impacts such as moderating temperatures during hot periods but it also resulted in more warming where evergreens were planted (Naudts et al. 2016). Moreover, the choice of male trees to reduce "messiness" of dropped fruits has resulted in an increased pollen load that plagues a significant portion of our urban populations. Exotic trees like eucalyptus also substantially increase fire risk, and most large trees left to grow for too many years can become serious fall hazards. In non-arboreal settings including the drylands, and even cities like San Francisco, urban afforestation has caused groundwater depletion, greater tree-fall risk, and increased flammability of the landscape in many places. Political ecologists like Gregory Simon have succeeded in exposing many of these problems in their work which takes narrating socio-nature over time seriously (Simon 2017).

The power of arborocentrism, though, is not limited to trees and forests. Deserts are in many ways portrayed as the evil twin of forests in the arborocentric imaginary. Deserts are where the trees are not and they have long been defined by their sparse vegetation, especially their lack of trees. The desertification narrative is intimately entwined with arborocentrism and has been used to "land grab" since at least the colonial period well into the twenty-first century.

It was quite common, for instance, during the colonial period to define land without trees as deforested and thus in need of "reforestation" even when there was no evidence of tree growth since the end of the last ice age. This was used by many colonial forest departments to acquire land for the state, from which the local population was dispossessed. Likewise, land without trees or other thick vegetation was frequently categorized as desertified and assumed to be ruined by deforestation, burning, or overgrazing.

The colonial logic then justified land expropriation based on the claim that the land had been "turned to desert" by Indigenous land use practices. Most often, the land was then provided to European colonists for "improved" production and land-use practices. I have elsewhere termed this process "accumulation by desertification" (Davis 2016). Such actions would not have been as easy without the environmental imaginary of arborocentrism with its twin narratives of deforestation and desertification.

Narrating Socio-Natures for Profit

Narrations of socio-natures, both discursive formations and imaginaries, nearly always include particular conceptualizations of socio-political economy, the excavation of which is key for political ecologists. In the case of the British enclosures, for example, it was not only that the landscape was being reordered for certain forms of "proper" agricultural production, it was equally significant that the values of "improvement" and profit generation were driving these dramatic changes and dispossessing peasants (Goldstein 2012). As Jesse Goldstein has shown, a key part of this transformation was narrating/representing "common," non-agricultural, lands as "waste[d]," which then justified or necessitated improvement with Anglo-European agricultural techniques for maximum productivity everywhere.

Analyzing arborocentrism in the midst of our climate change crisis allows us to examine the use of long-standing environmental narratives and imaginaries for political-economic gain and territorial control today. A particularly good example comes from the building of large solar farms as part of the energy transition to renewables. It is useful to consider the ways that narrating socio-natures is often used in several ways, in David Harvey's words, to "realize value" in capitalist systems (Harvey 2019).

Harvey, interpreting Marx, explains that the crucial point in the capitalist production process is not the manufacturing of goods, but the selling of goods since this is the point at which value is realized in the form of money. Harvey is primarily theorizing about the creation of new wants, needs and desires through the use of advertising, harnessing ideology, etc. to sell more manufactured goods, thereby realizing value, especially in urban settings.

In sunny dryland rural settings around the world, though, like Morocco and India, a related process is quietly taking place. Powerful actors, such as the World Bank and many governments, are presenting these regions as marginal, often degraded, and/or empty with an assemblage of maps, data sets, and environmental narratives that claim this land is available for investment in solar energy that will perform an environmental good, produce profit, and provide consistent and reliable growth (McCarthy and Thatcher 2019).

Digging deeper we find that most of these landscapes have been politically, scientifically and ideologically narrated as "desertified," "marginal," empty, and value-less, for at least two centuries, facilitating land expropriation by the (colonial) state, that is, accumulation by desertification (Baka 2013; Cantoni and Rignall 2019; Davis 2016; Rignall 2016; Stock and Birkenholtz 2021). The contemporary (neo)colonial environmental narratives being wielded by the likes of the World Bank to "sell" investment in solar farms in "marginal" drylands are based on these older environmental narratives, and therefore replicate the well-known errors of (neo)colonial dryland science and (neo)imperial ignorance about sustainable livelihoods in the drylands, leading to serious ecological and social problems where most of these solar installations are built.

There is a complicated nesting of environmental narratives and imaginaries involved which demonstrates not only the importance of narrating socio-ecological assemblages for the present moment, but also the fact that they can only be fully understood in their historical context. The "need" for a renewable energy transition is being driven by the meta-environmental narrative of a climate change "emergency," which depoliticizes and precludes discussions of potential alternatives like "de-growth," for instance. The desertification narrative is then being used to present sunny arid lands as empty, worthless/ruined and available for investments in solar energy. The deployment of these narratives may be seen, then, as similar to advertising in the role they play in the realization of value (Harvey 2019) when the decision is ultimately made to sign the leases and contracts to invest in constructing new solar farms. Political ecologists have examined some of these issues and documented the subsequent problems for local peoples and ecologies.

Future Socio-Natural Narratives

Arborocentrism is playing a central role in another area of climate change mitigation: forest carbon offsets. Despite the fact that other biomes, especially grasslands, can more reliably store carbon and are less vulnerable to catastrophic fire (Dass et al. 2018), planting trees has become the preferred way to "save the planet" in the global imaginary, thanks in large part to arborocentrism. In addition to the simplistic and seductive mania for planting trees in the name of carbon sequestration (as with the trillion tree campaign), there is the equally significant but less visible two billion dollar global market in voluntary carbon offsets (Greenfield 2023). The fact that this scheme hinges on future predictions of forest cover and forest loss that are themselves informed by estimates of deforestation reliant on (neo)colonial maps and data sets strongly influenced by arborocentrism is little understood.

These complicated markets essentially allow carbon generating companies to purchase "carbon offsets" in the form of allegedly protecting forests elsewhere in the world that would otherwise be deforested, theoretically storing an equal amount of carbon. Using organizations like Verra, the global leader in carbon standards for the voluntary offset market, clothing and airline companies, for example, can label their products and services as "carbon neutral," a new form of marketing that enhances sales and profits. Recent research has strongly suggested, however, that the baseline estimates of deforestation (to be remediated by these protections) used by Verra have been exaggerated by between 400% and 900% (Greenfield 2023; Guizar-Coutino et al. 2022; West et al. 2020; West et al. 2023). Therefore an estimated 90% of their rainforest carbon offsets appear to be worthless, nothing more, in many ways, than false advertising ... to realize value.

Tracing the situated histories of the problems tackled by political ecologists nearly always reveals the significance of the various ways we narrate socio-natures. The more we learn about the ways we as humans have represented socio-ecological assemblages and the more we excavate our environmental imaginaries, the better prepared we will be to try to create a world as we want it to be. With such understanding, we could let go of the shackles of arborocentrism, for instance, and embrace "good fire" and targeted livestock grazing to prevent catastrophic wildfires in our forests. We could stop planting problematic trees in our cities and stop building into the flammable wildlands. Before we can productively "imagine futures otherwise," however, we need to decolonize the problematic knowledge constructions from the past that continue to captivate our minds, our policies, and our actions. It is definitely worth the effort.

Notes

1 For an excellent discussion of environmental narratives and related terms such as discursive formations, in the context of the fuelwood crisis in India, "disingenuous natures" and the repurposing of multi-scalar narratives, see Simon and Peterson (2019).
2 See the appendix on methods in Klinger (2017) for a detailed iteration of some of these methodologies and their triangulation. This book won the AAG's Meridian Book Award and is an excellent example of the kind of political ecology discussed in this chapter.

References

Baka, J. (2013) The Political Construction of Wasteland: Governmentality, Land Acquisition and Social Inequality in South India. *Development and Change*, 44 (2):409–428.
Bassett, T. J., and K. B. Zuéli. (2003) The Ivorian Savanna: Global Narratives and Local Knowledge of Environmental Change. In *Political Ecology: An Integrative Approach to Geography and Environment-Development Studies*, edited by K. S. Zimmerer and T. J. Bassett. New York: Guilford.
Biehler, D. (2009) Permeable Homes: A Historical Political Ecology of Insects and Pesticides in US Public Housing. *Geoforum*, 40 (6):1014–1023.
Braun, B. (2002) *The Intemperate Rainforest: Nature, Culture, and Power on Canada's West Coast*. Minneapolis: University of Minnesota Press.
Bumpus, A. G., and D. M. Liverman. (2011) Carbon Colonialism? Offsets, Greenhouse Gas Reductions, and Sustainable Development. In *Global Political Ecology*, edited by R. Peet, P. Robbins, and M. J. Watts. London: Routledge.
Cantoni, R., and K. Rignall. (2019) Kingdom of the Sun: A Critical, Multiscalar Analysis of Morocco's Solar Energy Strategy. *Energy Research and Social Science*, 51 (1):20–31.
Carney, J. (2001) *Black Rice: The African Origins of Rice Cultivation in the Americas*. Cambridge, MA: Harvard University Press.
Connolly, C., P. Kotsila, and G. D'Alisa. (2017) Tracing Narratives and Perceptions in the Political Ecologies of Health and Disease. *Journal of Political Ecology*, 24 (1):1–11.
Cooper, M. H. (2017) Open Up and Say "Baa": Examining the Stomachs of Ruminant Livestock and the Real Subsumption of Nature. *Society and Natural Resources*, 30 (7):812–828.
Dass, P., B. Houlton, and Y. Wang. (2018) Grasslands May be More Reliable Carbon Sinks that Forests in California. *Environmental Research Letters*, 13 (7):1–8.
Davis, D. K. (2007) *Resurrecting the Granary of Rome: Environmental History and French Colonial Expansion in North Africa*. Athens: Ohio University Press.
Davis, D. K. (2011) Imperialism, Orientalism, and the Environment in the Middle East: History, Policy, Power, and Practice. In *Environmental Imaginaries of the Middle East and North Africa*, edited by D. K. Davis and E. Burke III. Athens: Ohio University Press.
Davis, D. K. (2015) Historical Approaches to Political Ecology. In *The Routledge Handbook of Political Ecology*, edited by T. Perreault, G. Bridge, and J. McCarthy. London: Routledge.
Davis, D. K. (2016) *The Arid Lands: History, Power, Knowledge*. Cambridge, MA: The MIT Press.
Davis, D. K. (2021) Deserting Arboreal (Bio)Politics. In *Manual for a Future Desert*, edited by I. Soulard, A. Meza, and B. El Baroni. Milan, Italy: Mousse Publishing.
Davis, D. K., and P. Robbins. (2018) Ecologies of the Colonial Present: Pathological Forestry from the *taux de boisement* to Civilized Plantations. *Environment and Planning E: Nature and Space*, 1 (4):447–469.
Ellis, E., M. Maslin, and S. Lewis. (2020) Planting Trees Won't Save the World. *New York Times*, February 12, www.nytimes.com/2020/02/12/opinion/trump-climate-change-trees.html
Fairhead, J., and M. Leach. (1995) False Forest History, Complicit Social Analysis: Rethinking Some West African Environmental Narratives. *World Development*, 23 (6):1023–1035.
Fairhead, J., and M. Leach. (1996) *Misreading the African Landscape: Society and Ecology in a Forest-Savanna Mosaic*. Cambridge: Cambridge University Press.
Fleischman, F., S. Basant, A. Chhatre, E. A. Coleman, and H. W. Fischer. (2020) Pitfalls of Tree Planting Show Why We Need People-Centered Natural Climate Solutions. *BioScience*, biaa094:1–4, DOI: 10.1093/biosci/biaa094
Forsyth, T. (2003) *Critical Political Ecology: The Politics of Environmental Science*. London: Routledge.

Freidberg, S. (2009) *Fresh: A Perishable History*. Cambridge, MA: Harvard University Press.
Galt, R. E. (2014) *Food Systems in an Unequal World: Pesticides, Vegetables, and Agrarian Capitalism in Costa Rica*. Tucson: University of Arizona Press.
Gandy, M. (2003) *Concrete and Clay: Reworking Nature in New York City*. Cambridge, MA: MIT Press.
Goldman, M. (2005) *Imperial Nature: The World Bank and Struggles for Social Justice in the Age of Globalization*. New Haven: Yale University Press.
Goldman, M. J. (2020) *Narrating Nature: Wildlife Conservation and Maasai Ways of Knowing*. Tucson, AZ: University of Arizona Press.
Goldman, M. J., P. Nadasdy, and M. D. Turner, eds. (2011) *Knowing Nature: Conversations at the Intersection of Political Ecology and Science Studies*. Chicago: The University of Chicago Press.
Goldstein, J. (2012) *Terra Economica*: Waste and the Production of Enclosed Nature. *Antipode*, 45 (2):357–375.
Gorostiza, S. (2015) "Liquid Power": An Interview with Erik Swyngedouw. *Undisciplined Environments*, July 2. https://undisciplinedenvironments.org/2015/07/02/
Greenfield, P. (2023) Revealed: More than 90% of Offsets Fail to Cut CO_2, Study Finds. *The Guardian*, January 27, 18–19.
Gregory, D. (2001) (Post)Colonialism and the Production of Nature. In *Social Nature: Theory, Practice, and Politics*, edited by N. Castree and B. Braun. Malden, MA: Blackwell Publishers.
Guizar-Coutino, A., J. P. Jones, and A. Balmford. (2022) A Global Evaluation of the Effectiveness of Voluntary REDD+ Projects at Reducing Deforestation and Degradation in the Moist Tropics. *Conservation Biology*, DOI: 10.111/cobi.13970
Guthman, J. (2004) *Agrarian Dreams: The Paradox of Organic Farming in California*. Berkeley: UC Press.
Guthman, J. (2019) *Wilted: Pathogens, Chemicals, and the Fragile Future of the Strawberry Industry*. Oakland: University of California Press.
Harvey, D. (2019) Realization Crises and the Transformation of Daily Life. *Space and Culture*, 22 (2):126–141.
Hecht, S., and A. Cockburn. (1989) *The Fate of the Forest: Developers, Destroyers, and Defenders of the Amazon*. Chicago: The University of Chicago Press.
Hecht, S. B., K. D. Morrison, and C. Padoch, eds. (2014) *The Social Lives of Forests: Past, Present and Future of Woodland Resurgence*. Chicago: The University of Chicago Press.
Hollander, G. M. (2008) *Raising Cane in the 'Glades: The Global Sugar Trade and the Transformation of Florida*. Chicago: The University of Chicago Press.
Huntsinger, L., and S. Barry. (2021) Grazing in California's Mediterranean Multi-firescapes. *Frontiers in Sustainable Food Systems*, 5 (715366)
Jarosz, L. (1993) Defining and Explaining Tropical Deforestation: Shifting Cultivation and Population Growth in Colonial Madagascar (1896–1940). *Economic Geography*, 69 (4):366–379.
Kirk, G. (In Press) Trains, Trees, and Terraces: Infrastructures of Settler Colonialism and Resistance in Southern Jerusalem. In *Gendered Infrastructures*, edited by Y. Truelove and A. Sabhlok. Morgantown: West Virginia University Press.
Klinger, J. M. (2017) *Rare Earth Frontiers: From Terrestrial Subsoils to Lunar Landscapes*. Ithaca: Cornell University Press.
Kosek, J. (2006) *Understories: The Political Life of Forests in Northern New Mexico*. Durham: Duke University Press.
Kull, C. A. (2004) *Isle of Fire: The Political Ecology of Landscape Burning in Madagascar*. Chicago: The University of Chicago Press.
Lave, R., C. Biermann, and S. N. Lane, eds. (2018) *The Palgrave Handbook of Critical Physical Geography*. Cham, Switzerland: Palgrave Macmillan.
Lave, R., and M. Doyle. (2021) *Streams of Revenue: The Restoration Economy and the Ecosystems it Creates*. Cambridge, MA: The MIT Press.
Lawrence, D., M. Coe, W. Walker, L. Verchot, and K. Vandecar. (2022) The Unseen Effects of Deforestation: Biophysical Effects on Climate. *Frontiers Forests and Global Change*, 5 (756115)
Leach, M., and R. Mearns, eds. (1996) *The Lie of the Land: Challenging Received Wisdom on the African Environment*. London: The International African Institute.
Mathavet, R., N. Peluso, A. Couespel, and P. Robbins. (2015) Using Historical Political Ecology to Understand the Present: Water, Reeds, and Biodiversity in the Camargue Biosphere Reserve, Southern France. *Ecology and Society*, 20 (4):17

McCarthy, J. (2001) States of Nature and Environmental Enclosures in the American West. In *Violent Environments*, edited by N. L. Peluso and M. Watts. Ithaca: Cornell University Press.

McCarthy, J., and J. Thatcher. (2019) Visualizing New Political Ecologies: A Critical Data Studies Analysis of the World Bank's Renewable Energy Resource Mapping Initiative. *Geoforum*, 102:242–254.

Naudts, K., Y. Chen, M. McGrath, J. Ryder, and A. Valade. (2016) Europe's Forest Management did not Mitigate Climate Warming. *Science*, 351 (6273):597–600.

Neumann, R. P. (1998) *Imposing Wilderness: Struggles over Livelihood and Nature Preservation in Africa*. Berkeley and Los Angeles: University of California Press.

Offen, K. H. (2004) Historical Political Ecology: An Introduction. *Historical Geography*, 32:19–42.

Ogden, L. A. (2011) *Swamplife: People, Gators, and Mangroves Entangled in the Everglades*. Minneapolis, MN: University of Minnesota Press.

Peluso, N. L. (1992) *Rich Forests, Poor People: Resource Control and Resistance in Java*. Berkeley: University of California Press.

Peluso, N. L. (2012) What's Nature Got to Do with It? A Situated Historical Perspective on Socio-Natural Commodities. *Development and Change*, 43 (1):79–104.

Popkin, G. (2019) How Much Can Forests Fight Climate Change? *Nature*, 565 (7739):280–282.

Rignall, K. E. (2016) Solar Power, State Power, and the Politics of Energy Transition in Pre-Saharan Morocco. *Environment and Planning A*, 48 (3):540–557.

Robbins, P. (2001a) Fixed Categories in a Portable Landscape: The Causes and Consequences of Land-Cover Categorization. *Environment and Planning A*, 33 (1):161–179.

Robbins, P. (2001b) Tracking Invasive Land Covers in India, Or Why Our Landscapes Have Never Been Modern. *Annals of the Association of American Geographers*, 91 (4):637–659.

Robbins, P. (2007) *Lawn People: How Grasses, Weeds, and Chemicals Make Us Who We Are*. Philadelphia: Temple University Press.

Robbins, P. (2020) *Political Ecology: A Critical Introduction*. Third ed. Oxford: Wiley-Blackwell.

Sayre, N. (2017) *The Politics of Scale: A History of Rangeland Science*. Chicago: The University of Chicago Press.

Sayre, N. F. (2008) The Genesis, History, and Limits of Carrying Capacity. *Annals of the Association of American Geographers*, 98 (1):120–134.

Schelenz, R. (2022) How the Indigenous Practice of "Good Fire" can Help our Forests Thrive. *University of California News*, April 6, 2022. https://universityofcalifornia.edu/news/how-indigenous-practice-good-fire-can-help-our-forests-thrive

Schroeder, R. (1993) Shady Practice: Gender and the Political Ecology of Resource Stabilization in Gambian Garden/Orchards. *Economic Geography*, 69 (4):349–365.

Senanayake, N. (2022a) Towards a Feminist Political ecology of Health: Mystery Kidney Disease and the Co-Production of Social, Environmental, and Bodily Difference. *Environment and Planning E: Nature and Space*, Pre-Print

Senanayake, N. (2022b) We Spray so We Can Live: Agrochemical Kinship, Mystery Kidney Disease, and Struggles for Health in Dry Zone Sri Lanka. *Annals of the American Association of Geographers*, 112 (4):1047–1064.

Showers, K. B. (2005) *Imperial Gullies: Soil Erosion and Conservation in Lesotho*. Athens: Ohio University Press.

Simon, G. L. (2017) *Flame and Fortune in the American West*. Berkeley: UC Press.

Simon, G. L., and C. Peterson. (2019) Disingenuous Forests: A Historical Political Ecology of Fuelwood Collection in South India. *Journal of Historical Geography*, 63 (1):34–47.

Sluyter, A. (2012) *Black Ranching Frontiers: African Cattle Herders of the Atlantic World, 1500–1900*. New Haven: Yale University Press.

St. George, Z. (2022) The Trouble with Trees. *The New York Times Magazine*, 24–31.

Stock, R., and T. Birkenholtz. (2021) The Sun and the Scythe: Energy Dispossessions and the Agrarian Question of Labor in Solar Parks. *The Journal of Peasant Studies*, 48 (5):984–1007.

Swyngedouw, E. (1997) Power, Nature, and the City: The Conquest of Water and the Political Ecology of Urbanization in Guayaquil, Ecuador, 1880–1990. *Environment and Planning A*, 29:311–332.

Swyngedouw, E. (2015) *Liquid Power: Contested Hydro-Modernities in Twentieth-Century Spain*. Cambridge, MA: MIT Press.

Temperton, V. M., N. Buchmann, E. Buisson, and G. Durigan. (2019) Step Back from the Forest and Step Up to the Bonn Challenge: How a Broad Ecological Perspective Can Promote Successful Landscape Restoration. *Restoration Ecology*, 27 (4):705–719.

Turner, M. D. (2003) Methodological Reflections on the Use of Remote Sensing and Geographic Information Science in Human Ecological Research. *Human Ecology*, 31 (2):255–279.

Veldman, J. W., J. C. Aleman, S. T. Alvarado, T. M. Anderson, and S. Archibald. (2019) Comment on "The Global Tree Restoration Potential". *Science*, 366 (6463):eaay7976

Veldman, J. W., G. E. Overbeck, D. Negreiros, and G. Mahy. (2015) Where Tree Planting and Forest Expansion are Bad for Biodiversity and Ecosystem Services. *BioScience*, 65 (10):1011–1018.

Vetter, S. (2020) With Power Comes Responsibility: A Rangeland Perspective on Forest Landscape Restoration. *Frontiers in Sustainable Food Systems*, 4 (549483), DOI: 10.3389/fsufs.2020.549483

Wainright, J. (2008) *Decolonizing Development: Colonial Power and the Maya*. Oxford: Wiley-Blackwell

Walker, P. A. (2004) Roots of Crisis: Historical Narratives of Tree Planting in Malawi. *Historical Geography*, 32 (89–109)

Watts, M. J. (2013) *Silent Violence: Food, Famine and Peasantry in Northern Nigeria*. Athens, GA: University of Georgia Press.

Werner, M., C. Berndt, and B. Mansfield. (2022) The Glyphosate Assemblage: Herbicides, Uneven Development, and Chemical Geographies of Ubiquity. *Annals of the American Association of Geographers*, 112 (1):19–35.

West, T. A., J. Borner, E. O. Sills, and A. Kontoleon. (2020) Overstated Carbon Emissions Reductions from Voluntary REDD+ Projects in the Brazilian Amazon. *PNAS*, 117 (39): 24188–24194.

West, T. A., S. Wunder, and E. O. Sills. (2023) Action Needed to Make Carbon Offsets from Tropical Forest Conservation Work for Climate Change Mitigation. Arxiv: 2301**0333354: https://arxiv.org/abs/2301.03354

Youdelis, M., J. Townsend, and J. Battacharrya. (2021) Decolonial Conservation: Establishing Indigenous Protected Areas for Future Generations in the Face of Extractive Capitalism. *Journal of Political Ecology*, 28 (1).

Zimmerer, K. S. (1996) *Changing Fortunes: Biodiversity and Peasant Livelihood in the Peruvian Andes*. Berkeley: University of California Press.

7 Valuing Nature

Constructing "Value" and Representing Interests in Environmental Decision Making

Marc Tadaki and Jim Sinner

Introduction

In Aotearoa New Zealand, depending on who you ask, brown trout (*Salmo trutta*) might be considered "river rats" (Holmes et al. 2021: 41) or "a most excellent thing" (Kós 2015). The value of nature is in the eyes of the beholder. Brown trout were introduced to New Zealand in the mid-1800s by British settlers who thought that local rivers had "no single fish worth the angler's catching" (Hursthouse 1857, cited in Knight 2019). Acclimatization societies raised trout in hatcheries and released them into streams and lakes, often without the consent of local Indigenous Māori. These societies also culled trout's key predators: freshwater eels (a major native fishery for Māori) and cormorants (Knight 2019). To enable acclimatization societies to recoup costs and control the fishery, the state prohibited anyone without a license—including Māori subsistence fishers—from catching trout or even from fishing for other species (such as eels) using methods that might catch trout by accident (White 1998).

Nowadays, debate over the value of trout has a distinct polarity. On the one hand, conservationists are increasingly asking why introduced salmonids (including trout) receive greater protection in law than New Zealand's many threatened endemic fish species (Weeks et al. 2016; Koolen-Bourke and Peart 2021). For some Māori, trout's presence is an open wound from colonization that, by displacing valued native fisheries, limits the ongoing transmission of Indigenous knowledge and cultural practices (Holmes et al. 2021). Furthermore, the *management* of trout—by excluding Māori values and voices—has weakened the ability of Māori to effectively fulfil their obligations as environmental custodians (see Tadaki et al. 2022).

On the other hand, some social groups and cultural practices are enlivened and sustained by trout. Angling can generate interest and care for the environment and can help build familial connections and responsibility for place (Holmes et al. 2021). Even for some Māori, such as those who have acquired management of trout fisheries, trout have become an important source of sustenance as well as a mechanism for transmitting cultural practices and knowledge (Tadaki et al. 2022).

The ecology of trout also complicates their management. In some places, removal of trout could cause native fish populations to decline further, as, for example, trout might be suppressing key competitors of at-risk native species. And since trout require cleaner water and deeper flow than most native fish, the removal of legal protections for trout—or simply giving native fish priority over trout—could lead to detrimental outcomes for the environment (an issue we take up in the conclusion).

DOI: 10.4324/9781003165477-10

Increasingly, decisions about river flow, pollution levels, or land use activities hinge on how society chooses to value trout and other fish through institutions of law and policy. The valuation of trout by different institutions draws on some people's aspirations, some types of evidence, and some types of reasoning (see also Alagona 2016; Perkins 2020 for an American example). The case of trout provides insight into how the practices of valuing nature are selective, normative, and contingent. In this chapter, we present a conceptual schema to help readers understand these valuation practices, what they mean, how they can be studied, and how practitioners can address their political effects.

Environmental Values in Political Ecology

From its roots in geography and cultural ecology, political ecology has had a longstanding interest in how people value their places and ecosystems (see Robbins 2012). Political ecologists have illuminated how culture, tradition, and ideology shape people's experiences of landscapes (Tuan 1977; Cosgrove and Petts 1990; Peet and Watts 1996; West 2016), and have drawn attention to how market structures shape practices of environmental use and management (Smith 2008; Blaikie and Brookfield 2015; Rocheleau 2008; Wyborn et al. 2021). An enduring theme has been the contest between "commensurable" ways of grasping how people relate to the environment (e.g., as conceived through monetary value) and the intangible, moral, and incommensurable character of people's relationships with particular environments (Bigger and Robertson 2017; Osborne et al. 2021).

The drive for growth and profit in the global economy creates pressure to exploit particular environments (e.g., harvest forests, abstract water) to generate a quantum of surplus that can be exchanged (Smith 2008; Robertson 2012). On the other hand, the "place values"—the moral worthiness of particular places and activities—are considered by some to be incommensurable with the logic of exchange (Gibbs 2010), and this immensurability often motivates communities to defend particular environments from exploitation (Ioris 2012). Political ecology has examined the relationship between the tendencies toward commensuration on the one hand, and the implications of incommensurability on the other (Bigger and Robertson 2017; Johnson 2017; Kay 2022).

Over the last three decades, political ecologists have revealed how attempts to subject the environment to the logics of capital and exchange (e.g., through monetization; see Chapter 10, this volume) have resulted in ecological degradation, social inequity, and cultural impoverishment (e.g., Castree 2008). By reducing the environment to concepts like "yield," "profit," or "carbon credits," governments, capitalists, and even academics have simplified ecosystems such as forests and rivers, ignoring and often then destroying other valued species, ecological functions, and forms of human experience (Scott 1998; Li 2007; see also Riechers et al. 2020).

At the same time, political ecologists have sought to make intangible place values visible and perhaps comparable—if not commensurable—with economic values, to provide greater parity in decision making (Gómez-Baggethun et al. 2010). Lansing et al. (1998), for example, challenged dam managers' claims that the "nonmarket" values of the Skokomish River could not be calculated and, by implication, could be assumed to be negligible. Drawing on tribal experiences and historical materials, Lansing et al. showed that the river acts as a source of "natural capital" for the tribe, and that the river's ongoing diversion by the utility company presents instantaneous losses (e.g., disappearance of habitat, fish, game, recreation), as well as progressive losses (e.g., weakening of community) to the tribe. By identifying and naming these categories of loss, political ecologists aim to make these relationships visible so that they can be advocated for through political struggle (see also Turner et al. 2008; Satterfield et al. 2013; Ioris 2012).

On the question of valuing nature, then, political ecology has provided (at least) two distinctive contributions. One is political ecology's critique of monetary and commensurable forms of value, and the other is political ecology's textured understandings of people, place, and landscape that is attentive to ideology, power relations, and political economy. To understand the politics of "valuing nature" we will draw from these contributions, but we will also need some conceptual tools from the environmental values literature.

Valuation as a Constructivist Process

The field of environmental values, and its applied cousin, environmental valuation, seeks to represent what is at stake for different people in environmental decisions. What valued ecosystem elements and relationships are affected by proposed new policies? Do gains exceed losses, and how are gains and losses distributed? The systematic inventorying and analysis of people's relationships with nature is called environmental valuation. Valuation is often imagined as a technical process that can help to "level the playing field" between different actors in environmental management by documenting what matters to different people, thereby enabling these values to be acknowledged by the wider community (see Jacobs et al. 2020).

As a *decision* tool, valuation can provide a basis for reasoning about which decisions are preferred by whom. As a *diagnostic* tool, valuation helps to identify which and whose relationships with nature have been given priority in decision making, thereby providing transparency and a democratic check on environmental governance. As an *advocacy* tool, valuation helps to bring certain environmental features and relationships into focus and prominence (e.g., a cultural keystone species, or the ecological functions of wetlands), allowing public discussion to be informed by and in response to these features. Often valuation advocacy aims to provide a counterweight to dominant and well-quantified values, such as monetary values.

In this chapter we focus on valuation as a social process in which researchers and practitioners construct "value" in a specific way and for a specific purpose. When we hear the term "values" mentioned in research and policy, we ask: what is being referred to here? Do these values reside within humans, like "transcendental values," or do values in this case refer to specific relationships between people and specific places? Is the singular "value" referred to, and if so, what does that mean?

In broad terms, one could think about valuing nature as what happens when decision makers become interested in learning about the "environmental values" of a place. We are less interested in whether values exist independently of human observers than we are in how human observers look for values, what they are searching for, what they do with what they find, and how that influences environmental decision making. Our starting point is that environmental valuation involves designating which phenomena have worth, which is a selective and power-laden process (see also Jacobs et al. 2020; Wyborn et al. 2021). We will identify key moments in the valuation process that have implications for both the results of the valuation exercise and the wider outcomes of environmental decision making.

To proceed, we describe four dominant concepts of value in the valuation literature and show how different research methods relate to these concepts. Braiding together work from environmental values and political ecology, we propose two foci for critical valuation research: the methodological politics of valuation, and the contextual politics of valuation. Finally, to demonstrate how these ideas can be applied, we interpret decision making about trout in Aotearoa through the lens of valuing nature.

Valuing Nature: Four Concepts of Value

When talking about values in environmental policy, people generally refer to one of four concepts, each involving particular assumptions (see Tadaki et al. 2017).

In mainstream economics, value is conceptualized as the "magnitude of preference" that an individual expresses for one outcome over another. Value is the price that individuals are willing to pay, e.g., to realize an outcome such as area of habitat saved from development, or improved water quality of a river. This value can be estimated through surveys that ask how much a person would pay for that outcome, whether directly (e.g., as a payment) or indirectly (e.g., through taxes). Value can also be inferred, for example, using proxy goods, such as the cost of travel to visit a national park (Young and Loomis 2014). Crucially, the "value" of nature expressed in monetary terms enables comparison with other things that we regularly pay for, making it theoretically possible to claim that the value of global ecosystem services in the mid-1990s was equivalent to approximately US$33 trillion per year (Costanza et al. 1997).

Second, value can be conceptualized as "contribution to a goal" (see Costanza 2000). Much biophysical science draws on this concept, as in, for example, efforts to quantify contributions of specific ecosystems (such as wetlands, farms) to the goals of "biodiversity" or "carbon sequestration" (e.g., Mokany et al. 2020; Tallis et al. 2021). Scientific metrics can be created to characterize environmental features using consistent criteria and then enable ranking of those features (e.g., rarity of species, habitat density). This makes it possible to talk about the relative "biodiversity value" of an ecosystem, for example.

Third, values are regularly invoked as "individual priorities" that structure how people act. Also described as held values (Brown 1984), this concept emerges from the field of psychology and seeks to characterize patterns in human motivation (Jones 2020). Environmental psychologists for example, might ask people which normative constructs such as "providing for family" or "freedom" matter most to them in their lives, and proceed to correlate these priorities with people's environmental attitudes and behaviors. When it is said that "people have different values" concerning an environmental issue, it often means that they have different individual priorities.

Fourth, values can be conceptualized as relations through which the environment becomes an object of moral concern for people (see O'Neill et al. 2008). Traditional food harvesting, for example, might be an object of moral concern not only because people need nutrition to survive, but also because food cultivation and harvesting rituals bind families and communities together and give meaning to their lives (Klain et al. 2014). The values of food harvesting extend beyond a willingness to pay for food, or beyond the contribution of an ecosystem to providing food, to include a constellation of moral reasons and relationships with a specific place-based ecosystem. These "relational values" (Chan et al. 2016, 2018) involve understanding how human relationships with nature contribute toward realizing a "good life."

Methods of Valuation

A variety of natural and social science methods are used to describe how nature matters to people (i.e., valuation). Readers interested in the conceptual underpinnings and practical issues relating to economic and non-monetary valuation methods can refer to reviews by Raymond et al. (2014), Scholte et al. (2015), Hirons et al. (2016), Tadaki et al. (2017), Harrison et al. (2018), and Arias-Arévalo et al. (2018).

Valuation methods all have their different foci, assumptions, and ways of applying expert judgement to represent the environment (Tadaki et al. 2017), which reflect a wider conceptual framework. Thus, in Table 7.1 we indicate common methods used for each

Table 7.1 Four concepts of value in environmental valuation

Value concept	*Common methods*
Value as a magnitude of preference	Revealed preference methods • Travel-cost pricing—quantitative calculations and modelling, using estimates of the cost of travel to a destination with known environmental qualities as a proxy for willingness to pay for that quality • Hedonic pricing—quantitative calculations and modeling, using a change in price of a commodity (e.g., housing) to approximate the price of a known environmental variable (e.g., distance to a stream or park) Stated preference methods • Contingent valuation—using surveys or interviews to ask people's willingness to pay for an environmental outcome • Choice experiments—using surveys to elicit willingness to pay by asking people to choose among bundles of environmental outcomes • Benefit transfer—application of prices from one study site/population to another • For a review of economic valuation methods see Tietenberg and Lewis (2018) and Young and Loomis (2014)
Value as contribution to a goal	• Biophysical modelling—experts model how different social and/or ecological scenarios contribute to different societally defined goals (e.g., clean water provision, access to nature) • Conservation prioritization—experts model the "goal" (e.g., biodiversity) as a function of ecological variables (e.g., species richness and abundance), and calculate this over space and time to identify the most cost-effective locations to manage • Participatory mapping—experts and/or residents identify locations that contribute to a "goal" such as sense of place, cultural tradition, and quantitatively evaluate a given location's relative contribution
Values as individual priorities	• Universal value item surveys—respondents use Likert scales to rate extent to which specified "priorities" matter to them • Q-method—interviews in which respondents organize statements along a spectrum of "most" to "least" agreement
Values as relations	• Interviews—eliciting how and why the environment matters to people, and the other elements (e.g., family, culture) that the environment is connected to • Participant observation—following the activities of people as they relate to and discuss the value of the environment in their day to day lives • Discourse analysis—tracing the discursive construction of "the environment" e.g., to identify which relationships are given prominence and priority in written and oral expression, and which are ignored or marginalized

concept of value. While methodological debates are certainly important, here we focus instead on discussing the assumptions and politics underpinning the valuation exercise as a whole.

Two Perspectives on the Politics of Valuation

We now return to political ecology to unpack how the valuation process unfolds. For present purposes, we identify two types of political choices in valuing nature: what we call methodological politics and contextual politics. By methodological politics we refer to the domain of conceptualization and methodology used in valuation exercises; here attention focuses on excavating the assumptions and ideological commitments of valuation, and the socio-ecological effects of these. By contextual politics, in contrast, we refer to how the valuation exercise is strategically shaped and interpreted by actors within historically grounded struggles for authority and agency. We also consider how to normatively evaluate the outcomes of valuation within context. We discuss each perspective in turn.

Methodological Politics of Valuation: Bounding and Measuring

Valuation practitioners make consequential choices about what values concepts to use, what methods to deploy, what entities or relations are worthy of moral concern, and how such entities should be measured. Political ecologists have helped to show how valuation frames the environment as an object of concern (i.e., "bounding"), and how scientific methods make aspects of the environment commensurable (i.e., "measuring"). The study by Lansing et al. (1998) effectively bounded two new objects of concern: instantaneous and progressive losses of "natural capital" from the ongoing diversion of the river. In our own research, we argue that categorizing sources of environmental value is a political act despite it often being considered a technical decision (Tadaki et al. 2015). Creating categories such as "recreation," "cultural values," or "instantaneous losses" funnels the diverse possible meanings of an environment into a small number of analytical containers and perceived political interests. In France, for example, the categories of "freshwater fish stock heritage" and "accidental pollution" have filtered public understanding of the river Seine and underpinned public judgements about whether changes in the river have been positive or negative (Bouleau 2017).

These bounded categories of "value" are products of people, place, and history (Ioris 2012) and have effects in the real world. Descriptions of people's relationships with wild fish could be split, for example, into fishing and consumption (e.g., harvesting), non-consumptive recreational (e.g., hiking experience), and cultural activities (e.g., traditional practices). Each category might invoke a different moral status in societal debate about river management, and these differences may be materially significant. For example, "river recreation" that is inclusive of picnicking might include concerns such as access, swimmability, and availability of greenspace, whereas a more specific category of "instream river recreation" might prioritize water quality and flow. However, splitting the categories in this way could also dilute a large pro-environment political constituency by disaggregating it into smaller pieces and possibly pitting one faction against the other. Thus, rather than merely describing an objective reality, valuation performs these categories of worth into existence, suggesting that these categories, and not others, are socially *valuable* (see also Bouleau 2017; Halffman 2019). Political ecologists look closely at these bounding processes and often focus on marginalized groups who may feel inadequately represented by such categories (see, e.g., Lansing et al. 1998).

Measurement often follows bounding. Measurement generally involves describing the quantum of value present in a given category, typically to allow comparison across space or time (Bigger and Robertson 2017). Measuring willingness to pay, ranking how highly people hold to certain normative ideals, or quantifying changes in biophysical parameters all provide a way to compare one situation with that of another. Measuring "conservation value" by reference to species richness and abundance, for example, enables analysis of conservation value as it has changed over time in a given place, or to compare how two places contribute to conservation value—as defined in this way (see e.g., Tadaki and Sinner 2014; Dempsey 2016).

By looking at how environmental value is bounded through categories and rendered comparable (or not) through measurement (see Figure 7.1), political ecologists expose the methodological politics of valuation for wider public scrutiny and deliberation (see also Lansing et al. 1998; Robertson 2012).

Contextual Politics of Valuation: Strategy and Normative Evaluation

There is also a need to understand the political work that valuation does in specific local contexts. Valuation doesn't influence the world in a mechanistic way; it is conducted and deployed by people who are seeking various ends, and who may or may not succeed. Some people (such as bureaucrats) might desire a "rational" calculation of costs and benefits to counter political influence, whereas others might seek to promote a key "value" category such as cultural heritage or recreational angling. Still others might want to promote specific valuation methods to remake the world in the image of a particular worldview (see e.g., Spash 2020). Furthermore, not only are the motivations for valuation mixed, but the effects are variable. For example, some economic valuation exercises can be used to legitimize efforts to privatize the environmental commons, whereas other

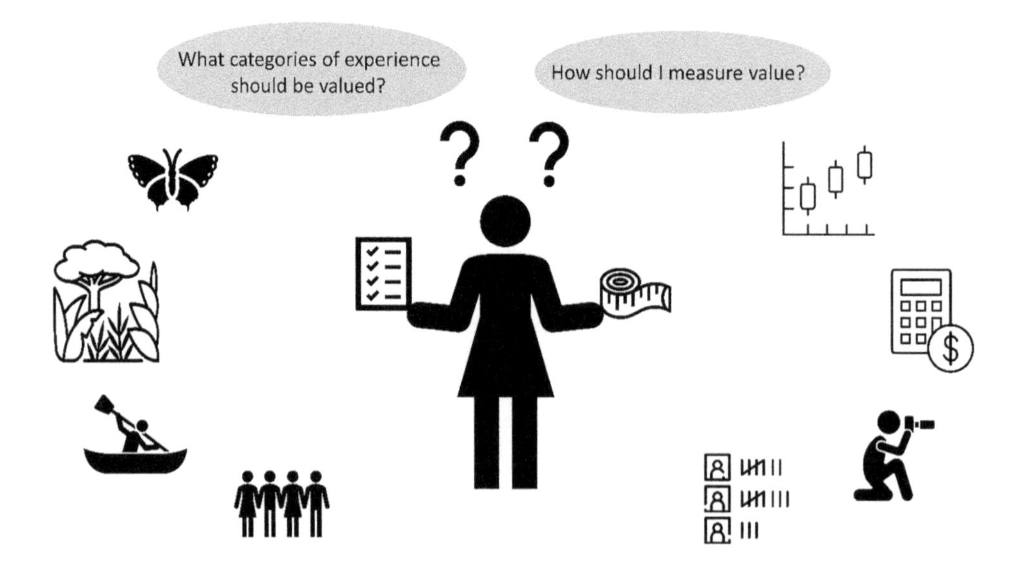

Figure 7.1 Methodological politics of valuation. Valuation practitioners make choices about what relationships and entities have value, and how that value should be measured and rendered comparable.

economic valuation exercises may empower marginalized groups to challenge dominant narratives of value (Kallis et al. 2013; Kolinjivadi et al. 2019). Explaining the diverse effects of valuation constitutes a major task and enduring challenge for political ecology (Castree 2008).

Political ecologists have proposed heuristics to help understand how valuation intervenes in local settings, by identifying the types of effects valuation can produce. Kallis et al. (2013) provocatively argue that the question is not whether to value using monetary methods; the question is in what contexts can specific valuation methods contribute to just and sustainable environmental outcomes? Kallis et al. propose four criteria to hold valuation practitioners to the ideals of justice while also allowing pragmatism in the selection and deployment of valuation methods. They argue that environmental valuation makes a positive contribution to environmental justice when it:

1. supports environmental improvement;
2. reduces inequality and redistributes power to disadvantaged interests;
3. preserves and enhances (rather than crowds out) multiple languages of value; and
4. confronts and/or prevents the enclosure of environmental commons.

In recent research, we elaborated these criteria to show more specifically *how* valuation intervenes to shape local environmental politics (Tadaki et al. 2021). We identified four mechanisms through which valuation affects and is affected by local politics:

- Valuation recognizes and strengthens particular forms of authority, for instance by granting a named government agency with comprehensive information about "values."
- Valuation reproduces regulatory categories of worth, while potentially transforming how these categories are interpreted.
- Valuation reallocates expertise about the environment by privileging certain types of knowledge and experience; for example, experts from "elsewhere," people with specific technical qualifications, local custodians, natural scientists.
- Valuation reworks or reinforces the uneven playing field of decision making by providing a common platform for values to be articulated, yet valuation cannot prevent powerful actors from trying to capture or subvert the valuation process, or from pursuing political goals through other means.

Thinking about the contextual politics of valuation helps political ecologists trace the strategic intent and normative impacts of valuation exercises (Figure 7.2). Valuation is not merely a technical method whose categories sometimes have consequences. Valuation is also wielded strategically, and its intent and outcomes can be about empowering particular ideologies, getting a particular result (e.g., for the environment, or a land use decision), strengthening political coalitions, or raising particular types of knowledge and expertise.

Applying the Framework: Valuing Trout in Aotearoa

Revisiting the case of introduced trout in Aotearoa, we can look at values in the decision-making process in concrete terms. A first point to note is that, in the real-world, decision making tends to consider only a narrow range of "values" and types of evidence. So we start by describing a decision process and then consider how the methodological and contextual politics of that process might be unpacked.

Figure 7.2 Contextual politics. Valuation practitioners make choices that affect contextual dynamics relating to legislated roles and policy categories, competing forms of knowledge, and power dynamics in the community.

In 2019, a collective of New Zealand farmers, the Lindis Catchment Group, applied to renew their permits to extract water from the Lindis River. Fish and Game New Zealand (F&G), which manages trout and salmon, contested this application in Environment Court. F&G presented evidence that the river required substantial flow if trout—highly valued in this river by the angling community—were to have sufficient food and grow to a sufficient size. They argued that a river flowing at pre-human "natural" levels was merited because of the high value of fish life in the river.

The Lindis Catchment Group argued, in contrast, that trout were an introduced species and therefore could not provide a justification for a "natural" flow level. In addition, they contended that, as an introduced species, trout's presence had reduced the ecological integrity of the Lindis, making it a low priority for flow restoration. In the end, the Court permitted the water extraction, reasoning that the presence of trout detracted from the river's natural character, so the river should not receive conservation priority.

In this court case, what we might call environmental values received little attention. The Court did not permit evidence or analysis of people's relationships with trout, their preferences for trout in the waterway, or how different people experienced the world. The decision focused almost exclusively on biophysical science, in which the value categories and associated measures of natural character, natural flow, and ecological integrity determined the future of the river and the fish in it.

As an instance of valuing nature, we can look at both the methodological and contextual politics of this valuation exercise. In terms of the methodological politics, political ecologists might focus on how "value" has been bounded and measured, and with what implications. Here we might home in on the concepts of natural character and ecological integrity, which were enshrined in legislation and therefore acted as "pre-bounded" entities of value. Scientists presented measures of ecological integrity and natural character, which are contested in both science and management spaces for their normative valence (e.g., see Low et al. 2010; Blue 2018). Not only are these categories and their definitions contested, but the concepts can also be measured in different ways. Some measures of ecological integrity count introduced species presence as a contribution toward the goal of biodiversity, whereas other measures define the goal of ecological integrity as excluding introduced fish or any human alteration (see e.g., Holmes et al. 2021).

Looking at the contextual politics of valuation, political ecologists might ask questions about strategic intent and the normative outcomes of the process. For instance, how did this valuation process privilege or marginalize certain types of expertise? Environment Court processes inherently privilege legal expertise and technical forms of evidence (see e.g., Jackson and Dixon 2007), which means that even if members of the community have strongly held views about what constitutes "ecosystem integrity" this does not feature into the Court's decision process. Indigenous Māori, for example, use concepts such as *mauri* (lifeforce) that are related but distinct from ecosystem integrity (Kitson and Cain 2022). Yet, the Court privileged only a narrow scientific definition and operationalization of these terms.

Stepping back to take a pragmatic look at the decision, we might answer the questions of Kallis et al. (2013) about the outcomes of this case:

- Is the environment better off? Not really, since further water extraction was permitted.
- Is inequality likely reduced? Not likely, since those who currently profit from extracting from the river are further allowed to do so.
- Are plural languages of value sustained? Not really, since legislation and the Court permits only biophysical evidence, so other forms of meaning are continually sidelined.
- Is the environment as a common sustained? Not really, since economic benefits are privatized and the continued flow extraction denies multiple alternative uses by human and animal communities.

In this case, this valuation exercise and decision-making process did not lead to an outcome that we consider just or sustainable. If this outcome could have been anticipated, then practitioners involved could ask themselves: can valuation be done differently to produce better outcomes?

Conclusion

From cultural ecology of the 1960s to the "valuing nature" initiatives of the present, political ecologists have sought to understand how and why people relate to their environments. As investments into valuing nature continue apace, understanding who wins and who loses from different ways of valuing nature remains an urgent task. To equip readers to grasp what is at stake with valuing nature, we provided a tour of concepts and methods in valuation, helping analysts to pinpoint the conceptual structure of the valuation exercise. Bringing together values research with political ecology, we proposed two heuristics—methodological and contextual politics—to help researchers identify key choices and diagnose the strategies and outcomes from valuing nature in decision making. Our example of introduced trout in Aotearoa showed that, despite this fish being valued in diverse ways by different people, in decision-making settings the types of values and forms of evidence considered can be sharply constrained. While political ecologists can and should contest this constraint, we also need to understand how these constraints are—and can be—worked within by actors seeking environmental justice and ecological sustainability. As interdisciplinary researchers and practitioners, political ecologists are well placed to develop a critical-yet-pragmatic approach to valuing nature.

References

Alagona PS. 2016. Species complex: Classification and conservation in American environmental history. *Isis*. 107(4):738–761.

Arias-Arévalo P, Gómez-Baggethun E, Martín-López B, Pérez-Rincón M. 2018. Widening the evaluative space for ecosystem services: A taxonomy of plural values and valuation methods. *Environmental Values*. 27:29–53.

Bigger P, Robertson M. 2017. Value is simple. Valuation is complex. *Capitalism Nature Socialism*. 28(1):68–77.

Blaikie P. and Brookfield H. 2015. *Land degradation and society*. Routledge.

Blue B. 2018. What's wrong with healthy rivers? Promise and practice in the search for a guiding ideal for freshwater management. *Progress in Physical Geography*. 42(4):462–477.

Bouleau G. 2017. A political ecology-based account of freshwater categorization. The French case. *L'Espace géographique*. 45:214–230.

Brown TC. 1984. The concept of value in resource allocation. *Land Economics*. 60(3):231–246.

Castree N. 2008. Neoliberalising nature: Processes, effects, and evaluations. *Environment and Planning A*. 40:153–173.

Chan KMA, Balvanera P, Benessaiah K, Chapman M, Díaz S, Gómez-Baggethun E, Gould R, Hannahs N, Jax K, Klain S et al. 2016. Why protect nature? Rethinking values and the environment. *Proceedings of the National Academy of Sciences USA*. 113(6):1462–1465.

Chan KMA, Gould RK, Pascual U. 2018. Editorial overview: Relational values: What are they, and what's the fuss about? *Current Opinion in Environmental Sustainability*. 35:A1–A7.

Cosgrove D, Petts G, editors. 1990. *Water, engineering and landscape*. London: Belhaven Press.

Costanza R. 2000. Social goals and the valuation of ecosystem services. *Ecosystems*. 3:4–10.

Costanza R, d'Arge R, de Groot R, Farber S, Grasso M, Hannon B, Limburg K, Naeem S, O'Neill RV, Paruelo J et al. 1997. The value of the world's ecosystem services and natural capital. *Nature*. 387:253–260.

Dempsey J. 2016. *Enterprising: Nature economics, markets, and finance in global biodiversity politics*. Malden, MA: John Wiley & Sons.

Gibbs LM. 2010. "A beautiful soaking rain": Environmental value and water beyond Eurocentrism. *Environment and Planning D: Society and Space*. 28:363–378.

Gómez-Baggethun E, De Groot R, Lomas, PL, Montes C. (2010). The history of ecosystem services in economic theory and practice: From early notions to markets and payment schemes. *Ecological Economics*. 69(6):1209–1218.

Halffman W. 2019. Frames: Beyond facts versus values. In: Turnhout E, Tuinstra W, Halffman W, editors. *Environmental expertise: Connecting science, policy, and society*. Cambridge: Cambridge University Press; pp. 36–57.

Harrison PA, Dunford R, Barton DN, Kelemen E, Martín-López B, Norton L, Termansen M, Saarikoski H, Hendriks K, Gómez-Baggethun E et al. 2018. Selecting methods for ecosystem service assessment: A decision tree approach. *Ecosystem Services*. 29:481–498.

Hirons M, Comberti C, Dunford R. 2016. Valuing cultural ecosystem services. *Annual Review of Environment and Resources*. 41(1):545–574.

Holmes R, Kitson J, Tadaki M, McFarlane K. 2021. *Diverse perspectives on the role of trout in Aotearoa New Zealand's biological heritage*. Cawthron Report No. 3691. Prepared for the Biological Heritage National Science Challenge. Available at www.cawthron.org.nz/our-news/trout-research/

Hursthouse C. 1857. *New Zealand, or Zealandia, the Britain of the south* (volume 1). London: Edward Standford.

Ioris AAR. 2012. The positioned construction of water values: Pluralism, positionality and praxis. *Environmental Values*. 21:143–162.

Jackson T, Dixon J. 2007. The New Zealand Resource Management Act: An exercise in delivering sustainable development through an ecological modernisation agenda. *Environment and Planning B: Planning and Design*. 34:107–120.

Jacobs S, Zafra-Calvo N, Gonzalez-Jimenez D, Guibrunet L, Benessaiah K, Berghöfer A, Chaves-Chaparro J, Díaz S, Gomez-Baggethun E, Lele S et al. 2020. Use your power for good: Plural valuation of nature—the Oaxaca statement. *Global Sustainability*. e8:1–7. https://doi.org/10.1017/sus.2020.1012

Johnson, L. 2017. What's beyond capitalist natures? The view from a camel's back. *Dialogues in Human Geography*. 7(3):319–322.

Jones RH. 2020. *The applied psychology of sustainability*. 2nd ed. New York: Routledge.

Kallis G, Gómez-Baggethun E, Zografos C. 2013. To value or not to value? That is not the question. *Ecological Economics*. 94:97–105.

Kay, K. 2022. "Performing Developability:" Generating threat and value in private land conservation. *Geoforum*. 128:37–45.

Kitson J, Cain A. 2022. Navigating towards Te Mana o te Wai in Murihiku. *New Zealand Geographer*. 78(1):92–97.

Klain SC, Satterfield TA, Chan KMA. 2014. What matters and why? Ecosystem services and their bundled qualities. *Ecological Economics*. 107:310–320.

Knight C. 2019. The meaning of rivers in Aotearoa New Zealand—Past and future. *River Research and Applications*. 35(10):1622–1628.

Kolinjivadi V, Van Hecken G, Almeida DV, Dupras J, Kosoy N. 2019. Neoliberal performatives and the "making" of Payments for Ecosystem Services (PES). *Progress in Human Geography*. 43(1):3–25.

Koolen-Bourke D, Peart R. 2021. *Conserving nature: Conservation reform issues paper*. Auckland: Environmental Defence Society.

Kós J. 2015. A most excellent thing: Ecological imperialism and the introduction of trout to Canterbury, New Zealand. *History Compass*. 13(4):159–170.

Lansing JS, Lansing PS, Erazo JS. 1998. The value of a river. *Journal of Political Ecology*. 5(1):1–22.

Li TM. 2007. *The will to improve: Governmentality, development, and the practice of politics*. Durham: Duke University Press.

Low JFB, Froude VA, Rennie HG. 2010. The nature of natural: Defining natural character for the New Zealand context. *New Zealand Journal of Ecology*. 34(3): 332–341.

Mokany K, Ferrier S, Harwood TD, Ware C, Di Marco M, Grantham HS, Venter O, Hoskins AJ, Watson JE. 2020. Reconciling global priorities for conserving biodiversity habitat. *Proceedings of the National Academy of Sciences*. 117(18):9906–9911.

O'Neill J, Holland A, Light A. 2008. *Environmental values*. London: Routledge.

Osborne T, Brock S, Chazdon R, Chomba S, Garen E, Gutierrez V, Lave R, Lefevre M, Sundberg J. 2021. The political ecology playbook for ecosystem restoration: Principles for effective, equitable, and transformative landscapes. *Global Environmental Change*. 70:102320.

Peet R, Watts M. 1996. Liberation ecology: Development, sustainability, and the environment in an age of market triumphalism. In: Peet R, Watts M, editors. *Liberation ecologies: Environment, development, social movements*. London: Routledge; pp. 1–45.

Perkins HA, 2020. Killing one trout to save another: A hegemonic political ecology with its biopolitical basis in Yellowstone's native fish conservation plan. *Annals of the American Association of Geographers*. 110(5):1559–1576.

Raymond CM, Kenter JO, Plieninger T, Turner NJ, Alexander KA. 2014. Comparing instrumental and deliberative paradigms underpinning the assessment of social values for cultural ecosystem services. *Ecological Economics*. 107:145–156.

Riechers M, Balázsi Á, Betz L, Jiren TS, Fischer J. 2020. The erosion of relational values resulting from landscape simplification. *Landscape Ecology*. 35(11):2601–2612.

Robbins P. 2012. *Political ecology*. 2nd ed. Chichester: John Wiley & Sons.

Robertson M. 2012. Measurement and alienation: Making a world of ecosystem services. *Transactions of the Institute of British Geographers*. 37(3):386–401.

Rocheleau DE. 2008. Political ecology in the key of policy: From chains of explanation to webs of relation. *Geoforum*. 39(2):716–727.

Satterfield T, Gregory R, Klain S, Roberts M, Chan KM. 2013. Culture, intangibles and metrics in environmental management. *Journal of Environmental Management*. 117:103–114.

Scholte SSK, van Teeffelen AJA, Verburg PH. 2015. Integrating socio-cultural perspectives into ecosystem service valuation: A review of concepts and methods. *Ecological Economics*. 114:67–78.

Scott JC. 1998. *Seeing like a state: How certain schemes to improve the human condition have failed*. New Haven: Yale University Press.

Smith N. 2008. *Uneven development: Nature, capital, and the production of space*. 3rd ed. Athens: University of Georgia Press.

Spash CL. 2020. A tale of three paradigms: Realising the revolutionary potential of ecological economics. *Ecological Economics*. 169:106518.

Tadaki M, Sinner J. 2014. Measure, model, optimise: Understanding reductionist concepts of value in freshwater governance. *Geoforum*. 51:140–151.

Tadaki M, Allen W, Sinner J. 2015. Revealing ecological processes or imposing social rationalities? The politics of bounding and measuring ecosystem services. *Ecological Economics*. 118:168–176.

Tadaki M, Sinner J, Chan KMA. 2017. Making sense of environmental values: A typology of concepts. *Ecology and Society*. 22(1):7. https://doi.org/10.5751/ES-08999-220107

Tadaki M, Sinner J, Šunde C, Giorgetti A, Glavovic B, Awatere S, Lewis N, Stephenson J. 2021. Four propositions about how valuation intervenes in local environmental politics. *People and Nature*. 3(1):190–203.

Tadaki M, Holmes R, Kitson J, McFarlane K. 2022. Understanding divergent perspectives on introduced trout in Aotearoa: A relational values approach. *Kōtuitui: New Zealand Journal of Social Sciences Online*.17(4): 461–478.

Tallis H, Fargione J, Game E, McDonald R, Baumgarten L, Bhagabati N, Cortez R, Griscom B, Higgins J, Kennedy CM. 2021. Prioritizing actions: Spatial action maps for conservation. *Annals of the New York Academy of Sciences*. 1505(1):118–141.

Tietenberg T, Lewis L. 2018. *Environmental and natural resource economics*. 11th ed. Boston: Pearson.

Tuan Y-F. 1977. *Space and place: The perspective of experience*. Minneapolis: University of Minnesota Press.

Turner NJ, Gregory R, Brooks C, Failing L, Satterfield T. 2008. From invisibility to transparency: Identifying the implications. *Ecology and Society*. 13(2):7. [online] URL: www.ecologyandsociety.org/vol13/iss12/art17/

Weeks ES, Death RG, Foote K, Anderson-Lederer R, Joy MK, Boyce P. 2016. Conservation Science Statement: The demise of New Zealand's freshwater flora and fauna: A forgotten treasure. *Pacific Conservation Biology*. 22(2):110–115.

West P. 2016. *Dispossession and the environment: Rhetoric and inequality in Papua New Guinea*. Columbia University Press.

White B. 1998. *Inland waterways: Lakes*. Waitangi Tribunal.

Wyborn C, Montana J, Kalas N, Clement S, Davila F, Knowles N, Louder E, Balan M, Chambers J, Christel L, Forsyth T. 2021. An agenda for research and action toward diverse and just futures for life on Earth. *Conservation Biology*. 35(4):1086–1097.

Young RA, Loomis JB. 2014. *Determining the economic value of water: Concepts and methods*. Routledge.

8 Enumerating Nature

Engaging Environmental Science and Data within Critical Nature-Society Scholarship

Lisa C. Kelley and Gregory L. Simon

Introduction

I am sitting at my desk in Denver, Colorado in late December, heavy snow falling onto city streets. With the aid of a growing geospatial industry, I am also one foot in eastern Sulawesi, Indonesia, opening maps that have been made to travel far from the places they purport to describe. The first map I open depicts the lands in and around Besulutu, Sulawesi as a series of gradations towards and away from what is deemed "high-integrity forest." This map draws my attention to a small patch of such forest, with the surrounding landscape depicted in different shades intended to indicate varying degrees by which lands had been anthropogenically diminished, detracting from forest integrity (Figure 8.1a). The second map classifies these lands as "agricultural" without further specification (Figure 8.1b). A third presents them as a more variegated mosaic, albeit one primarily comprised of "shrub/brush"—code in state maps for "underdeveloped" or "unused" land (Figure 8.1c).

It is difficult to reconcile these maps with one another. It is also difficult to reconcile them with the material landscapes of eastern Sulawesi, which defy neat categorization. Those areas deemed to be "forest" in the above maps, for instance, are comprised of diverse planted species, and act simultaneously as sites of watershed protection, collective vegetable, fruit, and timber harvest, and as gathering grounds and grazing lands. Nonetheless, the simplicity of the landscape renderings above is not a bug. It is a design feature. Classifying and compartmentalizing complex (socio)natural relations into discrete environmental units, domains, and statistical abstractions transmutes the landscape into a form "more susceptible to careful measurement and calculation" (Scott 2020: 12). The new management and control techniques that emerge in turn aid their makers in attempts to bring reality into closer accordance with the posited grid, whether this involves developing "idle," "unused," and "under-utilized" commons into state forest reserves (Peluso and Vandergeest 2001) or monotypic palm plantations (McCarthy et al. 2014).

Other commonalities unite these apparently contradictory landscape views. All, for instance, depend on an analytical "space of separation" between knower and known (Mitchell 2002: 100). All introduce such separation by drawing on the remotely sensed data to render the landscapes of Besulutu more abstract, and thus, commensurable with other lands regionally, nationally, and globally. All offer institutional or scientific narrations on themes prevalent within dominant conservation and development practice. All, despite technical distinctions and shifting refrains, reinforce familiar messages about the tropics: forests are being lost, agricultural expansion is the cause, and protecting some lands while rendering others more productive is the solution. Each map is not only an

DOI: 10.4324/9781003165477-11

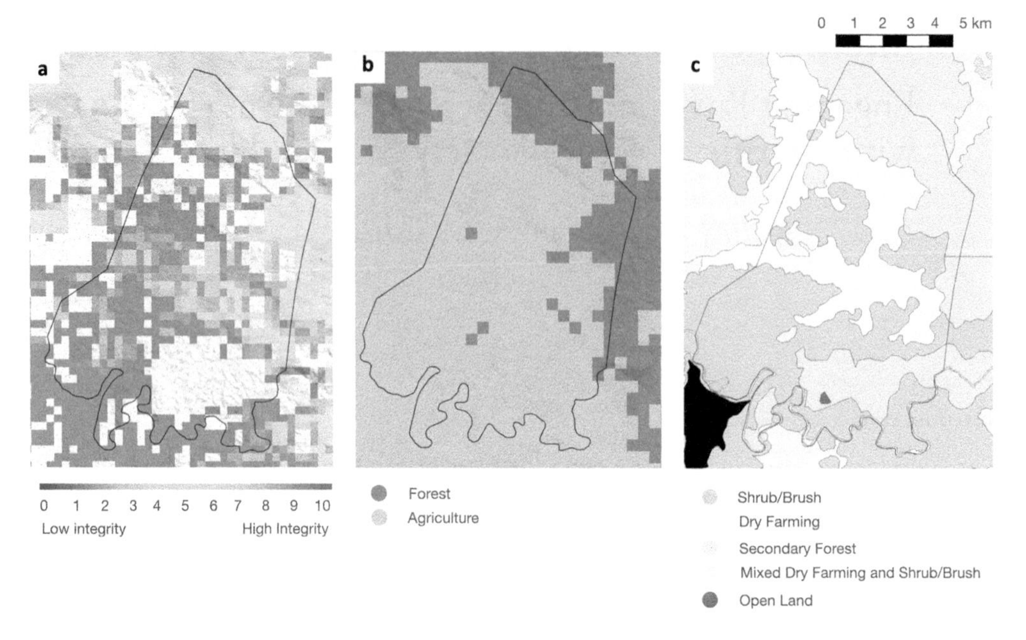

Figure 8.1 Contrasting landscape views in Sulawesi, Indonesia.

enumeration of nature but a claim about what nature is, how it came to be, and what its future should be. It would thus be stranger if the maps were the same as each reflects the distinct concerns, interests, and objectives of their makers. State maps code the landscape to better identify sites of prospective and existing commodity production, while conservation scientists describe forests as "high-integrity" due to their proposed higher biodiversity and carbon value relative to anthropogenically diminished lands.

These dynamics highlight how enumerating nature and human interactions with nature is political—a question of a priori ideological and theoretical commitments, methodological choices, and prevailing political and cultural ecologies infusing the collection, interpretation and representation of environmental data and information. This is particularly so given the Cartesian dualisms that structure the dominant scientific practices guiding many enumerations, including those separating nature from society, some members of society from others, and dominant scientific methods from other ways of knowing. Nonetheless, they do little to resolve the question of *how* critical nature-society scholars might engage with (or in, or against) such enumerations. This line of inquiry is vital given the colonial and imperial legacy of scientific knowledge production about the environment, the positivistic orientations of related datasets, and the instrumentalization of scientific practices to advance state and corporate interests (Forsyth 2015).

To address these concerns, we break this chapter into three parts, beginning with a brief overview of the many ways political ecologists have engaged social-environmental data and science historically. We then turn to a review of several key insights from political ecology, critical physical geography, and science and technology studies surrounding the politics and partialities of scientific ways of knowing and enumerating nature. Finally, we examine how scholars in these fields are carving new terrain through more explicitly place-based, justice-based, and transdisciplinary ways of enumerating nature.

Political Ecological Engagements with Socio-environmental Data and Science

In an early conceptualization, Blaikie and Brookfield (1987: 17) suggested that "regional political ecology ... combines the concerns of ecology and a broadly defined political economy." This formulation extended the emphasis on integrating biophysical, social, and biological data already integral to cultural ecology and the hazards school (e.g., Sauer 1952) while situating human-environment relations more fully in a capitalist political economy through attention to increasingly globalized flows of capital, deepening market integration, and dynamics of state resource control and territorialization (Blaikie and Brookfield 1987; Peluso 1992). Such work often drew on both environmental data and theoretical toolkits from neo-Marxist development theory and peasant studies to counter then-dominant narratives of deforestation, soil erosion, and other environmental changes as products of poverty and population. Correcting the "naive view" of functional homeostasis built into some strands of cultural ecology, political ecologists also drew on physical geographical and ecological methods to interrogate the "naturalness" of even iconic forms of wilderness (e.g., the tropical forest; Fairhead and Leach 1996), the supposed "backwardness" and "wastefulness" of peasant livelihoods (Dove 1983; Hecht 1993), and the social origins and ontology of "natural hazards" (Watts 1989).

Writing 25 years after Blaikie and Brookfield (1987), Robbins (2019 [2012]: 3) defines political ecology as "empirical, research-based explorations in the condition and change of social/environmental systems, with explicit consideration of relations of power." This definition largely parallels Blaikie and Brookfield's original formulation. As Walker (2006: 391) argues, however, the "the range of 'explorations' and 'linkages' in social/environmental systems under the label of political ecology is [now] vast," informed by an embrace of post-structuralism from the 1990s onward. Scholars self-identifying as political ecologists draw upon, *inter alia*, feminist development studies, discourse theory, critical environmental history, postcolonial theory, actor network theory, and science and technology studies. Such approaches, while disparate, tend to emphasize the "institutional nexus of power, knowledge, and practice" (Walker 2005: 75) and "a sort of political-ecological thick description" that "matches the nuanced, richly textured beliefs and practices of the world" (Peet and Watts 1996: 38). Political ecologists are also broadly united in their embrace of post-positivist science, emphasizing the partiality of all knowledge claims (i.e., accepting "truth options"—Castree 2016; and "truth tensions"—Büscher 2020) and the value-laden nature of scientific practice (Haraway 2020; Harding 2015).

These shifts in political ecology closely inform how the field now engages the enumeration of social and environmental data. On the one hand, the embrace of post-structuralism (Braun and Castree 1998; Castree and Braun 2001) has generated approaches to critically analyzing environmental datasets and scientific practices with the goal of more fully illuminating the a priori suppositions and interests, institutional norms, disciplinary orientations, and value-laden methodological choices—scholarship we review below. This body of work has explored how nature is understood, defined, and measured scientifically; how and in what specific contexts particular environmental constructs come to make sense and gain validity; and whether or not such constructs engage with nature's complexity and materiality (Forsyth 2015).

Simultaneously, the post-structural turn has been marked by an ever-diversifying range of engagements between critical human geography, physical geography, and the environmental sciences. While a growing awareness of the politics of enumerating nature has led some scholars to adopt an "ambivalent, if not wary, relationship to the simple models of materiality and measurability that undergird expert practices and method" (Robbins 2015: 90), emerging bodies of work have also reflexively engaged such critiques to fashion new ways of enumerating nature. The final section reviews several of these key works. Among other things, such work has adopted alternative forms of subject narration (Harris 2021; Goldman 2020); "multi-perspectival" or pluralistic enumerations of nature (e.g., Bennett et al. 2022); and anti-colonial, (trans)queer, and participatory scientific practices (e.g., Goodling 2020; Hazard 2022; Liboiron 2021; Wesner 2019).

The Politics of Enumerating Nature: Why Count? Who Counts? And What Counts?

In the critical physical and environmental social sciences, research examining the data collection process typically begins from a constructivist perspective and acknowledges that enumerating nature is shaped by theoretical commitments, power-laden choices and a priori assumptions that a researcher, or team of researchers, bring to the scientific process (Rhoads and Thorn 1996). Nearly half a century ago, Latour and Woolgar (1979) argued for the importance of understanding "laboratory life" and, more generally, science as "social practice." This is because institutions, identities, norms and relationships fundamentally shape what questions are asked, what methods are used, and what research outcomes are produced. Indeed, our ability to solve problems in nature is always complicated because "we ourselves are a part of the mystery that we are trying to solve" (Planck 1932: 217).

For political ecologists, this means unpacking the "black box" of environmental knowledge production (King and Tadaki 2018) to problematize the lineages, channels and relationships through which scientific statements arise and representations of nature come to make sense (Peluso 2012). This work has considered not only the methods used to construct those arguments (Sayre 2017) or what gets invisibilized by such mainstream socio-natural renderings (Hess 2020; Birkenholtz and Simon 2022), but also what methods and epistemologies (Malone and Polyakov 2020) and data governance approaches (Nost and Goldstein 2022; Dillon et al. 2019) may be productively employed to reassess engrained scientific practices and understandings.

Deconstructing the Assumptions Shaping Environmental Knowledge and Science

Political ecology has a long tradition of exploring the outsized influence of certain individuals, institutions, and epistemic communities in fashioning commonsense scientific enumerations of nature (Mountz et al. 2015; Lave 2015; Lane 2017). It also commonly engages the power-laden ways governments, corporations, and other powerful groups shape the political rationale, financial means, conceptual parameters, and structures of inclusion/exclusion for studying, knowing, and managing nature. Such work, among other things, deconstructs the "intuitive conceptual obviousness" of dominant constructs structuring rangeland science and management (Sayre 2008), and other claims about nature accepted as authoritative (Castree 2014) leveraged within instrumental uses of science (Turnhout et al. 2014) and which become hegemonic within associated management practices (Lave 2012).

Political ecologists continue to highlight and question the development and use of these "environmental orthodoxies" (Forsyth 2003), referring to such norms and beliefs in other contexts as "received wisdoms" (Leach and Mearns 1996), "knowledge regimes"

(Adger et al. 2001), "storylines" (Hajer 1995), or "discursive formations" (Peet and Watts 1996), among other descriptors. This work often blends multiple analytical traditions including: critical science studies exploring the tacit and taken-for-granted knowledge underlying nature's enumeration; Feminist, Indigenous, and Black traditions of thought that highlight how processes of scientific knowledge production repress the perspectives and situated knowledges of marginalized groups; and political economic analysis of the financial and political forces shaping scientific knowledge production.

A number of studies exemplify this mode of environmental inquiry. These studies interrogate simplistic and familiar science paradigms permeated with "tragedy of the commons" logic (e.g., Fairhead and Leach 1996; Davis 2004); examine colonial accounts of environmental decline that erase Indigenous knowledges and histories (Whyte 2018; Curley 2019); and engage the science of green development projects where neoliberal logic shapes how climate problems and solutions proceed (Bigger et al. 2018). Such work also challenges the Eurocentric notions of time, nature, and resource scarcity that lead to (neo)Malthusian, colonial, and universalizing explanations of environmental decline (Davis and Todd 2017; Mehta et al. 2019; Ojeda et al. 2020). Cognate work questions the marginalization of Black and Indigenous thought within political ecology that also begets Eurocentric erasures within counter-narratives of environmental extraction and decline (Davis et al. 2019; Curley and Smith 2023).

A similarly robust body of scholarship has also questioned the scientific and political strategies that aggregate and disaggregate the environment analytically. Such research highlights how socio-environmental reductions and delineations reflect methodological choices and therefore the tools, priorities and biases of scientists and resource managers. Kelley (2018), for example, examines the variegated patterns and diverse causes of cacao expansion in Southeast Sulawesi, Indonesia. Kelley describes how remote sensing techniques and data may produce environmental classifications that capture macroscale land use processes and support particular explanations of environmental change. However, descriptions of those same areas constructed through ethnographic methods will likely lead to a mosaic of land categories that reflect richer and more variable (yet also complementary) understandings of local social-ecological change. This work also highlights how nature, in both its specificity and variability, defies the neat and often convenient imposition of environmental categories and classifications by those occupying positions of power (e.g., Robbins 2001; Simon 2010; Davis 2016; Duvall et al. 2018).

Knowledge Co-production and its Politics

For many environmental researchers, a fundamental consideration shaping the enumeration process is that scientific knowledge of the environment is co-produced. This somewhat slippery term has at least three, albeit overlapping, variations—material-constructivist, dialogical and dialectical. First, co-production has been understood as a social-material phenomenon where society influences its surrounding environments while, all the meanwhile, the biophysical world exerts its influence on humans (Ogden 2011). For example, as Robbins (2012) argued, we construct lawns and their toxic mono-ecologies while at the same time, lawns make us who we are (particularly White, middle-class suburban residents in the United States). This material-constructivist version of co-production is relevant to the politics of enumeration because it highlights the always situated nature of efforts to enumerate the environment. Research choices and science trajectories are influenced by the things and places we study; often as a consequence of their opacity, complexity, temporality, intractability or locality (Kroepsch and Clifford 2022). Indeed, while we are charged with making sense of the environments under our culturally loaded frames of investigation, those same biophysical conditions actively influence our research possibilities, including the questions we ask and the data we collect (Biermann 2016). Using this relational lens, some have argued for an approach to environmental research that forefronts deep human-nature mutualism and symbiosis (Haraway 2016; Eyster et al. 2023), leading scholars to advocate for a more care- and kinship-based form of scientific research (Lave, Biermann, and Lane 2018; Liboiron 2021; Hazard 2022).

Second, in a more dialogical sense, the co-production of scientific knowledge can arise when multiple research contributors produce knowledge and put findings to practice through collaborative, interactive and shared engagement (Chambers et al. 2021). This form of co-production, which underpins many transdisciplinary projects and can differentiate participatory action research from more extractive approaches of data collection (Klenk et al. 2017), provides space to evaluate (or imagine) how different groups may co-frame research questions, integrate data, blend methods (i.e., from the qualitative and quantitative environmental sciences), hybridize theories and epistemologies, and co-generate management and policy responses (Miller and Wyborn 2018). (Innovations and critiques of this dialogical mode of co-production, which includes forms of community-based participatory research and citizen science, are outlined in subsequent sections.)

Third, the idiom of co-production highlights how knowledge of the environment informs scientific research questions, practices, technologies and outputs, while also being shaped by those very same research components (Jasanoff 2004). This dialectical view of science underscores how environmental knowledge is influenced by an array of circulating data, choices and practices that members of society (as producers, purveyors and recipients of science) simultaneously draw from and actively contribute to (Beck et al. 2021). For example, environmental scientists don't just measure differences, boundaries and thresholds, we produce them, legitimate them, and learn and act *through* them. This dialectic form of co-production is particularly furtive for political ecologists because it suggests that how we research, know and eventually manage nature connects with the ways powerful interests seek to render the environment researchable, knowable and manageable, such as for industry lobbyists (Michaels 2008), authoritarian regimes (McCarthy 2019), deregulatory advocates (Mansfield 2021) and science skeptics (Hess 2020). This creates conditions where "the exercise of power perpetually creates knowledge and, conversely knowledge constantly induces effects of power" (Foucault 1980: 52). This is of particular concern where efforts to enumerate nature produce and reinforce biased, incomplete and disingenuous environmental renderings that may harm or disempower certain groups (Simon 2022).

The Political Work in/of Knowledge Blind Spots

With an abundance of scholarship exploring the messy politics of knowledge production, we seem to know a lot less about subject ignorance and uncertainty (Proctor 2008). A smaller yet growing body of research has begun to take the subject of knowledge-absence in environmental research more seriously (Senanayake and King 2021; Birkenholtz and Simon 2022). Indeed, examining uncounted and ambiguous natures can be as (if not more) revelatory than the study of natures we have already successfully enumerated (Scoones and Stirling 2020). Although ignorance and uncertainty may stifle efforts to enumerate nature, these "non-knowledges" often remain circulative, affective and political in their very absence—often becoming influential through intentional acts of science obfuscation, suppression and distortion (Murphy 2006; Kleinman and Suryanarayanan 2013; McGoey 2019).

To address these knowledge gaps, critical scholars of the environment have studied factors that influence what research questions are deemed fundable and what questions are not; what social-environmental conditions are studied and what are not; and what data is eventually enumerated and what is not (Goldstein, Paprocki, and Osborne 2019; Clifford and Travis 2021; Sedell 2021). Studies have shown how these informational shortcomings may persist due to a lack of will, interest or desire by key actors, thus stifling research and informed decision making. In these contexts, we know what we don't know yet still choose not to improve upon current understandings due to countervailing political and economic priorities. These outcomes produce a form of "undone science" where information gaps persist and may stifle efforts to study and combat issues of critical importance (Hess 2007; Frickel et al. 2010).

Undone science often emerges from the imbalances of knowledge production found within colonial, patriarchal and racist science structures that privilege certain voices and knowledge contributions over others. Systemic inequities, injustices, and exclusions in science occur in a range of academic settings related to, for example, citational (Hawthorne 2019) and editorial (Crane et al. 2023) practices that curtail knowledge participation from certain scientists, and also research design considerations around whose knowledge of the environment (e.g., research participants) should count and be considered of "expert" quality when enumerating nature (de Santos 2018; D'Ignazio and Klein 2020). In contemporary research settings, forms of undone science can thus also be traced to systemic patterns of marginalization that have limited scientific engagements by diverse and multiple positionalities, thereby further entrenching other scientific identities, perspectives and methodologies (Tuhiwai Smith 2012).

It is also the case that the complex and often confounding nature of biophysical environments will produce forms of uncertainty for scientists and resource users and managers alike. "Inscrutable natures," as Kroepsch and Clifford (2022) call them, have produced uncertainty and stymied efforts to manage the environment in contexts ranging from groundwater (Birkenholtz 2018; Kroepsch 2018), subsurface gas (Kinchy and Schaffer 2018), and rangelands (Sayre 2017). Others have described how blind spots and analytical gaps lead to forms of compounding and accumulating ignorance that occur *from* knowledge blind spots (Hess 2020; Shattuck 2021). These "regimes of imperceptibility" (Murphy 2006) highlight conditions where ignorance persists because we don't know our own knowledge deficiencies, leading to a retrograde cycle of ignorance and inaction.

Research has also demonstrated the utility and strategic purpose of producing knowledge shortcomings where knowledge deficiencies can be a valuable resource used to obscure, suppress, discount and dispute scientific findings in order to sow confusion, increase ambiguity, cause controversy or delay regulatory action (Kuehn 2004; Neimark et al. 2019; Dillon et al. 2019; Hesse 2021). Focusing on the "for whom?" of ignorance, this research demonstrates the value of non-knowledge for powerful interests in relation to a range of topics including climate change (Oreskes and Conway 2010; Freudenburg and Muselli 2013), synthetic pregnancy hormones (Langston 2010), sugar (Goldman et al. 2014), fracking chemicals (Kinchy and Schaffer 2018), fluoride (Richter et al. 2018), and radiation risk (Greene 2017), to name a few. Furthermore, uncertainty and ignorance may prove exculpatory (Hall 2011; Kopack 2019) by reducing liability for industry scientists if research findings fail to demonstrate causality. Meanwhile, maintaining scientific unpredictability may actually generate continued opportunities for groups to justify further research and the acquisition of more research funding (McGoey 2019). As these examples suggest, examining the politics of enumeration requires reorienting our conception of "epistemology" to include the production of knowledge gaps and their furtive influence.

Alternative Practices of Enumerating Nature

Political ecologists have long critiqued approaches to enumerating nature within the environmental sciences. Efforts to unsettle engrained scientific practices and "truths," however, are limited by a growing distance between political ecology and direct engagements with methodological protocols and practices from the biophysical sciences (Lave et al. 2014; Walker 2005; but see also Turner 2016). Political ecology, like geography more broadly, is an historically White (Bruno and Faiver-Serna 2022; Pulido 2002) and cis (Gieseking 2023) discipline. These histories not only shape geographical imagination, but also how political ecology has engaged the politics, partialities, and potential of engaging nature and its enumerations (Meehan et al. 2023). Emerging bodies of work are addressing these limitations to fashion a more inclusive, integrative, and explicitly emancipatory praxis of enumerating nature within political ecology, including through more politically engaged scholarship with diverse "publics" (Osborne 2017). This section engages with work in this spirit, grounding a review of relevant literatures in illuminating examples of each of three trends we note.

Engaging Mixed-Methods to Reveal Scientific Knowledge Instabilities

A first strand of emerging work draws upon approaches to enumerating nature to better understand how discursive and material natures are co-produced. Work within the field of critical physical geography, for instance, is integrating insights from political ecology and methods from physical geography and the environmental sciences (Biermann et al. 2020) to document dynamics ranging from the socio-biophysical ramifications of the US Farm Bill (Malone and McClintock 2023) and stream restoration policies (Lave and Doyle 2021) to the "socio-ecology of fear" shaping wildlife management (Anderson et al. 2023). This work and related scholarship within critical remote sensing (Bennett et al. 2022; Bennett and Faxon 2021) evidences the value of engaging biophysical and geospatial methods with a "care for the subject" (Schuurman and Pratt 2002) and by "standing in faith" (TallBear 2014) with biophysical scientists.

Christine Biermann's (2018) analysis of the practices and politics of tree-ring science in Great Smoky Mountains National Park, for instance, illustrates how deep engagements with biophysical research methods can contribute to a transformation of dominant science from within. Biermann first leverages 390 core samples from 245 trees and historical climate data within a moving correlation analysis to document the temporally unstable relationship between climate variables and pine growth—a finding that challenges the assumption of a consistent, linear relationship between tree growth and climate variables so central to dendroclimatological reconstructions of historical environmental conditions. Biermann then draws on 48 interviews to understand how the fallacy of this uniformitarian assumption is being engaged by tree-ring scientists. While Biermann identifies a tendency within tree-ring science to view uncertainty as a threat to authoritative claims on climate, she also highlights how uncertainty is being embraced by a subset of scientists to shed light on the "dynamic interactions among climate, landscape, trees, and people" (Biermann 2018: 220) that beget temporally unstable tree growth-climate relations.

Tianna Bruno's (2023) scholarship also draws on dendrochronological analysis, albeit in this case, to speak to how dynamic socio-ecological interactions are shaped by and reflect the "biophysical afterlives of slavery." To theorize such dynamics, Bruno draws upon dendrochronology and elemental chemical analysis to document the cumulative histories of pollution and contamination inscribed in tree growth histories, highlighting how such archives document a more substantial legacy of contamination than do the "conventional colonial and hegemonic archives" of environmental agencies such as EPA. Bruno then integrates these understandings with archives of Black ecological life ranging from oral traditions and community members' oral histories and lived experiences to theorize how Black people and trees have been mutual witnesses and survivors of both anti-Black environmental violence and the living ideological and sociopolitical legacies of slavery.

These approaches, in other words, not only reveal the internal and external politics of enumerating nature (King and Tadaki 2018). They also illuminate how alternative practices of enumerating and representing nature are already taking root within and using methods from the biophysical sciences. In the case of Biermann and Bruno's scholarship, integrating these practices with qualitative methodologies and perspectives from feminist science and technology studies or Black geographies and environmental justice also provide a means of theorizing the complex land use histories and the dynamics of racism and colonialism that shape landscapes, both socially and biophysically.

Reimagining the "Who," "What," and "How" of Dominant Scientific Approaches

Recent work in political ecology, anti-colonial environmental science, and critical physical geography also questions who is authorized to speak and be heard within critical nature-society accounts (Ojeda et al. 2022), expanding historical notions of "who counts" and "what counts" within nature's enumerations. Such work problematizes the norms, politics, and institutional bases begetting "regimes of imperceptibility" and problematic environmental orthodoxies (Murphy 2006). These studies also evidence the varied valid means of "truth" telling that can be deployed to enumerate nature (Harris 2022), including analytical traditions within Black and Indigenous thought and reimagined methodological approaches from the biophysical sciences (McGuire and Mawyer 2023). By highlighting diverse ontologies and pluriversal ways of experiencing and knowing nature (Escobar 2018) this research reflects broader efforts to decolonize the scientific enumeration process (Smith 2021).

Cleo Wölfle Hazard's (2022) scholarship on river policy and science is one powerful example of this work. As Hazard argues, logics of rationality, control, and White supremacy not only threaten rivers, they also pervade river science. Hazard draws on his own lived experiences, queer and trans thought, Indigenous studies, and critical physical geography to model an alternative approach to studying and transforming rivers. Among other subjects, Hazard engages queer and trans strategies to foster collectivities that celebrate bodily and affective transformation as a model of the "politics of solidarity in multiplicity" that can transform riverine relations. Hazard shows how approaches like this can reveal how the excess and unruliness of rivers, the agency of other-than-human beings, and the feelings and affects of scientists form political and ecological "underflows" which are already transfiguring cisheteronormative settler-colonial water politics and land relations.

Max Liboiron's (2021) *Pollution is Colonialism* similarly explores ways for navigating the "complex and compromised" terrain of dominant science, in this case, by developing feminist and anti-colonial scientific approaches to study how plastic affects fish populations significant to culturally embedded food systems in Newfoundland. Liboiron begins by denaturalizing the assumptions of assimilative capacity, universalism, and colonial entitlement to land so central to dominant ways of defining and managing pollution, using these insights as the basis for reflecting upon how "good Land relations [can be made] integral to dominant scientific practices." For Liboiron and the CLEAR laboratory they lead, embedding science with such relations involves practices and protocols ranging from community peer review to not wearing earbuds and reflecting on sample origins when dissecting fish ("it's rude to tune out your relations"). These practices involve a range of strategies including engaging models of generalization (e.g., Tuck and McKenzie 2014) while also embracing critical appraisals of place to avoid, and perhaps counter, claims of universalism.

Towards a More Liberatory Praxis of Engaging Nature's Enumerations

Political ecologists have done more than draw attention to the political and cultural exclusions that mediate epistemic exclusions. Scholars have also pushed back against such omissions (Wesner 2019), including by elevating the situated knowledges and knowledge production practices of diverse and marginalized communities as no less "scientific" than those of science "experts" (Wylie et al. 2017). Such work within political ecology attempts to realize the field's "liberatory potential" (Peet and Watts 1996) by more fully engaging with traditions and innovations within participatory action research and direct action politics (Fortmann 2009; Heynen and Van Sant 2015; Rocheleau 2003) and efforts to counter dominant ideas and their cartographic renderings (Harris and Hazen 2006; Dalton and Stallmann 2017; Sletto et al. 2020). These bodies of work seek to overcome entrenched knowledge gaps and "undone" science while simultaneously advancing more inclusive understandings of the multiple ways nature is already and effectively being enumerated outside the academy (Arancibia and Motta 2019). Davies (2018: 1549), for instance, shows how the "slow observations" of long-term residents in Louisiana's Cancer Alley capture diverse temporalities of violence shaped by "unexpected infrastructural expansion, nocturnal toxic smells, or sudden earth tremors" while highlighting the multiple relevant barometers for tracking petrochemical pollution that already exist (e.g., fruit trees, alligators, and family injuries). As part of the political ecology project, however, community-based participatory research itself can be the subject of reflexive analysis by exploring the uneven power dynamics (Vincent 2022) and neoliberal extractivist research practices (Haverkamp 2021) that have a way of influencing even the most well-intentioned co-production process. This might include imbalanced contributions, skewed representation, and inadequate compensation of certain research participants in the design and implementation of research projects (Turnhout et al. 2020; Lemos et al. 2018; Castree et al. 2014).

Efforts to develop more inclusionary science must also reckon with questions about who science enumerations are ultimately for. An increasing number of studies seek to demonstrate approaches to translating and communicating science in service of more open and inclusive knowledge participation. This includes justice-based and interdisciplinary environmental pedagogy (Malone 2021) and multiple modes and means of engagement, including social media and museum exhibits (Arce-Nazario 2016; Siti et al. 2019).

Political ecologist Tracy Osborne (2017: 845) highlights how such work can be the basis for a more explicitly public political ecology "that joins public debates on environmental change and strives to shape the consciousness of civil society towards an integrated view of Earth Stewardship." Forging closer connections between people and institutions inside and outside academia is the basis for fostering such a community of praxis and more just sustainabilities (Sze 2018). This work goes beyond acknowledgement and recognition into the realms of direct action and assumed risks, as geographer Scott Warren's work to provide food and water for migrants at the US border illustrates.

Conclusion

Enumerating nature is political. From the underlying theoretical commitments shaping our research questions to the methodological choices influencing data collection, researching socio-environments is always value-laden and power-infused. These judgements and decisions determine what we study, how we conduct research, and whose interests our research ultimately serves. This chapter has reviewed key themes and concepts in political ecology, science and technology studies, critical physical geography and beyond to highlight ongoing critiques of conventional scientific practices and the knowledge it produces. While political ecologists have generated countless insights, we focus primarily on three here: excavating the underlying assumptions informing our enumeration efforts; underscoring the origins, purpose and influence of knowledge and its blind spots; and highlighting the features and (dis)advantages of using various forms of knowledge co-production to explain the development of environmental information and expertise.

The case to begin this chapter, describing maps from Besulutu, Sulawesi, reminds us that scientific choices inform not only the character of our data but also how that data narrates the environment back to us. Socio-environmental science lives far beyond its creation. For example, data substantiating and reinforcing land descriptors like "unproductive" or "wasteland" can inform and actuate the land management goals of powerful interests, which may include the pursuit of commodity production, biodiversity conservation, or carbon sequestration. But science is also a realm of imagination, creativity and alternative data collection and analysis strategies. More reflexive, inclusive, and participatory engagements with socio-environmental data and science continue to take shape and point to promising directions for further work in this space.

References

Adger, W. N., Benjaminsen, T. A., Brown, K., and Svarstad, H. (2001). Advancing a political ecology of global environmental discourses. *Development and Change*, *32*(4), 681–715.

Anderson, R. M., Charnley, S., Epstein, K., Gaynor, K. M., Martin, J. V., and McInturff, A. (2023). The socioecology of fear: A critical geographical consideration of human-wolf-livestock conflict. *The Canadian Geographer/Le Géographe Canadien*, *67*(1), 17–34.

Arancibia, F., and Motta, R. (2019). Undone science and counter-expertise: Fighting for justice in an Argentine community contaminated by pesticides. *Science as Culture*, *28*(3), 277–302.

Arce-Nazario, J. A. (2016). Translating land-use science to a museum exhibit. *Journal of Land Use Science*, *11*(4), 417–428. https://doi.org/10.1080/1747423X.2016.1172129

Beck, S., Jasanoff, S., Stirling, A., and Polzin, C. (2021). The governance of sociotechnical transformations to sustainability. *Current Opinion in Environmental Sustainability*, *49*, 143–152.

Bennett, M. M., Chen, J. K., Alvarez León, L. F., and Gleason, C. J. (2022). The politics of pixels: A review and agenda for critical remote sensing. *Progress in Human Geography*, *46*(3), 729–752.

Bennett, M. M., and Faxon, H. O. (2021). Uneven frontiers: Exposing the geopolitics of Myanmar's borderlands with critical remote sensing. *Remote Sensing*, *13*(6), 1158.
Biermann, C. (2016). Securing forests from the scourge of chestnut blight: The biopolitics of nature and nation. *Geoforum*, *75*, 210–219.
Biermann, C. (2018). Shifting climate sensitivities, shifting paradigms: Tree-ring science in a dynamic world. *The Palgrave Handbook of Critical Physical Geography*, 201–225. Palgrave.
Biermann, C., Kelley, L. C., and Lave, R. (2020). Putting the Anthropocene into practice: Methodological implications. *Annals of the American Association of Geographers*, *111*(3), 808–818.
Bigger, P., et al. (2018). Reflecting on neoliberal natures: An exchange. *Environment and Planning E: Nature and Space*, *1*(1–2), 25–75.
Birkenholtz, T. (2018). Commentary: STS of the underground. *Engaging Science, Technology, and Society*, *4*, 155–164.
Birkenholtz, T., and Simon, G. (2022). Introduction to themed issue: Ignorance and uncertainty in environmental decision-making. *Geoforum*, *132*, 154–161.
Blaikie, P., and Brookfield, H. (1987). *Land Degradation and Society*. Routledge.
Braun, B., and Castree, N. (1998). *Remaking Reality. Nature at the Millenium*. Routledge.
Bruno, T. (2023). Ecological memory in the biophysical afterlife of slavery. *Annals of the American Association of Geographers*, *113*(7), 1543–1553.
Bruno, T., and Faiver-Serna, C. (2022). More reflections on a white discipline. *The Professional Geographer*, *74*(1), 156–161.
Büscher, B. (2020). *The Truth about Nature: Environmentalism in the Era of Post-Truth Politics and Platform Capitalism*. University of California Press.
Castree, N. (2014). *Making Sense of Nature*. Routledge.
Castree, N. (2016). Geography and the new social contract for global change research. *Transactions of the Institute of British Geographers*, *41*(3), 328–347.
Castree, N., Adams, W., Barry, J. et al. (2014). Changing the intellectual climate. *Nature Climate Change*, *4*, 763–768. https://doi.org/10.1038/nclimate2339
Castree, N., and Braun, B. (2001). *Social Nature: Theory, Practice and Politics*. Oxford: Basil Blackwell.
Chambers, J. M., et al. (2021). Six modes of co-production for sustainability. *Nature Sustainability*, *4*, 983–996. https://doi.org/10.1038/s41893-021-00755-x
Clifford, K. R., and Travis, W. R. (2021). The new (ab)normal: Outliers, everyday exceptionality, and the politics of data management in the Anthropocene. *Annals of the American Association of Geographers*, *111* (3), 932–943.
Crane, N. J., Ergler, C., Griffin, P., Holton, M., Rhiney, K., Robinson, C., and Simon, G. (2023). Whose geography do we review? *Geography Compass*, *17*(2), p. e12676.
Curley, A. (2019). Unsettling Indian water settlements: The Little Colorado River, the San Juan River, and colonial enclosures. *Antipode*, *53*(3), 705–723.
Curley, A., and Smith S. (2023). The cene scene: Who gets to theorize global time and how do we center indigenous and black futurities? *Environment and Planning E: Nature and Space*, Online First.
D'Ignazio, C., and Klein, L. F. (2020). *Data Feminism*. MIT Press.
Dalton, C. M., and Stallmann, T. (2017). Counter-mapping data science. *The Canadian Geographer/ Le Geographe Canadien*, *62*, 93–101.
Davies, T. (2018). Toxic space and time: Slow violence, necropolitics, and petrochemical pollution. *Annals of the American Association of Geographers*, *108*(6), 1537–1553. https://doi.org/10.1080/24694452.2018.1470924
Davis, D. K. (2004). Desert "wastes" of the Maghreb: Desertification narratives in French colonial environmental history of North Africa. *Cultural Geographies*, *11*(4), 359–387.
Davis, D. K. (2016). *The Arid Lands: History, Power, Knowledge*. MIT Press.
Davis, H., and Todd, Z. (2017). On the importance of a date, or decolonizing the Anthropocene. *ACME: An International E-Journal for Critical Geographies*, *16*(4), 761–780.
Davis, J., Moulton, A. A., Van Sant, L., and Williams, B. (2019). Anthropocene, capitalocene ... plantationocene?: A manifesto for ecological justice in an age of global crises. *Geography Compass*, *13*(5), e12438.
Dillon, L., Lave, R., Mansfield, B., Wylie, S., Shapiro, N., Chan, A. S., and Murphy, M. (2019). Situating data in a Trumpian era: The environmental data and governance initiative. *Annals of the American Association of Geographetrs* *109*(2), 545–555.

Dove, M. R. (1983). Theories of swidden agriculture, and the political economy of ignorance. *Agroforestry Systems*, *1*, 85–99.

Duvall, C. S., Butt, B., and Neely, A. (2018). The trouble with savanna and other environmental categories, especially in Africa. *The Palgrave Handbook of Critical Physical Geography*, 107–127.

Escobar, A. (2018). *Designs for the Pluriverse: Radical Interdependence, Autonomy, and the Making of Worlds*. Duke University Press.

Eyster, H. N., Satterfield, T., and Chan, K. M. (2023). Empirical examples demonstrate how relational thinking might enrich science and practice. *People and Nature*, *5*(2), 455–469.

Fairhead, J., and Leach, M. (1996). *Misreading the African Landscape: Society and Ecology in a Forest-Savanna Mosaic*. Cambridge University Press.

Forsyth, T. (2003). *Critical Political Ecology: The Politics of Environmental Science*. London: Routledge.

Forsyth, T. (2015). Integrating science and politics in political ecology. *The International Handbook of Political Ecology*. London: Edward Elgar Publishing, 103–116.

Fortmann, L. (2009). *Participatory Research in Conservation and Rural Livelihoods: Doing Science Together*. John Wiley & Sons.

Foucault, M. (1980). *Language, Counter-memory, Practice: Selected Essays and Interviews*. Cornell University Press.

Freudenburg, W. R., and Muselli, V. (2013). Reexamining climate change debates: Scientific disagreement or Scientific Certainty Argumentation Methods (SCAMs)? *The American Behavioral Scientist* *57*(6), 777–795.

Frickel, S., Gibbon, S., Howard, J., Kempner, J., Ottinger, G., and Hess, D. (2010). Undone science: Social movement challenges to dominant scientific practice. *Science, Technology & Human Values* *35*(4), 444–473.

Gieseking, J. J. (2023). Reflections on a cis discipline. *Environment and Planning D: Society and Space*, 02637758231191656.

Goldman, G., Carlson, C., Bailin, D., Fong, L., and Phartiyal, P. (2014). *Subtracted Science: How Industry Obscures Science and Undermines Public Health in Sugar Research*. Union of Concerned Scientists, Cambridge, MA.

Goldman, M. J. (2020). *Narrating Nature: Wildlife Conservation and Maasai Ways of Knowing*. University of Arizona Press.

Goldstein, J. E., Paprocki, K., and Osborne, T. (2019). A manifesto for a progressive land-grant mission in an authoritarian populist era. *Annals of the American Association of Geographers*, *109*(2), 673–684.

Goodling, E. (2020). Intersecting hazards, intersectional identities: A baseline critical environmental justice analysis of US homelessness. *Environment and Planning E: Nature and Space*, *3*(3), 833–856.

Greene, G. (2017). *The Woman Who Knew Too Much, Revised Ed.: Alice Stewart and the Secrets of Radiation*. University of Michigan Press.

Hajer, M. A. (1995). *The Politics of Environmental Discourse: Ecological Modernization and the Policy Process*. Oxford University Press.

Hall, S. S. (2011). Scientists on trial: At fault? *Nature* *477*(7364), 264–269.

Haraway, D. (2016). *Staying with the Trouble: Making Kin in the Chthulucene*. Duke University Press.

Haraway, D. (2020). Situated knowledges: The science question in feminism and the privilege of partial perspective. In *Feminist Theory Reader* (pp. 303–310). Routledge.

Harding, S. (2015). Objectivity for sciences from below. *Objectivity in Science: New Perspectives from Science and Technology Studies*, 35–55.

Harris, D. M. (2021). Storying climate knowledge: Notes on experimental political ecology. *Geoforum*, *126*, 331–339.

Harris, L. M. (2022). Towards enriched narrative political ecologies. *Environment and Planning E: Nature and Space*, *5*(2), 835–860.

Harris, L. M., and Hazen, H. D. (2006). Power of maps: (Counter) mapping for conservation. *ACME: An International E-Journal for Critical Geographies* *4*(1): 99–130.

Haverkamp, J. (2021). Where's the love?: Recentring Indigenous and feminist ethics of care for engaged climate research. *Gateways: International Journal of Community Research and Engagement*, *14*(2), 1–15.

Hawthorne, C. (2019). Black matters are spatial matters: Black geographies for the twenty-first century. *Geography Compass*, *13*(11), e12468.
Hazard, C. W. (2022). *Underflows: Queer Trans Ecologies and River Justice*. University of Washington Press.
Hecht, S. B. (1993). The logic of livestock and deforestation in Amazonia. *Bioscience*, *43*(10), 687–695.
Hess, D. (2007). *Alternative Pathways in Science and Industry*. MIT Press.
Hess, D. (2020). The sociology of ignorance and post-truth politics. *Sociological Forum 35*(1), 241–249.
Hesse, A. (2021). Geographies of uncertainty and negotiated responsibilities of occupational health. *Geoforum 123*, 184–193.
Heynen, N., and Van Sant, L. (2015). *Political Ecologies of Activism and Direct Action Politics*. Routledge.
Jasanoff, S. (Ed.) (2004). *States of Knowledge: The Co-Production of Science and the Social Order*. Routledge.
Kelley, L. C. (2018). The politics of uneven smallholder cacao expansion: A critical physical geography of agricultural transformation in Southeast Sulawesi, Indonesia. *Geoforum*, *97*, 22–34.
Kinchy, A., and Schaffer, G. (2018). Disclosure conflicts: Crude oil trains, fracking chemicals, and the politics of transparency. *Science, Technology & Human Values 43*(6), 1011–1038. https://doi.org/10.1177/0162243918768024
King, L., and Tadaki, M. (2018). A framework for understanding the politics of science (core tenet #2). *The Palgrave Handbook of Critical Physical Geography*, 67–88.
Kleinman, D. L., and Suryanarayanan, S. (2013). Dying bees and the social production of ignorance. *Science, Technology & Human Values 38*(4), 492–517.
Klenk, N., Fiume, A., Meehan, K., and Gibbes, C. (2017). Local knowledge in climate adaptation research: Moving knowledge frameworks from extraction to co-production. *Wiley Interdisciplinary Reviews: Climate Change*, *8*(5), e475.
Kopack, R. A. (2019). Rocket wastelands in Kazakhstan: Scientific authoritarianism and the Baikonur Cosmodrome. *Ann. Am. Assoc. Geograph. 109*(2), 556–567. https://doi.org/10.1080/24694452.2018.1507817
Kroepsch, A. (2018). Groundwater modeling and governance: Contesting and building (sub)surface worlds in Colorado's northern San Juan basin. *Engaging Science, Technology, and Society 4*, 43–66.
Kroepsch, A. C., and Clifford, K. R. (2022). On environments of not knowing: How some environmental spaces and circulations are made inscrutable. *Geoforum*, *132*, 171–181.
Kuehn, R. R. (2004). Suppression of environmental science. *American Journal of Law & Medicine 30* (2–3), 333–369.
Lane, S. N. (2017). Slow science, the geographical expedition, and critical physical geography. *The Canadian Geographer/Le Géographe Canadien*, *61*(1), 84–101.
Langston, N. (2010). *Toxic Bodies*. Yale University Press.
Latour, B., and Woolgar, S. (1979). *Laboratory Life: The Social Construction of Scientific Facts*. Princeton University Press.
Lave, R. (2012). Bridging political ecology and STS: A field analysis of the Rosgen Wars. *Annals of the Association of American Geographers 102*(2), 366–382.
Lave, R. (2015). The future of environmental expertise. *Annals of the Association of American Geographers 105*(2), 244–252.
Lave, R., Biermann, C., and Lane, S. N. (2018). Introducing critical physical geography. *The Palgrave Handbook of Critical Physical Geography*, 3–21. Palgrave.
Lave, R., and Doyle, M. (2021). *Streams of Revenue: The Restoration Economy and the Ecosystems it Creates*. MIT Press.
Lave, R., et al. (2014). Intervention: Critical physical geography. *The Canadian Geographer/Le Géographe Canadien*, *58*(1), 1–10.
Leach, M., and Mearns, R. (1996). *The Lie of the Land: Challenging Received Wisdom on the African Environment*. London: James Currey.
Lemos, M. C. et al. (2018). To co-produce or not to co-produce. *Nature Sustainability 1*, 722–724. https://doi.org/10.1038/s41893-018-0191-0
Liboiron, M. (2021). Pollution is colonialism. In *Pollution Is Colonialism*. Duke University Press.

Malone, M. (2021). Teaching critical physical geography. *Journal of Geography in Higher Education*, *45*(3), 465–478.
Malone, M., and Polyakov, V. (2020). A physical and social analysis of how variations in no-till conservation practices lead to inaccurate sediment runoff estimations in agricultural watersheds. *Progress in Physical Geography: Earth and Environment*, *44*(2), 151–167.
Malone, M., and McClintock, N. (2023). A critical physical geography of no-till agriculture: Linking degraded environmental quality to conservation policies in an Oregon watershed. *The Canadian Geographer/Le Géographe Canadien*, *67*(1), 74–91.
Mansfield, B. (2021). Deregulatory science: Chemical risk analysis in Trump's EPA. *Social Studies of Science 51* (1), 28–50.
McCarthy, J. (2019). Authoritarianism, populism, and the environment: Comparative experiences, insights, and perspectives. *Ann. Am. Assoc. Geograph. 109*(2), 301–313.
McCarthy, J. F., Vel, J. A., and Afiff, S. (2014). Trajectories of land acquisition and enclosure: Development schemes, virtual land grabs, and green acquisitions in Indonesia's Outer Islands. In *Green Grabbing: A New Appropriation of Nature* (pp. 295–324). Routledge.
McGoey, L. (2019). *The Unknowers: How Strategic Ignorance Rules the World*. Zed Books.
McGuire, G., and Mawyer, A. (2023). Cultivating the unseen: Pa'akai and the role of practice in coastal care. *Ethnobiology Letters*, *14*(2), 22–36.
Meehan, K., Gergan, M. D., Mollett, S., and Pulido, L. (2023). Unsettling race, nature, and environment in geography. In *Annals of the American Association of Geographers* (pp. 1–8). Taylor & Francis Group.
Mehta, L., Huff, A., and Allouche, J. (2019). The new politics and geographies of scarcity. *Geoforum 101*, 222–230.
Michaels, D. (2008). *Doubt is Their Product: How Industry's Assault on Science Threatens your Health*. Oxford University Press.
Miller, C. A., and Wyborn, C. (2018). Co-production in global sustainability: Histories and theories. *Environmental Science & Policy 113*, 88–95.
Mitchell, T. (2002). *Rule of Experts: Egypt, Techno-politics, Modernity*. Univ. of California Press.
Mountz, A. et al. (2015). For slow scholarship: A feminist politics of resistance through collective action in the neoliberal university. *ACME: An International Journal for Critical Geographies*, *14*(4), 1235–1259.
Murphy, M. (2006). *Sick Building Syndrome and the Problem of Uncertainty: Environmental Politics, Technoscience, and Women Workers*. Duke University Press.
Neimark, B., Childs, J., Nightingale, A. J., Cavanagh, C. J., Sullivan, S., Benjaminsen, T. A., Batterbury, S., Koot, S., and Harcourt, W. (2019). Speaking power to "post-truth": Critical political ecology and the new authoritarianism. *Ann. Am. Assoc. Geograph. 109*(2), 613–623.
Nost, E., and Goldstein, J. E. (2022). A political ecology of data. *Environment and Planning E: Nature and Space*, *5*(1), 3–17.
Ogden, L. A. (2011). *Swamplife: People, Gators, and Mangroves Entangled in the Everglades*. University of Minnesota Press.
Ojeda, D., Nirmal, P., Rocheleau, D., and Emel, J. (2022). Feminist ecologies. *Annual Review of Environment and Resources*, *47*, 149–171.
Ojeda, D., Sasser, J. S., and Lunstrum, E. (2020). Malthus's specter and the Anthropocene. *Gender Place Culture 27*(3), 316–332.
Oreskes, N., and Conway, E. M. (2010). *Merchants of Doubt*. Bloomsbury Press.
Osborne, T. (2017). Public political ecology: A community of praxis for earth stewardship. *Journal of Political Ecology*, *24*(1), 843–860.
Peet, R., and Watts, M. (1996). *Liberation Ecologies: Environment, Development, Social Movements*. Psychology Press.
Peluso, N. L. (1992). *Rich Forests, Poor People: Resource Control and Resistance in Java*. University of California Press.
Peluso, N. L. (2012). What's nature got to do with it? A situated historical perspective on socio-natural commodities. *Development and Change 43*(1), 79–104.
Peluso, N. L., and Vandergeest, P. (2001). Genealogies of the political forest and customary rights in Indonesia, Malaysia, and Thailand. *The Journal of Asian Studies*, *60*(3), 761–812.
Planck, M. (1932). *Where is Science Going? The Universe in the Light of Modern Physics*. W. W. Norton and Company.

Proctor, R. N. (2008). Agnotology: A missing term to describe the cultural production of ignorance (and its study). In *Agnotology: The Making and Unmaking of Ignorance*, (pp. 1–33). Stanford University Press.
Pulido, L. (2002). Reflections on a white discipline. *The Professional Geographer*, *54*(1), 42–49.
Rhoads, B. L., and Thorn, C. E. (1996). Toward a philosophy of geomorphology. *The Scientific Nature of Geomorphology*, (pp. 115–143). John Wiley & Sons.
Richter, L., Cordner, A., and Brown, P. (2018). Non-stick science: Sixty years of research and (in)action on fluorinated compounds. *Social Studies of Science*, *48*(5), 691–714.
Robbins, P. (2001). Fixed categories in a portable landscape: The causes and consequences of land-cover categorization. *Environment and Planning A*, *33*(1), 161–179.
Robbins, P. (2012). *Lawn People: How Grasses, Weeds, and Chemicals Make Us Who We Are*. Temple University Press.
Robbins, P. (2015). The trickster science. *The Routledge Handbook of Political Ecology*, 89–101. Routledge.
Robbins, P. (2019). *Political Ecology: A Critical Introduction*. John Wiley & Sons.
Rocheleau, D. (2003). Participation in context: What's past, what's present, and what's next. *Managing Natural Resources for Sustainable Livelihoods: Uniting Science and Participation*, (pp.169–183). Routledge.
de Santos, B. S. (2018). *The End of the Cognitive Empire: The Coming of Age of Epistemologies of the South*. Duke University Press.
Sauer, C. O. (1952). *Agricultural Origins and Dispersals*. American Geographical Society.
Sayre, N. F. (2008). The genesis, history, and limits of carrying capacity. *Annals of the Association of American Geographers*, *98*(1), 120–134.
Sayre, N. F. (2017). *The Politics of Scale: A History of Rangeland Science*. University of Chicago Press.
Schuurman, N., and Pratt, G. (2002). Care of the subject: Feminism and critiques of GIS. *Gender, Place and Culture: A Journal of Feminist Geography*, *9*(3), 291–299.
Scoones, I., and Stirling, A. (2020). *The Politics of Uncertainty: Challenges of Transformation*, (p. 196). Taylor & Francis Group.
Scott, J. C. (2020). *Seeing Like a State: How Certain Schemes to Improve the Human Condition Have Failed*. Yale University Press.
Sedell, J. K. (2021). No fly zone? Spatializing regimes of perceptibility, uncertainty, and the ontological fight over quarantine pests in California. *Geoforum*, *123*, 162–172.
Senanayake, N., and King, B. (2021). Geographies of uncertainty. *Geoforum*, *123*, 129–135.
Shattuck, A. (2021). Toxic uncertainties and epistemic emergence: Pesticide use and imperceptibility in Lao PDR. *Annals of the American Association of Geographers*, *111*(1), 216–230.
Simon, G. L. (2010). The 100th meridian, ecological boundaries, and the problem of reification. *Society and Natural Resources*, *24*(1), 95–101.
Simon, G. L. (2022). Disingenuous natures and post-truth politics: Five knowledge modalities of concern in environmental governance. *Geoforum*, *132*, 162–170.
Siti, M., Hoover, E., Ekowati, D., Owen, A., Elmhirst, R., and Heath, L. (2019). *Extracting Us: Looking Differently: Feminism, Politics and Coal Extraction*. Exhibition.
Sletto, B. et al. (2020). *Radical Cartographies: Participatory Mapmaking from Latin America*. Austin, TX: University of Texas Press.
Smith, L. T. (2021). *Decolonizing Methodologies: Research and Indigenous Peoples*. Bloomsbury Publishing.
Sze, J. (2018). *Sustainability: Approaches to Environmental Justice and Social Power*. NYU Press.
TallBear, K. (2014). Standing with and speaking as faith: A feminist-indigenous approach to inquiry. *Journal of Research Practice*, *10*(2), N17.
Tuck, E., and McKenzie, M. (2014). *Place in Research: Theory, Methodology, and Methods*. Routledge.
Tuhiwai Smith, P. L. (2012). *Decolonizing Methodologies: Research and Indigenous Peoples*. Zed Books.
Turner, M. D. (2016). Political ecology II: Engagements with ecology. *Progress in Human Geography*, *40*(3), 413–421.

Turnhout, E., Metze, T., Wyborn, C., Klenk, N., and Louder, E. (2020). The politics of co-production: Participation, power, and transformation. *Current Opinion in Environmental Sustainability*, *42*, 15–21. https://doi.org/10.1016/j.cosust.2019.11.009

Turnhout, E., Neves, K., and Lijster, E. (2014). "Measurementality" in biodiversity governance: Knowledge, transparency, and the Intergovernmental Science-Policy Platform on Biodiversity and Ecosystem Services (IPBES). *Environment & Planning A*, *46*(3), 581–597.

Vincent, K. (2022). Development geography I: Co-production. *Progress in Human Geography*, *46*(3), 890–897. https://doi.org/10.1177/03091325221079054

Walker, P. A. (2005). Political ecology: Where is the ecology? *Progress in Human Geography*, *29*(1), 73–82.

Walker, P. A. (2006). Political ecology: Where is the policy? *Progress in Human Geography*, *30*(3), 382–395.

Watts, M. J. (1989). *Silent Violence: Food, Famine, and Peasantry in Northern Nigeria* (Vol. *15*). University of Georgia Press.

Wesner, A. (2019). Messing up mating: Queer feminist engagements with animal behavior science. *Women's Studies*, *48*(3), 309–345.

Whyte, K. P. (2018). Indigenous science (fiction) for the Anthropocene: Ancestral dystopias and fantasies of climate change crises. *Environment and Planning E: Nature and Space*, *1*(1–2), 224–242.

Wylie, S., Shapiro, N., and Liboiron, M. (2017). Making and doing politics through grassroots scientific research on the energy and petrochemical industries. *Engaging Science, Technology, and Society*, *3*, 393–425.

Part III

Capital, Colonialism, and Political Economy

9 Globalizing Nature

Long-Standing Structures and Contemporary Processes

Fernanda Rojas-Marchini and Jessica Dempsey

Introduction: Planting Chile's Green Gold

During the summer of 2015, one of the authors of this chapter was asked to collaborate on mapping a water pipe that would bring water to a group of houses near Valdivia, located on Mapuche unceded lands in Chile. It was a wet afternoon and I was getting used to having leeches climbing my legs while using the GPS. With a member of the rural water committee, we walked the pipe's projected trail, crossing through a dense mixture of native forest trees. The smell of *canelos*, *tepas*, and *melis* was so strong we could quickly forget the forest plantation monoculture next to the pipe's trail. The committee asked the municipality to provide fiscal water infrastructure to deal with the seasonal drought; the map I was tasked with helping the committee in their demand (Figure 9.1).

The decrease in rainfall has become more frequent in this region due to climate change, and forest plantations consume most of the water table that would otherwise supply the houses during summer. The pipe projected by the committee took the water from a stream inside one of these plantations, a plantation owned by a Chilean firm, and it crossed through another plantation owned by a Japanese firm. In one of the meetings organized by the committee, a state agent encouraged the members to ask the forest companies for water and tanks. After years of work and following the state agent's suggestion, the committee got the state funding and built the pipe. Yet, despite all the efforts, houses still experience water shortages.

The scenario described above is the result of a profound transformation in Chile. A century after the military and settler invasion of Mapuche lands, forests were transformed into globalized nature. One environmental geographer observed that in just over a generation "Chile has created one of the world's most competitive forest resources" (Clapp 1995). This was not an act of new resource "discovery;" rather it required enormous amounts of state effort, coordination and force. After Salvador Allende's CIA-backed overthrow in 1973, reforms conducted by the military state prepared the terrain to attract national and foreign capital: Flexibilization of taxes, free trade agreements, state subsidies, new labor and water Acts,[1] and the enclosure and privatization of public and Indigenous lands. The amount of capital mobilized through foreign investments increased nearly 500% between 1985 and 1989 in the country (Moguillansky 1999: 45).[2] The vertiginous increase in foreign investment cannot be explained simply as a trend but as a change in the state economic structure.

Forest economies like the ones described above prove how hard it is to define globalizing nature. What does it mean to say something is globalizing? Where does the state end and globalization begin? As Barnes and Christophers (2018) write, there is no singular

DOI: 10.4324/9781003165477-13

Figure 9.1 Trail where the water pipeline was projected.

thing called globalization, rather, it refers to an ongoing process composed of events, logics, and processes that involve increasing economic and political interconnectivities. Globalization involves complexity and contradiction, top-down discipline and self-regulation, state and free market, and fast-paced reforms. It is not a natural process of diffusion, but rather a making "deeply infused with the exercise of power" (Hart 2002, quoted in Barnes and Christophers 2018).

For many scholars, globalization is a process tightly entwined with neoliberalism, a set of policies and practices that took hold in the 1980s wherein states re-regulated to produce better conditions for global capital mobility, for creating competition between jurisdictions for investment: who constructs the coziest home for capital to reproduce itself? Who provides the lowest taxes, the least red tape, the most flexible labor, the least stringent and most flexible regulations, the most security against land defenders? Yet capital's free(er) movement[3] is constantly resisted and reformed by communities and globalized movements. And too the era of neoliberalism corresponds with the construction of a global apparatus of concern about the environmental and social impacts of hundreds of years of industrialization and economic development, processes advancing more quickly under the most recent globalization regime. All of these more recent processes layer onto long-standing structures of colonialism, imperialism, racialization, and othering (Roberts and Mahtani 2010), which have contributed to consolidating a globalized and uneven growth-based model. In this chapter we review how political ecologists have understood and conceptualized these transformations across three major themes. First, we point to those who study a much longer history of "globalization," namely, the long arc of colonial and imperial relations. Then, we move into what is more traditionally considered "globalizing," focused on re-regulations in the neoliberal era. We then focus on the emergence of neoliberal green governance.

Colonialism and the Expansion of (Neo)extractivism

Globalization has long transformed nature. Recent waves of economic globalization amply studied by political ecologists are preceded by a longer arc of extractivism that produced and continues to sustain a web of global relations. Such processes are trackable to the colonization of the Americas. Alfred Crosby's argument about the "Columbian exchange" is central—and is a common point of departure—to political ecological work in the area of globalization and nature. Crosby argues that widespread colonization post-1492 triggered an enormous shift as living things from different biomes (plants, animals, microorganisms including viruses) began moving around the globe at a fast pace, with devastating consequences for, in particular, Indigenous Peoples of the Americas (Crosby and Von Mering 1972; Crosby 1986). A more recent wave of political ecology research argues that the human and nonhuman exploitation that took place with the colonial encounter defines today's Anthropocene (Davis and Todd 2017). As Heather Davis and Zoe Todd argue, the beginning of the geological epoch defined as the Anthropocene must be placed in 1492 as a counter strategy to the universal claim that the inevitable "human nature" is the root cause of the ecological crisis—one of the main arguments pushed by the mainstream view of the Anthropocene (p. 763). Janae Davis and co-authors (Davis et al. 2019) take the argument further to destabilize the useful—but still limited—view of the Plantationocene developed by Donna Haraway (2015), Anna Tsing (2015), and their interlocutors. The Plantationocene is a concept that transfers the universality of the Anthropocene to the plantation; however, as Davis et al., argue, such a movement needs to have a dialogue with Black Geographies and place colonialism, racism and enslavement at the center of these accounts:

> we acknowledge the need for a radical awareness of the plantation's role in producing global environmental change. However, an interest in ecological ethics must not overshadow attention to the dynamics of power (racial, gender, sexual, or otherwise).
>
> (Davis et al. 2019: 10)

Global exchange processes triggered by the Columbian exchange and based on extractivism are part and parcel of more contemporary processes of globalization. The term "extractivism" alludes to a mode of accumulation that emerged alongside the European colonization of the Americas, Africa, and Asia, in one of the first eras of globalization. According to Alberto Acosta (2013), extractivism was established "on a massive scale 500 years ago" (ibid.: 62). The accumulation commenced in 1492 with the Spanish arrival to the Caribbean and the Columbian exchange, materialized by commercial relations driven by the extraction of minerals. Around 18,000 tonnes of silver were taken from America to Spain between 1503 and 1660 (Dussel 1998: 10). Extractivism was facilitated by racism, the structure used by Europeans to codify the difference between conquerors and conquered people based on the idea of race (Quijano 2008: 182). Racism allowed

Europeans to naturalize the domination of entire human groups and nonhuman natures (ibid.). One of the first racial structures was the hierarchy system imposed between Indians, Mestizos, and Black people. Authorized by such a structure, since 1520 and until the final stage of slavery in the nineteenth century, around 14 million people had been trafficked as plantation slaves (Dussel 1998: 12). The magnitude of these numbers reflects extractivism's profound and ongoing origins.

Extractivism emerged from these racial interstices as a structural piece of global capitalism, a prolongation of what Anibal Quijano defined as "the constitution of a new structure of control of labor and its resources and products" (Quijano 2008: 182). It provided imperial metropolises in Europe with the substance to accumulate wealth while impoverishing colonies. For instance, the Spanish crown was able to win wars and expand its dominion because of the wealth took from America; 25 years after the discovery of the mines of Potosí in Peru and Zacatecas in Mexico, Spain was able to finance an armada to defeat the Turks in 1571 in Lepanto, which led to the control of Mediterranean routes (Dussel 1998: 10). In the words of political ecologist Emiliano Terán Mantovani (2016: 257), extractivism represents "not only an economic structure but also an ecological-political regime on the fabric of life, territories, flows of energy and matter, bodies, institutions, and cultural webs."[4] From soy monocultures in Argentina to trawler fishing in West Africa to coal mines in Indonesia, extractivism configures a geography of exploitation that drains out territories' wealth and transfers this wealth to imperial spaces. This arrangement is composed by institutions doing the extracting, e.g., industrial firms, states and supranationals (Andrade 2022), and people living and suffering extractivism's concrete outcomes, namely, racialized communities, First Nations, rural communities, and marginalized cities and neighborhoods, especially women, children, and LGBTQI+ people (Bolados and Sánchez 2017; Valenzuela-Fuentes et al. 2021).

The concept of extractivism has also been deployed to highlight the last 40 years of capitalist accumulation driven by neoliberal governance schemes and global finance. Some scholars refer to this mode as the "extractivism economy" (e.g., Achiume 2019), while others have added the "neo" to emphasise global finance and the expansion of capital flows (e.g., Svampa 2019; Brand et al. 2016) including those enabled by free trade agreements signed by Latin American countries in the 1990s and 2000s (see Dingemans and Ross 2012). Maristella Svampa (2019: 6) understands neo-extractivism as

> a way of appropriating nature and a development model based on the over-exploitation of natural goods, largely non-renewable, characterized by its large scale and its orientation toward export, as well as by the vertiginous expansion of the borders of exploitation to new territories, which were previously considered unproductive or not valued by capital.

In a similar engagement, Tendayi Achiume (2019) includes in extractivism all the "industries, actors, and financial flows, as well as the economic material and social processes and outputs associated with the globalized extraction of natural resources." (ibid.: 2). Both views emphasise the global and financial side of this late trajectory; however, while Svampa stresses the political economy of extractivism, its impacts and the organized resistance carried by social movements, Achiume seeks to explain the relationship between extractivism and racism. Legal devices such as the doctrine of discovery contributed to embed racism into the extractivism economy: "[t]he overwhelming material and social benefits of the colonial extractivism economy accrued along racial lines" (Achiume 2019: 7).

It is important to keep in sight that neo-extractivism has been associated with progressive and left-wing versions of extractivism characterized by states' greater involvement in the economy, which, nevertheless, does not avoid the destruction of nature (e.g., Gudynas 2009; Andrade 2022). However, as Brand et al. (2016) argue, and also following Svampa's work, we see value in mobilizing the concept of "neo-extractivism" to refer to a model that emphasises a "dynamic global capitalist economy" (ibid.: 128), including but not only referring to the more progressive variants. Neo-extractivism is a part of "the current phase of globalizing capitalism" (ibid.: 129) while also being a prolongation of extractivism because it feeds from racial and geopolitical hierarchies established with colonization.

In sum, neo-extractivism manifests in multiple modes with global and local impacts, some of which are rather visible while others still unknown. Here we outline three interrelated features. First, neo-extractivism depends on long-lasting logics of exploitation to access territories and bodies, logics initiated with the colonization of the Americas. Maria Mies's (1986) work is illuminating here; she observed the ways in which those cast as "closer to nature" or not quite fully human are primary for accumulation processes (Sundberg et al. 2020; also see Figure 9.2). The rise of neo-extractivism actualizes the dehumanization shown by Mies (1986), a dehumanization that is fundamental for the geography of exploitation.

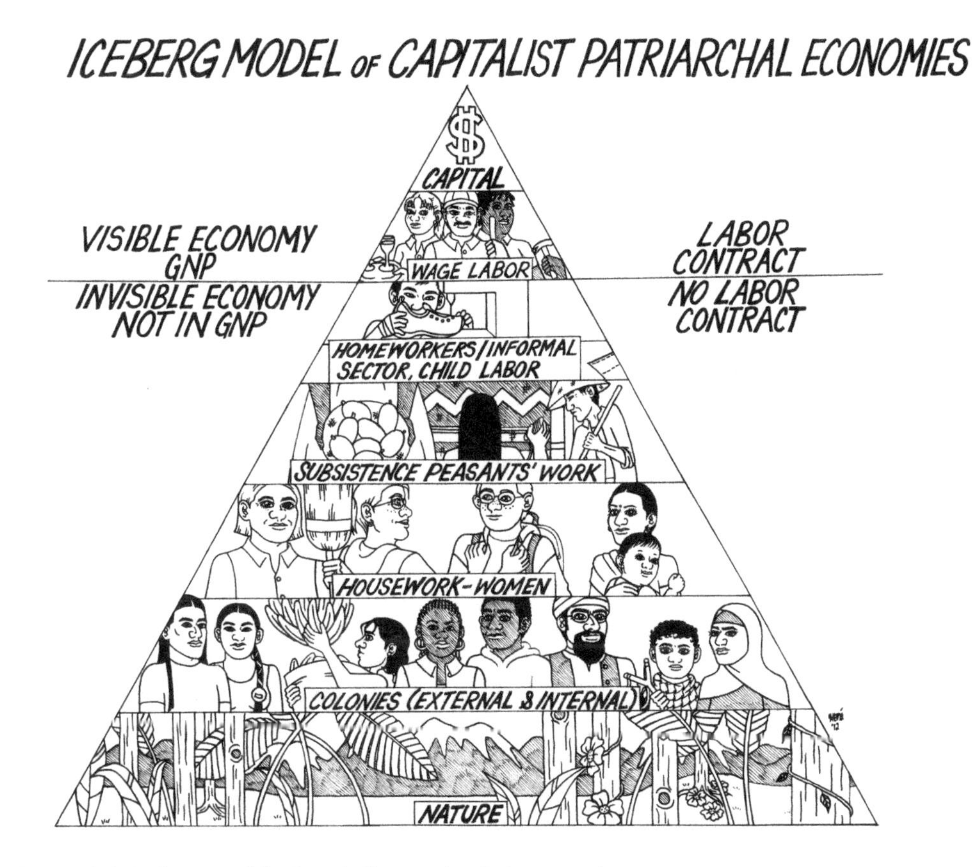

Figure 9.2 Iceberg model of capitalist patriarchal economies.

Source: From Ecosocialist Horizons, https://ecosocialisthorizons.com/ecosocialist-images/

Second, in order to access these bodies/territories, neo-extractivism seeks to produce a geography of exploitation: the creation of conditions of possibility for capital circulation and accumulation, re-regulations that we address in the following section. Third, the most notorious materialization of neo-extractivism are sacrifice zones, polluted spaces "overwhelmingly inhabited by racialized peoples" (Sundberg et al. 2020). These zones go hand in hand with the production of a geography of exploitation—cheap natures (Patel and Moore 2018) and territories are necessary to feed this system of accumulation. According to Maristella Svampa and Enrique Viale, sacrifice zones are spaces where environmental injustices are intensified, where bodies and lives become expendable and sacrificial (Svampa and Viale 2014: 84). Hierarchies and devaluation processes taking place in sacrifice zones are applied to the lives and livelihoods that do not align with the hegemonic economy (ibid.).

Box 9.1 Querying globalization and neoliberalism

Most geography graduate students of the late 1990s and 2000s (and hopefully beyond) would have encountered the intellectual tour de force of J. K. Gibson-Graham's *The End of Capitalism as We Know It*. The book dropped a feminist and post-structural bomb into dominant framings of capitalism and globalization, arguing that Marxist dominated political economic approaches award too much power to capital as a material force determining the shape of everything it encounters. Drawing from post structural approaches, Gibson-Graham emphasizes how representations shape political possibilities, and that representations of capitalism as an always dominating, expansive, monolithic, hegemonic force contributed to shoring up its power; "the project of understanding the beast has produced a beast," (Gibson-Graham 1996: 1). They also critique feminist political economic theorizations (like Mies's iceberg model), as falling into this capitalocentrism, by which they mean a tendency to define all other economic forms in reference to capitalism (for a review of Mies and Gibson-Graham, see Collard and Dempsey 2020). In the chapter "Querying Globalization," Gibson-Graham examines academic and progressive representations of globalization, finding representations that picture a one-way street of capital always "penetrating" noncapitalist or other-than-capitalist communities and spaces. Rather, Gibson-Graham argues that it is essential to see economic relations as varied and diverse, to cut capitalism and globalization down to size, so to speak, and to open up a much wider range of political possibilities beyond "waiting for the revolution" (1996: 251), as the final chapter of the book is titled.

These post-structural, feminist, and Marxist debates have influenced political ecologists studying neoliberalism: to what extent are our studies and analyses narrowly constrained by conceptual frameworks and broader theories that emphasize overwhelming power of structural forces? In the 2007 landmark edited collection *Neoliberal Environments*, Larner suggests that the authors tend to gloss over "contradictions and inconsistencies" in taking neoliberalism as their starting point, and through this desired coherence, suggests that they might be falling into the patterns identified by Gibson-Graham. Contra this view, in a 2008 review essay Noel Castree argues that the focus on case studies and "diverse investigations" risks obscuring "the common 'logics' and processes operating within or between otherwise different spatiotemporal settings" (Castree 2008a: 137). In other words, the hyper attention to the local and the attention to variegation occludes wider generalization, theorization, and also political potential. This tension shows up in a wide variety of political ecological debates, including more recent ones regarding payments for ecosystem services (PES).

Neoliberal Re-regulation for Extractivism

Political ecologists have also approached contemporary globalization and globalizing nature by carefully tracing the processes associated with neoliberalism. But, what is neoliberalism? A contested term with multiple beginnings and definitions (Hayter and Barnes 2012: 201; Peck 2013), in this section our focus is on a set of governance and governmental policies and practices—re-regulations—entwined with increasing globalization of trade and investment (e.g., trade between high income countries increased 15 times between 1950 and 1990, cited in Barnes and Christophers 2018).[5] Rather than a thing or a simple recipe, political ecologists emphasize the importance of understanding neoliberalism as a process of neoliberalization that unfolds heterogeneously and "entails shifts in human–environment relations at a range of scales" (Kay 2018).

Political ecologists have been at the forefront of studying the heterogeneous "re-regulations" of nature during the neoliberal era (e.g., Heynen et al. 2007; Castree 2008a, 2008b). A key pattern identified by this body of work is that environmental re-regulation was a part of creating a smoother space for capital to flow throughout the globe. This includes through multilateral and bilateral agreements that emerged in the 1990s, which often worked to protect corporate investments and logics of domination between rich and non-rich economies. In an examination of NAFTA, James McCarthy (2004) argued that even when trade agreements incorporate environmental provisions such as the creation of new environmental institutions and regulations, the evidence indicates these mechanisms contribute to the degradation they seek to remedy and can also privatize the conditions of production. A more recent example of this is the "Trans-Pacific Partnership," or TPP11, a multilateral trade agreement that seeks to extra-strengthen corporate protections in the face of institutional and regulatory changes in the Global South. For José Gabriel Palma (2020: 1003), TPP11 is an instrument to maintain corporate power, a salvage device for neoliberalism. And, its environmental implications are profoundly destructive. In the case of Chile, TPP11 proposes tax cuts in forestry, large-scale agriculture, and fishing, activities that already present deductions of 100% with Australia and Peru, and of 90% with the rest of party countries except Vietnam (Ahumada 2019).

As this work on investment agreements and tax re-regulations in Chile shows, political ecologists have conducted extensive research to identify how countries are reforming policies, laws and governance systems—including environmental—to produce conditions that are more attractive for foreign investments. Researchers have focused on understanding the impacts of structural adjustment conditionalities (e.g., currency devaluations, tax reforms, austerity), investment, and extraction. For example, in Bolivia such policies, along with technological innovations, led to increased foreign actors and increased deforestation, as well as unequal land distribution (Hecht 2005). In Nigeria, the impacts of structural adjustment programs led to a misogynist campaign to reduce the birth rates, which prompted Silvia Federici to develop a historical analysis on the exploitation of women in capitalist societies. That is part of the origins of "Caliban and the Witch," a major hallmark for feminist political economy and ecology (Federici 2004; see also the rich feminist discussion of overpopulation narratives related to climate change in Hendrixson et al. 2020).

In a still relevant 1994 publication, Laura Pulido points to how increasingly globalized economies pit regulatory environments against one another. She describes how the expansion of environmental rights in Los Angeles came up against another set of forces, namely neoliberal restructuring, which smoothed the pathways for polluting industries to move to jurisdictions with not only less environmental rights, but also lower labor costs—across the border to Mexico. While grassroots groups began organizing themselves transnationally to oppose what she calls "pollution flight," this opposition had "limited power against the forces of capital flight" (Pulido 1994: 931). Pulido links efforts to stop capital flight to the erosion of environmental rights in places like California, in a process sometimes termed the race to the bottom.

Often drawing from Marxist, Polanyian and power and knowledge approaches (see Castree 2008a, 2008b; Bigger and Dempsey 2018), political ecologists show that these neoliberal re-regulations can lead to new waves of enclosure, privatization and dispossession, aiming "to identify specific winners and losers in such reforms" (McCarthy and Prudham 2004). For example, Emel and Huber (2008) examine Tanzanian mining restructuring in line with IMF and World Bank neoliberal strategies. Responding to widespread narratives that Tanzania (and other Global South countries) were high-risk investment jurisdictions, the country reconstructed its mining royalty and tax laws to be more attractive to transnational mining companies and investors. And foreign investment did follow, growing from US $370 million in 1998 to one billion in 2002. Despite being a so-called "risky investment," profits also followed, and as Emel and Huber show these restructured tax and royalty schemes meant that the profits largely flowed to large companies like Barrick Gold, headquartered in Canada (Emel and Huber 2008). Looking across the globe, Gavin Bridge (2007: 85) outlines how in a short period over 90 states adopted new mining laws in an effort to, as he colorfully describes, produce the "underground as a site for the circulation of international capital."

Political ecologists are also studying the financialization of agriculture and land, particularly as it became a desired investment asset class after the 2008 financial crisis (Kish and Fairbairn 2017). Here again concern is with who benefits and who loses from such shifts in ownership and control, the logics underpinning the rush for land, the geographical areas where it concentrates, and the outcomes produced, including forced displacement of communities (Peluso and Lund 2011), transformation of labor (Li 2011), and control over water access (Mehta et al. 2012). As Saturnino Borras and co-authors point out (2012), one key finding from the land grabbing debate is a variegated role for the state. Borras et al. (2012) suggest that in some cases states allow land purchases by the agribusiness sectors, finance and banking because it lacks the legal frameworks to control this process. In other cases, states actively pursue private land acquisitions and the consequent "foreignization" of land.

Responses to the macro-structural perspective on land grabbing did not take long to appear, with political ecologists addressing more situated dimensions of this phenomenon. Diana Ojeda argues (2018) that land grabbing should be analyzed over the long arc of dispossession. A multiscalar and complex process, land grabbing can take place by small purchases, as opposed to the common notion of single land purchases of vast tracts of land. Ojeda further argues that more substantial focus on historical dimensions can enhance our understanding of land grabbing, addressing pre-existing social relations shaping rural spaces where land deals occur (Ojeda 2018: 401; Araya 2017). In a similar spirit, Stefan Ouma (2014) argues that the literature on land grabbing may greatly benefit from addressing the situated financial dimensions entailed in this phenomenon. As he observes, among the reasons behind land purchases is a buyers' desire to diversify investment portfolios during economic crises, which is why the lands acquired under this regime are considered to be an "alternative asset class" (p. 163).

Of course neoliberal re-regulations do not play out evenly and outcomes can be unpredictable. Looking across a wide body of research, Liverman and Vilas's (2006) review of NAFTA concludes that "local and historical factors mediate the effects of neoliberal processes and how the withdrawal of national controls can open up new forms of control and regulation at the local level." And if there is one thing political ecologists agree on it is that resistance is everywhere. None of the patterns we have outlined above in this chapter go uncontested; reflecting Karl Polanyi's "double movement," many of these neoliberal reforms led to pushback and new forms of the national and international alliances. And finally, even if a general outcome of the neoliberal globalizing processes has increased economic interconnection and entwinement, a smoother space for capital to flow through, there is no reason to believe that will always be the case. Some scholars suggest we are emerging into a more "geoeconomic" world order, with economic policy driven more by security and geopolitical interests, particularly as tensions rise between China and the US (Roberts et al. 2019) and also related to climate change (Riofrancos 2022).

Neoliberal "Green" Global Governance to "Green Extractivism"

The neoliberal re-regulation era was also an era of resistance, with increasingly transnational movements contesting the legitimacy of states and supra-national institutions (see Box 9.2). And political ecologists have been at the forefront of examining the discursive, institutional and geo-political economic efforts to remake neoliberalism, economic development and capitalism writ large as sustainable and green. Political ecologist Kathleen McAfee (1999) uses the term "green developmentalism" to describe a governance

Box 9.2 Globalizing resistance

Social movements and civil society flourished in the 1990s to resist the globalization of nature. Concentrated mostly in the Global South, these movements denounced the neoliberal reforms conducted under programs like the structural adjustments pushed by the United States. One emblematic example is Brazil, which underwent a radical transformation in the 1990s, with massive waves of privatizations of state industries and utilities. These transformations were faced by civil society and social movement mobilization already configured during the agrarian reform started in the 1970s. The growth of the Movement for Rural Landless Workers, MST (*Movimiento do Los Trabajadores Rurales Sin tierra*) resulted from that period (Wolford 2003; Pinto 2020) and it continues to resist the takeover of peasant agriculture by agribusiness, which continues to grow in Brazil and other countries (Edelman and Borras 2016; Pinto 2020).

Another example is one of the world's largest social movement, *La Vía Campesina* (LVC). Focused on defending peasant and small-scale farmers and the advancement of food sovereignty, LVC brings together "millions of peasants, landless workers, Indigenous people, pastoralists, fishers, migrant farmworkers, small and medium-size farmers, rural women and peasant youth from around the world."[6] LVC formed in 1993 out of concern for the neoliberal policies and the future of small farming, particularly focused on the agricultural reforms that pursued trade liberalization, considered in the GATT[7] negotiations (which would transform into the World Trade Organization, WTO). These reforms would open up agricultural land to international competition, reducing farmer support and food sovereignty, with a consequent increase of precarity (Edelman and Borras 2016). Through a series of international meetings, LVC opened up to include other movements besides farmers, all concerned with food sovereignty and the risks of the expansion of large scale, concentrated and increasingly corporate and financialized food production.

LVC and its member organizations were central to the rise of the Global Justice movement, also known as the "anti-globalization" movement. Growing in the 1990s, this movement contested neoliberal globalization tendencies such as the rise of corporate and financial sector power in light of trade and investment agreements. Considered a "movement of movements," the anti-globalization coalition involved labor, environmental, peasant, antiwar, social movements and organizations from the Global North and South, expressing their opposition at major gatherings of states like the World Trade Organization held in Seattle in 1999 (the WTO meetings ultimately failed in part due to the 50,000 plus people who convened in Seattle). In 2001 the World Social Forum held its first meeting in Brazil under the tag "another world is possible," opening a space for learning and solidarity building among these resistance movements, many of whom were bound together in their opposition to global neoliberal reforms.

formation that emerged with the Convention on Biological Diversity, where environmental goals and poverty alleviation are to be met largely through established economic and political structures with heavy doses of science, technology and, above all, markets to drive efficiency, "selling nature to save it" (McAfee 1999; see also Bernstein 2002). As climate change challenges the legitimacy of institutions (including states), Wainwright and Mann (2018) similarly characterize the emergence of "Climate Leviathan," a political formation that includes capitalist elites and states cooperating toward the dream of "planetary sovereignty," alongside a prerogative to deem "what and who must be sacrificed in the interests of life on Earth" (p. 15). Carbon markets, green finance, and carbon capture and storage are strategies pushed by this ruling block as they attempt to "stabilize their position amidst planetary crises" (p. 15). Political ecologists and allied scholarship have spilled much critical ink to better understand how these hegemonic formations define the terrain of the politically possible pushing more transformative political and economic demands to the side (e.g., Asiyanbi and Lund 2020; Cavanagh and Benjaminsen 2017; Osborne 2015; Carton 2014).

Taking an institutional ethnographic approach, Michael Goldman (2005) set out to understand how, in light of massive criticism for environmental and social harms, the World Bank managed to remake itself, maintaining and expanding its power and influence. *Imperial Nature* lays out how the World Bank constructed a powerful discursive framework—green neoliberalism—one that directly responded to and in many ways absorbed criticisms. In this way, green neoliberalism can be understood as a reactionary, or counterrevolutionary governing strategy, a point made compellingly by Sara Nelson (2014), whose historical research focuses on the emergence of ecosystem services. Elizabeth Povinelli (2011) suggests that "late liberalism" is precisely a mode of governing attentive to its exclusions and contradictions, a form that excels at absorbing and incorporating opposition.

Another characteristic of green neoliberalism is the increasing activity of NGOs and other quasi-state actors, including the private sector, in global environmental governance since Rio. Political ecologists describe this trend as a "shift from government to governance" (Bigger and Dempsey 2018), pointing to the reconfiguration of regulatory functions to include non- or quasi-state actors. What is common to these strategies, from forest to organic to marine certification (Guthman 2007) to investment principles and coalitions (Clapp 2017; Cohen et al. 2021), is that they are often voluntary forms of environmental governance and that, for the most part, achieving "sustainable" outcomes relies on consumer choice and market signals; compliance is not through state fiat (Cashore et al. 2004). Geographers have also studied the emergence of risk disclosure approaches to deal with unsustainable, climate and biodiversity degrading capital flows. Such approaches posit that once firms assess and disclose risks from climate and other environmental changes (like water scarcity or ecosystem degradation), the market will discipline accordingly. Christophers's (2017) analysis shows that financial and corporate sectors strongly prefer these approaches over regulated approaches, others point to how these frameworks are largely shaped by multinational firms and finance (Knox-Hayes and Levy 2014), a form of regulatory capture (Kedward et al. 2020).

As the above literature suggests, political ecologists and allied scholars are interested in understanding the power relations through which these green neoliberal formations, initiatives, and projects emerge, what effects they have in the world, and who benefits and loses from them. More generally, political ecological work draws attention to the way these mechanisms like the Paris Agreement, carbon trading or climate risk disclosure frameworks are ineffective in scaling down emissions rapidly or lessening impacts for the most marginalized populations (Carton 2014; Kama 2014; Christophers 2017). Going further, Wainwright and Mann (2018) argue that these failures should not be understood as bad policy, but rather as the logical or "rational" end point of a political and economic system replete with contradictions and limitations "that cause them to 'fail'" (p. 38).

Political ecologists have also been at the forefront of analyzing the colonial nature of many global efforts to halt climate change and the decimation of nonhuman forms of life. McAfee, for her part, notes that green developmentalism, in its focus on commodifying genetic resources in the service of wider biodiversity saving efforts, "abstracts nature from its spatial and social contexts and reinforces the claims of global elites to the greatest share of the earth's biomass and all it contains." (McAfee 1999: 133). That is, political ecologists point to how green neoliberalism excels at bracketing out concentrated wealth and power amassed through imperial relations that produced the "third world" in the first place, rarely considering redistribution or even the unsustainability of growth on a constrained planet. There is also an established body of work examining how green discourses perpetuate colonial ideologies of nature (e.g., Sullivan 2010; Alimonda 2011; Carpinatti 2019; Millaleo Millaleo 2019; Baldwin 2009) and facilitate new rounds of accumulation through processes such as "green" grabbing (Fairhead et al. 2012; Benjaminsen and Bryceson 2012).

The energy transition is also implicated in green colonialism and in structures of racialization and patriarchy (Sultana 2014). Going back to themes identified in the first section of this review, Zografos (2022) implicates the Global North Green New Deal in failing to address the impacts of transition outside the North, particularly due to demands for critical minerals like cobalt and lithium needed for renewables and batteries, which he argues are producing "green sacrifice zones." Drawing from theorist Lisa Lowe (2015), Zografos connects lines between colonial structures of race and social difference that are often obscured by the emphasis on just transitions for fossil fuel dependent communities in the Global North. Similarly, Voskoboynik and Andreucci (2021) use the term "green extractivism" to explain the mineral-intensive development model proposed in Latin America to fulfill the energy transition, examining the lithium triangle region covering Bolivia, Chile, and Argentina. Across the Atlantic, DeBoom (2021) studies Chinese state-based investments in uranium mining in Namibia, and, using Mbembe (2019) as her guide, proposes "climate necropolitics" as a theoretical framework for analyzing the discourses, practices, and logics through which violence is rendered legitimate in the service of combating climate change.

As with plain old extractivism, struggle permeates this green variant. See Normann's (2021) study of the Saami People's struggle to halt wind energy expansion in Norway, or the recent constitution of the Plurinational Observatory of Andean Salt Flats to defend these natural formations and Indigenous communities from lithium mining (Jeréz Henríquez 2021), as well as Indigenous resistance to small-scale hydroelectric development in Chile (Hernando-Arrese and Rasch 2022). But not all communities oppose green extractivism. Through ethnographic and historical methods, Lorca et al. (2022) examine how some Indigenous communities in Chile's Salar de Atacama region negotiate with lithium mining companies to increase the benefits they receive from mining, while also experiencing negative impacts such as water scarcity and community tensions. Importantly, this work unsettles the notions of Indigenous people as "uniform and unified actors," and also complicates mining (or extractivism) as pushed only by external actors or institutions (see also Atleo 2021).

Across these works on green developmentalism to green extractivism or "eco-extractivism" (Núñez et al. 2020), there is a tension about how to understand the global and local dynamics, how to understand the relationship between structural forces like capitalism, colonialism, racialization and the agency of communities—as outlined in Box 9.1. This played out in a recent debate over analytical approaches to payments for ecosystem services (PES) in the journal *Ecological Economics*. On one side, Fletcher and Büscher (2017: 224) argue that PES scholarship tends to focus on the "micropolitics," which occludes an understanding of what they term the "PES conceit," namely that "the approach implicitly accepts neoliberal capitalism as both the problem and the solution to the ecological crisis." On the other side, Van Hecken (2018) suggests that understanding the "variegated and hybrid forms of what actually emerges from PES interventions" through their complicity in neoliberalism is simplistic while it silences the agency of those resisting, adapting and proposing "divergent PES ontologies" (p. 317).

Conclusions

In this chapter we reviewed the colonial bedrock that underpins the globalization of nature, tracing extractivism's varied genealogies and how political ecologists have engaged these processes alongside global connections and local impacts. An entire web of processes involving commerce, legal arrangements, and war comprise the structures of early extractivism, an extractive era that connects with the rise of more recent economic globalization. The latter cannot be properly understood without paying attention to these long-standing, still very present structures.

The trajectory of forest plantations in Chile—how they settled and expanded through the years—illustrate well the three phases presented here. First, colonization by the Spanish crown made visible Indigenous ancestrally occupied lands to global, mostly European imperial metropoles. After Chile achieved its independence from the crown in 1810, a phase of Indigenous land dispossession took place and Mapuche lands concentrated massive amounts of native forest. The state of Chile invaded these lands (1861–1883) and established a new system of land ownership, reducing Mapuche lands from 10 million hectares to approximately 600,000 hectares (Marimán et al. 2006). Forty years later, the Pinochet dictatorship supported by Chilean economists trained at the Chicago School of Economics imposed a set of neoliberal reforms characterized by massive waves of privatization, elimination of price controls, and incentives to increase foreign investments. Added to the suspension of the agrarian reform started in 1962, the military *junta* divided the last patches of Indigenous common lands and promoted forest plantations through state subsidies, lower import tariffs, and trade agreements (Rojas Marchini 2022). While national elites quickly showed up to benefit from these reforms, foreign investors arrived in a second phase from Canada, Japan, New Zealand, Switzerland, and the United States (Clapp 1995). The investment boom attracted small landowners to sell their property to forestry companies (Rojas Marchini and Folch 2018). In a more recent trend, various Chilean governments have tried to remake forest plantations as carbon offsets, arguing that they contribute to the country's carbon neutrality (Hoyos-Santillan et al. 2021). However, these arguments brush aside the fact that forest plantations release carbon dioxide once they are cut, a process that occurs every 12–18 years (ibid.). As this story shows, the long-standing structures of extractivism combine with contemporary processes whereby forests are remade and reframed by investors and states interested in maintaining the nature-based, export-oriented, and uneven global economy.

Political ecologists have contributed greatly—and will keep doing so—to understanding the processes associated with the globalization of nature. With the intention of proposing future directions and opportunities, we invite readers to think about the following themes and questions: First, as the transition to green energy becomes increasingly paramount there are new geopolitical dynamics emerging as states aim to secure access to critical minerals like lithium used to produce batteries. In a recent article Riofrancos (2022) points to a new protectionism emerging as states aim to "onshore" mining or processing of these critical minerals as a national security imperative. Globalizing nature has always been tied up in geopolitical power relations and more work can be done at this intersection, along the lines of geopolitical ecology (see Bigger and Neimark 2017; DeBoom 2017).

Second, the transformative, radical change needed to halt biodiversity loss and adapt to climate change has put pressure on states the world over. States in the Global South deemed by development institutions as biodiversity rich but economically poor (McAfee 1999) are particularly pressured by the transformative imperative. What are the possible paths for these states to transform themselves toward a green and just future? Centuries of uneven development complicate this answer. How can political ecologists analyze such a challenge?

And finally, what kinds of theoretical frames have, or should political ecologists engage to think about capitalism, globalization, and neoliberalism? Four years ago,

Bigger and Dempsey (2018) demonstrated with a bibliometric analysis that the field of neoliberal natures was dominated by a narrow range of people and deployed a narrow range of theoretical frameworks. There is much to be gained by diversifying who is in the field and expanding conceptual perspectives and approaches.

Notes

1 The Labor Act was designed to weaken unions, while the Water Act established water as a private good separated from the land.
2 We arrived at this number by analyzing the curve of the graph elaborated by Moguillansky (1999), which shows the increase of approximately 200 million dollars to 1000 million dollars between 1985 and 1989, a rise of 500%.
3 The free movement of capital is different from the free movement of people, which does not increase due to globalization—although flows of migration do increase because of globalization.
4 All quotes from Spanish-written sources have been translated by the authors.
5 Neoliberal logics involve more than the economic. They also entail economic thinking embedded into social spheres of living (Brown 2015; Gago 2014; Larner 2007).
6 See https://viacampesina.org/es/la-via-campesina-la-voz-las-campesinas-los-campesinos-del-mundo
7 The General Agreement on Tariffs and Trade (GATT) is a multilateral trade agreement conceived in 1947 to promote trade liberalization.

References

Achiume, T. (2019). *Global Extractivism and Racial Equality. Report of the Special Rapporteur on Contemporary Forms of Racism, Racial Discrimination, Xenophobia and Related Intolerance* (UN Special Rappporteur A/HRC/41/54; p. 20). United Nations.

Acosta, A. (2013). Extractivism and neoextractivism: Two sides of the same curse. In M. Lang and D. Mokrani (Eds.), *Beyond Development: Alternative Visions from Latin America* (pp. 61–87). Amsterdam: Netherlands, Transnational Institute / Rosa Luxemburg Foundation.

Ahumada, J. M. (2019, January 16). El TPP-11 y el desarrollo en Chile: ¿aliados o adversarios? *CIPER Chile*. www.ciperchile.cl/2019/01/16/el-tpp-11-y-el-desarrollo-en-chile-aliados-o-adversarios/

Alimonda, H. (2011). La colonialidad de la naturaleza. Una aproximación a la Ecología Política Latinoamericana. In H. Alimonda (Ed.), *La Naturaleza Colonizada. Ecología Política y Minería en América Latina* (pp. 21–60). Buenos Aires.

Andrade, A. (2022). Neoliberal extractivism: Brazil in the twenty-first century, *The Journal of Peasant Studies*, 49(4), 793–816, DOI: https://doi.org/10.1080/03066150.2022.2030314

Araya, A.L., (2017). Domesticando el despojo: Palma africana, acaparamiento de tierras y género en el Bajo Aguán, Honduras. *Revista Colombiana de Antropología*, 53(1), 151–185.

Asiyanbi, A., and Lund, J. F. (2020). Policy persistence: REDD+ between stabilization and contestation. *Journal of Political Ecology*, 27(1), preprint. https://doi.org/10.2458/v27i1.23493

Atleo, C. (2021). Between a rock and a hard place: Canada's carbon economy and Indigenous ambivalence. In *Regime of Obstruction: How Corporate Power Blocks Energy Democracy*, (Ed.), W. Carroll. AU Press.

Baldwin, A. (2009). Carbon nullius and racial rule: Race, nature and the cultural politics of forest carbon in Canada. *Antipode*, 41(2), 231–255. https://doi.org/10.1111/j.1467-8330.2009.00671.x

Barnes, T. J., and Christophers, B. (2018). *Economic Geography: A Critical Introduction*. Wiley-Blackwell.

Benjaminsen, T. A., and Bryceson, I. (2012). Conservation, green/blue grabbing and accumulation by dispossession in Tanzania. *The Journal of Peasant Studies*, 39(2), 335–355. https://doi.org/10.1080/03066150.2012.667405

Bernstein, S. (2002). Liberal environmentalism and global environmental governance. *Global Environmental Politics*, 2(3), 1–16. DOI: https://doi.org/10.1162/152638002320310509

Bigger, P., and Dempsey, J. (2018). Reflecting on neoliberal natures: An exchange. *Environment and Planning E: Nature and Space*, 1(1–2), 25–75. https://doi.org/10.1177/2514848618776864

Bigger, P., and Neimark, B. D. (2017). Weaponizing nature: The geopolitical ecology of the US Navy's biofuel program. *Political Geography*, 60, 13–22. https://doi.org/10.1016/j.polgeo.2017.03.007

Bolados García, P., and Sánchez Cuevas, A. (2017). Una ecología política feminista en construcción: El caso de las "Mujeres de zonas de sacrificio en resistencia", Región de Valparaíso, Chile. *Psicoperspectivas*, 16(2), 33–42. https://doi.org/10.5027/psicoperspectivas-vol16-issue2-fulltext-977

Borras, S. M., Kay, C., Gómez, S., and Wilkinson, J. (2012). Land grabbing and global capitalist accumulation: Key features in Latin America. *Canadian Journal of Development Studies/Revue Canadienne d'études Du Développement*, 33(4), 402–416. https://doi.org/10.1080/02255189.2012.745394

Brand, U., Dietz, K., and Lang, M. (2016). Neo-extractivism in Latin America—One side of a new phase of global capitalist dynamics. *Ciencia Política*, 11(21). https://doi.org/10.15446/cp.v11n21.57551

Bridge, G. (2007). Acts of enclosure: Claim staking and land conversion in Guyana's gold fields. In Heynen, J., McCarthy, J., Prudham, S., Robbins, P. (Eds). *Neoliberal Environments: False Promises and Unnatural Consequences*. Routledge.

Brown, W. (2015). *Undoing the Demos: Neoliberalism's Stealth Revolution*. Zone Books.

Carpinatti, B. (2019). *Colonialismo verde: ecología política de la conservación de la Naturaleza en Guinea Ecuatorial* (T). Universidad Nacional de Misiones. Retrieved from https://rid.unam.edu.ar/handle/20.500.12219/2150?show=full

Carton, W. (2014). Environmental protection as market pathology?: Carbon trading and the dialectics of the "double movement". *Environment and Planning D: Society and Space* 32(6), 1002–1018.

Cashore, B. W., Auld, G., and Newsom, D. (2004). *Governing Through Markets: Forest Certification and the Emergence of Non-state Authority*. Yale University Press.

Castree, N. (2008a). Neoliberalising nature: The logics of deregulation and reregulation. *Environment and Planning A* 40(1), 131–152.

Castree, N. (2008b). Neoliberalising nature: Processes, effects, and evaluations. *Environment and Planning A* 40(1), 153–173.

Cavanagh, C., and Benjaminsen, T. (2017). Political ecology, variegated green economies, and the foreclosure of alternative sustainabilities. *Journal of Political Ecology*, 24, 200–216.

Christophers, B. (2017). Climate change and financial instability: Risk disclosure and the problematics of neoliberal governance. *Annals of the American Association of Geographers*, 107(5), 1108–1127. Retrieved August 9, 2022 from https://doi.org/10.1080/24694452.2017.1293502

Clapp, J. (2017). Responsibility to the rescue? Governing private financial investment in global agriculture. *Agric Hum Values* 2017(34), 223–235.

Clapp, R. A. (1995). Creating competitive advantage: Forest policy as industrial policy in Chile. *Economic Geography*, 71(3), 273. https://doi.org/10.2307/144312

Cohen, D., Nelson, S., and Rosenman, E. (2021). Reparative accumulation? Financial risk and investment across socio-environmental crises. *Environment and Planning E*. 5(4), 2356–2382.

Collard, R.-C., and Dempsey, J. (2020). Two icebergs: Difference in feminist political economy. *Environment and Planning A: Economy and Space*, 52(1), 237–247. https://doi.org/10.1177/0308518X19877887

Crosby, A. W., (1986). *Ecological Imperialism: The Biological Expansion of Europe, 900–1900*. Cambridge University Press.

Crosby, A. W., and Von Mering, O. (1972). *The Columbian Exchange: Biological and Cultural Consequences of 1492*. Greenwood Press.

Davis, H., and Todd, Z. (2017). On the importance of a date, or decolonizing the Anthropocene. *ACME*, 16(4), 761–780.

Davis, J., Moulton, A. A., Van Sant, L., and Williams, B. (2019). Anthropocene, Capitalocene, … Plantationocene?: A Manifesto for Ecological Justice in an Age of Global Crises. *Geography Compass*, 13(5), e12438. https://doi.org/10.1111/gec3.12438

DeBoom, M. J. (2017). Nuclear (geo)political ecologies: A hybrid geography of Chinese investment in Namibia's uranium sector. *Journal of Current Chinese Affairs*, 46(3), 53–83. https://doi.org/10.1177/186810261704600303

DeBoom, M. J. (2021). Climate necropolitics: Ecological civilization and the distributive geographies of extractive violence in the Anthropocene. *Annals of the American Association of Geographers*, 111(3), 900–912.

Dingemans, A., and Ross, C. (2012). Free trade agreements in Latin America since 1990: An evaluation of export diversification. *CEPAL Review*, 2012(108), 27–48. https://doi.org/10.18356/0cc7e81b-en

Dussel, E. (1998). Beyond Eurocentrism: The world system and the limits of modernity. In *The Cultures of Globalization*, edited by Fredric Jameson and Masao Miyoshi. Duke University Press. https://doi.org/10.1215/9780822378426

Edelman, M., and Borras, S. M. (2016). *Movimientos agrarios transnacionales: Historia, organización y políticas de lucha*. La Paz: Icaria editorial.

Emel, J., and Huber, M. T. (2008). A risky business: Mining, rent and the neoliberalization of "risk". *Geoforum*, 39(3), 1393–1407. https://doi.org/10.1016/j.geoforum.2008.01.010

Fairhead, J., Leach, M., and Scoones, I. (2012). Green grabbing: A new appropriation of nature? *Journal of Peasant Studies*, 39(2), 237–261. https://doi.org/10.1080/03066150.2012.671770

Federici, S. (2004). *Caliban and the Witch: Women, the Body and Primitive Accumulation*. New York: Autonomedia.

Fletcher, R., and Büscher, B. (2017). The PES conceit: Revisiting the relationship between payments for environmental services and neoliberal conservation. *Ecological Economics*, 132, 224–231.

Gago, V. (2014). *La razón neoliberal: Economías barrocas y pragmática popular*. Tinta Limón Ediciones.

Gibson-Graham, J. K. (1996). *The End of Capitalism As We Know It*. Wiley-Blackwell.

Goldman, M. (2005). *Imperial Nature: The World Bank and Struggles for Social Justice in the Age of Globalization*. Yale University Press.

Gudynas, E. (2009). Diez tesis urgentes sobre el nuevo extractivismo. *Extractivismo, política y Sociedad*, 187, 187–225.

Guthman, J. (2007). The Polanyian way? Voluntary food labels as neoliberal governance. *Antipode*, 39(3), 456–478. https://doi.org/10.1111/j.1467-8330.2007.00535.x

Haraway, D. (2015). Anthropocene, Capitalocene, Plantationocene, Chthulucene: Making kin. *Environmental Humanities*, 6(1), 159–165. https://doi.org/10.1215/22011919-3615934

Hart, G. P. (2002). *Disabling Globalization: Places of Power in Post-Apartheid South Africa*. University of California Press.

Hayter, R., and Barnes, T. J. (2012). Neoliberalization and its geographic limits: Comparative reflections from forest peripheries in the global north. *Economic Geography*, 88(2), 197–221.

Hecht, S. B. (2005). Soybeans, development and conservation on the Amazon frontier. *Development and Change*, 36(2), 375–404. https://doi.org/10.1111/j.0012-155X.2005.00415.x

Hendrixson, A., Ojeda, D., Sasser, J. S., Nadimpally, S., Foley, E. E., and Bhatia, R. (2020). Confronting populationism: Feminist challenges to population control in an era of climate change. *Gender, Place and Culture*, 27(3), 307–315. https://doi.org/10.1080/0966369X.2019.1639634

Hernando-Arrese, M., and Rasch, E. D. (2022). The micropolitical life of energy projects: A collaborative exploration of injustice and resistance to small hydropower projects in the Wallmapu, Southern Chile. *Energy Research and Social Science*, 83, 102332. https://doi.org/10.1016/j.erss.2021.102332

Heynen, J., McCarthy, J., Prudham, S., and Robbins, P. (Eds.) (2007). *Neoliberal Environments: False Promises and Unnatural Consequences*. Routledge, Taylor & Francis Group.

Hoyos-Santillan, J., Miranda, A., Lara, A., Sepulveda-Jauregui, A., Zamorano-Elgueta, C., Gómez-González, S., Vásquez-Lavín, F., Garreaud, R. D., and Rojas, M. (2021). Diversifying Chile's climate action away from industrial plantations. *Environmental Science and Policy*, 124, 85–89. https://doi.org/10.1016/j.envsci.2021.06.013

Jeréz Henríquez, B. (2021). La colonialidad de la minería del litio sobre los salares altoandinos: Conflictos socioambientales para la electromobilidad verde del norte global. In P. López and M. Betancourt Santiago (Eds.), *Conflictos territoriales y territorialidades en disputa* (pp. 371–390). CLACSO. https://public.ebookcentral.proquest.com/choice/PublicFullRecord.aspx?p=6782470

Kama, K. (2014). On the borders of the market: EU emissions trading, energy security, and the technopolitics of "carbon leakage". *Geoforum* 51, 202–212.

Kay, K. (2018). Financialization, adaptable assets and the evolution of neoliberal environments. In Bigger, P., and Dempsey, J. *Reflecting on Neoliberal Natures: An Exchange. Environment and Planning E: Nature and Space*, 1(1–2), 25–75. https://doi.org/10.1177/2514848618776864

Kedward, K., Ryan-Collins, J., and Chenet, H. (2020). Managing nature-related financial risks: A precautionary policy approach for central banks and financial supervisors. *SSRN Electronic Journal*. https://doi.org/10.2139/ssrn.3726637

Kish, Z., and Fairbairn, M. (2017). Investing for profit, investing for impact: Moral performances in agricultural investment projects. *Environment and Planning A: Economy and Space*, 50(3), 569–588. https://doi.org/10.1177/0308518X17738253

Knox-Hayes, J., and Levy, D. (2014). The political economy of governance by disclosure: Carbon disclosure and nonfinancial reporting as contested fields of governance. In A. Gupta and M. Mason (Eds.), *Transparency in Global Environmental Governance: Critical Perspectives*. The MIT Press. https://doi.org/10.7551/mitpress/9780262027410.003.0009
Larner, W. (2007). Neoliberal governmentalities. In Nik Heynen, James McCarthy, Scott Prudham, and Paul Robbins (Eds.), *Neoliberal Environments*, 217–220. Abingdon: Routledge.
Li, T. M. (2011). Forum on global land grabbing: Centering labor in the land grab debate. *Journal of Peasant Studies* 38(2): 281 –298.
Liverman, D. M., and Vilas, S. (2006). Neoliberalism and the environment in Latin America. *Annual Review of Environment and Resources*, 31(1), 327–363. https://doi.org/10.1146/annurev.energy.29.102403.140729
Lorca, M., Olivera Andrade, M., Escosteguy, M., Köppel, J., Scoville-Simonds, M., and Hufty, M. (2022). Mining indigenous territories: Consensus, tensions and ambivalences in the Salar de Atacama. *The Extractive Industries and Society*, 9, 101047, ISSN 2214–790X, https://doi.org/10.1016/j.exis.2022.101047
Lowe, L. (2015). *The Intimacies of Four Continents*. Duke UP.
Marimán, P., Caniuqueo, S., Millalén, J., and Levil, R. (Eds.). (2006). *—Escucha, winka—!: Cuatro ensayos de Historia Nacional Mapuche y un epílogo sobre el futuro*. Santiago: LOM Ediciones.
Mbembe, A. (2019). *Necropolitics* (S. Corcoran, Trans.). Duke University Press.
McAfee, K. (1999). Selling nature to save it? Biodiversity and green developmentalism. *Environment and Planning D: Society and Space*, 17(2), 133–154. https://doi.org/10.1068/d170133
McCarthy, J. (2004). Privatizing conditions of production: Trade agreements as neoliberal environmental governance. *Geoforum*, 35, 327–341.
McCarthy, J., and Prudham, S. (2004). Neoliberal nature and the nature of neoliberalism. *Geoforum*, 35, 275–283.
Mehta, L., Veldwisch, G., and Franco, J. (2012). Introduction to the special issue: Water grabbing? Focus on the (re)appropriation of finite water resources. *Water Alternatives*, 5(2), 193–207.
Mies, M. (1986). *Patriarchy Accumulation on a World Scale: Women in the International Division of Labour*. Zed Books.
Millaleo Hernández, S. (2019). Colonialismo, Racismo Ambiental y Pueblo Mapuche. *Revista Anales*, 16, 269–282.
Moguillansky, G. (1999). *La Inversión en Chile: ¿El Fin de un Ciclo de Expansión?* Santiago de Chile: Fondo de Cultura Económica y CEPAL.
Nelson, S. H. (2014). Resilience and the neoliberal counter-revolution: From ecologies of control to production of the common. *Resilience*, 2(1), 1–17.
Normann, S. (2021). Green colonialism in the Nordic context: Exploring Southern Saami representations of wind energy development. *Journal of Community Psychology*, 49(1), 77–94.
Núñez, A., Benwell, M. C., and Aliste, E. (2020). Interrogating green discourses in Patagonia-Aysén (Chile): Green grabbing and eco-extractivism as a new strategy of capitalism? *Geographical Review*, (August), 1–19.
Ojeda, D., (2018). Landgrabbing in Latin America: Sedimented spaces of dispossession. In Cupples, J., Palomino-Schalscha, M., and Prieto, M. (Eds.). *The Routledge Handbook of Latin American Development* (1st ed.). Routledge. https://doi.org/10.4324/9781315162935
Osborne, T., (2015). Tradeoffs in carbon commodification: A political ecology of common property forest governance. *Geoforum*, 67, 64–77.
Ouma, S. (2014). Situating global finance in the Land Rush Debate: A critical review. *Geoforum*, 57, 162–166.
Palma, J. G. (2020). América Latina en su "Momento Gramsciano": Las limitaciones de una salida tipo "nueva socialdemocracia europea" a este impasse. *El Trimestre Económico*, 87(348), 985–1031. https://doi.org/10.20430/ete.v87i348.1146
Patel, R., and Moore, J. W. (2018). *A History of the World in Seven Cheap Things: A Guide to Capitalism, Nature, and the Future of the Planet*. University of California Press.
Peck, J. (2013). Explaining (with) Neoliberalism. *Territory, Politics, Governance*, 1(2), 132–157. https://doi.org/10.1080/21622671.2013.785365
Peluso, N. L., and Lund, C. (2011). New frontiers of land control: Introduction. *Journal of Peasant Studies*, 38(4), 667–681. https://doi.org/10.1080/03066150.2011.607692
Pinto, L. H. (2020). Movimiento de los Trabajadores Rurales Sin Tierra (MST). (Brasil, 1984-2015). In A. Salomón and J. Muzlera (Eds.), *Diccionario del Agro Iberoamericano*. Teseopress.com.

Povinelli, E. A. (2011). *Economies of Abandonment: Social Belonging and Endurance in Late Liberalism*. Duke University Press.

Pulido, L. (1994). Restructuring and the contraction and expansion of environmental rights in the United States. *Environment and Planning A*, 26, 915–936.

Quijano, A. (2008). Coloniality of power, Eurocentrism, and Latin America. In M. Moraña, E. Dussel, and C. A. Jáuregui (Eds.), *Coloniality at Large: Latin America and the Postcolonial Debate* (pp. 181–224). Duke University Press.

Riofrancos, T. (2022). The security–sustainability nexus: Lithium onshoring in the Global North. *Global Environmental Politics*, 1–22. https://doi.org/10.1162/glep_a_00668

Roberts, A., Choer, H., and Ferguson, V. (2019). Toward a geoeconomic order in international trade and investment. *Journal of International Economic Law*, 22(4,December), 655–676. https://doi.org/10.1093/jiel/jgz036

Roberts, D. J., and Mahtani, M. (2010). Neoliberalizing race, racing neoliberalism: Placing "race" in neoliberal discourses. *Antipode*, 42(2), 248–257.

Rojas Marchini, M. F. (2022). *Making the Environmental State in Chile: Knowledge, Markets, and Legal Frameworks for Biodiversity Conservation* (T). University of British Columbia. Retrieved from https://open.library.ubc.ca/collections/ubctheses/24/items/1.0421413

Rojas Marchini, F., and Folch, T. (2018). Stepping into the forest: Forest economy, rural livelihoods, and socionatures in Chile. In J. Hutton, D. Ibañez, and K. Moe (Eds.), *Wood Urbanism: From the Molecular to the Territorial* (pp. 244–251). Actar D.

Sullivan, S. (2010). "Ecosystem service commodities"—a new imperial ecology? Implications for animist immanent ecologies, with Deleuze and Guattari. *New Formations*, 69(Spring 2010), 111–128.

Sultana, F. (2014). Gendering climate change: Geographical insights. *The Professional Geographer*, 66(3), 372–381.

Sundberg, J., Dempsey, J., and Marchini, F. R. (2020). Nature–culture. In *International Encyclopedia of Human Geography* (pp. 315–324). Elsevier. Retrieved from https://doi.org/10.1016/B978-0-08-102295-5.10889-3

Svampa, M. (2019). *Neo-extractivism in Latin America: Socio-environmental Conflicts, The Territorial Turn, and New Political Narratives*. (1st ed.). Cambridge University Press. Retrieved from https://doi.org/10.1017/9781108752589

Svampa, M., and Viale, E. (2014). *Maldesarrollo: La Argentina del extractivismo y el despojo*. Buenos Aires, Argentina: Katz.

Terán Mantovani, E. (2016). Las nuevas fronteras de las commodities en Venezuela: Extractivismo, crisis histórica y disputas territoriales. *Ciencia Política*, 11(21). https://doi.org/10.15446/cp.v11n21.60296

Tsing, A. L. (2015). *The Mushroom at the End of the World: On the Possibility of Life in Capitalist Ruins*. Princeton: Princeton University Press.

Valenzuela-Fuentes, K., Alarcón-Barrueto, E., and Torres-Salinas, R. (2021). From resistance to creation: Socio-environmental activism in Chile's "sacrifice zones". *Sustainability*, 13, 3481. https://doi.org/10.3390/su13063481

Van Hecken, G. (2018). Silencing agency in payments for ecosystem services (PES) by essentializing a neoliberal "monster" into being: A response to Fletcher and Büscher's "PES Conceit". *Ecological Economics*, 122, 314–318.

Voskoboynik, D.M., and Andreucci, D. (2021). Greening extractivism: Environmental discourses and resource governance in the "Lithium Triangle". *Environment and Planning E: Nature and Space*. https://doi.org/10.1177/25148486211006345

Wainwright, J., and Mann, G. (2018). *Climate Leviathan: A Political Theory of Our Planetary Future*. London: Verso Books.

Wolford, W. (2003). Producing community: The MST and land reform settlements in Brazil. *Journal of Agrarian Change*, 3(4), 500–520. https://doi.org/10.1111/1471-0366.00064

Zografos, C. (2022). The contradictions of Green New Deals: Green sacrifice and colonialism. *Soundings*, 80(Spring 2022), 37–50.

10 Monetizing Nature

From Resource-Making to Financialization

Kelly Kay

It was still dark out when I started the walk along Florida State Road A1A from my motel to the Ritz Carlton. Since there were no sidewalks, no buses, and no rideshares, the only way to get to the conference without a car of my own was to walk three miles along the shoulder of the road. At times, I walked through the front lawns of vacation homes—built on former paper company land—happy to be off the road but worried that I might not blend in at the conference if I arrived with wet shoes from walking through the grass. As I finally reached the outer gates of the conference hotel and straightened my blazer, several private jets flew right over my head. They landed at the airstrip on the other side of the highway, presumably to attend the same event as me: the timberland investment conference. After grabbing my nametag, I walked to the exhibit area, passing by tables for companies that specialize in forest carbon credits, land sales, recreational leasing, and geospatial technologies to help with timberland valuation. Dotted around the entry hall were groups of men in suits, excited to see each other, catch up, and trade stories. I settled into a chair and opened the conference program, which was prefaced with a note about insider trading, an important reminder of why most of the conference participants were here: to do deals. Many people remarked that it's unusual to see someone from UCLA at a forestry conference, though as I came to realize as the weekend went on, most of them assumed I worked for the university's endowment.

This conference, and others like it that I've attended in recent years, is an important node in a burgeoning industry: timberland investment. While forests have long been commodified, a growing interest from pension funds and other institutional investors in land and other "real assets" (Fairbairn 2021) has meant that forests are being remade once more to ensure that they produce financial returns and other ecosystem services and monetizable property rights (Gunnoe and Gellert 2011), in addition to fast-growing trees suited to paper and lumber production (Boyd et al. 2001). Broadly, this chapter is focused on monetization, by which I mean the conversion of the biophysical world into money. In it, I consider a range of ways that capital has remade environments and non-human life in the service of generating profits or rents. These include enclosures and the creation of new property rights, the making of new resources and commodities, as well as through marketization—or the creation of markets for goods and services where they had not previously existed. The chapter also engages significantly with neoliberalism—the "economic and political philosophy that questions, and in some versions entirely rejects, government interventions into the market…and eschews social and collective controls over the behaviors and practices of firms, the movement of capital, and the regulation of socio-economic relationships" (Heynen et al. 2007, 3) and financialization—the "process

DOI: 10.4324/9781003165477-14

of ontological reconfiguration through which different qualities of nature and resource-based production are translated into a financial value form to be traded on specialized markets" (Ouma et al. 2018, 501).

While financialization is newer, the ongoing and expanding commodification, monetization, and marketization of nature have been enduring concerns of political ecologists for more than three decades. Further still, since the field's very inception, political ecologists have been concerned with how environments come to be enrolled in capitalist value circuits, transforming them as a result, and the implications of those structural transformations (e.g., Blaikie and Brookfield 1987; Watts 2013). In this chapter, I trace the arc of literature on nature's monetization, beginning with several classic texts and key concepts that political ecologists have drawn on, then moving briefly into a discussion of how property and resources are made. From there, I review the voluminous literature on neoliberal natures and the financialization of nature, before closing with some reflections on future directions for scholarship on the monetization of nature.

Political Economy of the Environment

Historically, much of the literature on the political economy of the environment has been centered around the making of resources or property, as well as considering some of the negative externalities that have emerged from the interface between capitalism and nature. Ecosocialist thinkers, like John Bellamy Foster (1999, 2000) and James O'Connor (1991), for example, have played a critical role in pulling environmental thought out of the work of Marx and other foundational thinkers in heterodox political economy. These concepts have become central for many political ecologists whose work is steeped in political economic thought. O'Connor's pioneering work on the "second contradiction of capitalism" aimed to extend Marx's ideas about labor exploitation and capitalism's internal crisis tendencies to argue that there may be a second road to socialism, and it stems from capitalism's tendency to destroy its own conditions of production, including the environment and infrastructure. Similarly, Karl Polanyi's *The Great Transformation* (2001 [1944]) has been another foundational text for many political ecologists. In it, Polanyi traces the rise of free market capitalism, arguing that one of its internal contradictions is the tendency to treat key inputs as commodities even though they were not produced for sale on the market. These "fictitious commodities" include land, labor, and capital. In Polanyi's formulation, treating these materials as commodities within a highly marketized society is likely to lead to social uprising and resistance, an outcome that he describes as the "double movement."

In his work on metabolism and the metabolic rift, John Bellamy Foster also engages with key concepts from heterodox political economy, aiming to highlight the relevance of Marx's thought for environmental scholarship. Foster draws on Marx's critiques of capitalist agriculture and agricultural chemistry to build the foundations for an ecological Marxism. In particular, he centers the concept of metabolism, or *Stoffwechsel*, which is rooted in his understanding of the labor process, which Marx describes as "a process between man and nature, a process by which man, through his own actions, mediates, regulates and controls the metabolism between himself and nature. He confronts the materials of nature as a force of nature" (Foster 1999, 380). For Foster, metabolism is a complex and interdependent process that links people and natures, and can be easily thrown out of balance when there is a rupture in nutrient cycling between the city and rural areas, or between humans and nature more generally. Metabolism has been a major anchoring concept for development studies and environmental studies, as well as for the field of urban political ecology (Heynen et al. 2006). Beyond metabolism, Marxian scholarship has continued to play a guiding role in political ecological research. In recent years, scholars have suggested a closer engagement with value theory, for example, as a means of cutting across distinct case studies to understand in a more holistic sense how capitalism operates vis-à-vis nature (Robertson and Wainwright 2013; Kay and Kenney-Lazar 2017).

Beyond the examples of foundational thought in political economy considered above, political ecologists have long had an interest in how the environment is turned into property, resources, and commodities. Guided by an understanding that resources are "made not born" and must emerge from distinct systems of "resource making" (Kama 2020), scholars in political ecology and related fields like critical resource geography have long been interested in understanding how nature comes to be monetized through its conversion to resources, and how those resources eventually circulate as commodities like gold (Hartwick 1998), wildlife (Lunstrum and Massé 2021), unconventional fossil fuels (Kama 2020), or even living parts of human bodies (Fannin 2021). Banoub et al. (2021), for example, highlight this dynamic in their work on commodity frontiers. Through case studies of gold mining, aquaculture, and plantation forestry, they demonstrate how capital "works through nature" to reshape the temporal, spatial, and material characteristics of the biophysical world to make them more amenable to commodity and resource production. This process includes both the "widening" (or spatial expansion) and "deepening" (or enhancing of productivity) of extant commodity production, as well as the wholesale transformation of non-human natures and reconstitution of existing commodity forms. Additionally, rooted in scholarship about the primacy of property-making to the formation of modern capitalism, political ecologists have also had an enduring concern with how environments are remade as property—private and excludable property in particular—as well as in resisting private property through engagement with concepts like common property regimes (Ostrom 1990) and commoning (Turner 2017). Kevin St Martin's (2001) work on community resource management, for example, interviews and conducts mapping exercises with New England fisherpeople as a means of pushing back on the notion that private property rights are required to prevent overfishing. While for many mainstream scientists and policymakers fisheries are often understood to be an unregulated common pool resource in need of intervention, St Martin's findings indicate active cooperation between fishers as a means of trying to prevent depletion of the resource upon which they all rely.

Neoliberal Natures: From Commodities to Conservation

In 1999, when Kathleen McAfee wrote her germinal piece "Selling Nature to Save It? (McAfee 1999)," it would have been hard to predict how strongly her concerns about the rise of "green developmentalism" and market-based environmental governance would come to ring true. McAfee's article, and many works that came after by political ecologists and scholars in adjacent fields, comprise a voluminous body of work on what came to be termed "neoliberal natures" (Heynen et al. 2007; McCarthy and Prudham 2004; Bakker 2010; Castree 2008a, 2008b). Scholarship on the neoliberalization of nature, while highly variegated, aimed to understand how post-Fordist economic transformations interface with environmental politics and governance, transforming nature-society relations as a result. While much environmental regulation in the 1960s and

1970s was centered on command-and-control approaches, the 1980s ushered in a new era focused around voluntary and market-based solutions to environmental governance which relied heavily on market mechanisms like cap and trade programs to resolve an ever-growing range of socio-ecological issues. For this "self-regulating market" (Polanyi 2001 [1944]) to function, it would have to become "increasingly wide in its geographic scope, comprehensive as the governing mechanism for allocating all goods and services, and central as a metaphor for organizing and evaluating institutional performance. This of course requires the deeply problematic 'commodification of everything'" (McCarthy and Prudham 2004, 276). For political ecologists, this sea change in the political-economic order of the day ushered in a rethinking of nature-state-society relations.

In a meta-review of the neoliberal natures literature, Noel Castree identifies a range of ideal-type processes that scholars have identified across individual cases as being central to nature's neoliberalization (2008a). These processes include: privatization, marketization, deregulation (or regulatory "roll-back"), reregulation (or regulatory "roll-out"), and the growing power and presence of "flanking mechanisms" in civil society, including charities and NGOs. Key to understanding each of these processes is a close engagement with the state. States are central to the "roll out" (Peck and Tickell 2002) of neoliberal environmental governance, as they play a critical role in creating and regulating new markets in ecosystem services and in creating and enforcing new property rights over formerly non-marketized ecologies. Paradoxically, neoliberal environmental governance is also understood to be anchored by deregulation (Castree 2008a) and the hollowing out and retreat of the state, as well as the rescaling of governance—including devolution to more localized scales and the scaling up of decision making and action to supranational institutions such as the UN Framework Convention on Climate Change. Scholars understand these "roll out" and "roll back" processes to be tightly interlinked (Peck and Tickell 2002), as deregulation creates space for new policies, and new rounds of marketization and privatization to emerge. Anchored by this understanding of the state's shifting function in environmental governance, ample case studies have been conducted in a range of geographies on (1) nature's enclosure/privatization and (2) the creation of new markets and environmental commodities.

Through the enclosure of existing commons, the creation and enforcement of new property rights, and the transfer of state assets to private hands, privatization has been central to the neoliberal project. Harkening back to Marx's work on primitive accumulation and the violent enclosure of the English commons—a critical precursor to modern capitalism—geographer David Harvey (2003) coined the term "accumulation by dispossession" to describe the transfer of wealth and property from the public to private sector resulting from neoliberal policies favoring privatization. We see the primacy of privatization as an accumulation strategy in a range of works by political ecologists. Becky Mansfield, for example, considers how neoliberal approaches to environmental management have meant a "property revolution" in managing regional fisheries, rationalizing and enclosing oceans using technologies like individual transferable quotas (2004, 323). James McCarthy (2004), in his work on NAFTA, is also concerned with enclosure and new property rights, arguing that trade agreements, and in particular the investor protections built into them, should be understood as the privatization—or primitive accumulation—of the conditions of production. Finally, in her work on the privatization of the water supply in England and Wales, Karen Bakker (2004, 2005) captures another example of the successful privatization of a previously public resource. Her work offers more than another case study of neoliberalization, however, as it underscores the need to delineate between distinct processes of market-based governance. In her case, while the ownership of the water system was successfully transferred from a nationalized monopoly to a privately-owned and publicly-traded company, privatization did not automatically mean successful commercialization or commodification. This highlights the importance of conceptual precision and the fact that monetizing nature will always be partial and incomplete.

Scholars have also conducted extensive research on marketization—or the creation of markets where they did not previously exist—as well as the rise of new environmental commodities. As Bakker reminds us above, this is a related but distinct process from privatization. Shifting ownership is just one piece in a complex puzzle of management, pricing, and reregulation. Also, critically, the creation of new environmental commodities differs from other more historic forms of nature's commodification. In a piece in the *Socialist Register*, entitled "Nature as Accumulation Strategy," the late Neil Smith extends his work on the production of nature (Smith 2007; see also Chapter 2, this volume) to analyze the slew of new environmental commodities that came to emerge from this decades-long process of nature's neoliberalization. He describes them as follows:

> Whereas the traditional commodification of nature generally involved harvesting use values as raw materials for capitalist production—wood for tables, oil for energy, iron ore for steel, various corns for bread—this new generation of ecological commodities is different. Whether they do or do not become the raw material for future production is incidental to their production. Instead, these commodities are simultaneously excavated (in exchange-value terms) from pre-existing socio-natural relations and as part of their production they are reinserted or remain embedded in socialized nature—the more "natural" the better.
>
> (Smith 2007, 17)

Smith (2010) raises an important distinction between harvesting use values from nature and the excavation of new environmental commodities from existing natures. Much of the literature in political ecology on payments for ecosystem services and other incentives and financing mechanisms for conservation take these types of new ecological commodities as their subject of concern. As an umbrella topic, much of the literature on new environmental commodities rely on the notion of payments for ecosystem services—compensation for the goods and services that "intact" natures are able to provide. In a review of the concept, Gómez-Baggethun et al. (2010) demonstrate that notions like natural capital and ecosystem services arose in the 1970s and 1980s primarily as a "pedagogic" concept intended to communicate to policy-makers that nature has value in and of itself. During the 1990s, more serious calculations began to be done, again, primarily for heuristic or policy purposes, to keep the environment on the agenda as public policy became much more concerned with cost-benefit analyses and other forms of quantification. From there, heuristic valuations gave way to markets, with an array of new environmental commodities arising. On this front, political ecologists have studied an array of tradable permit systems (Bigger 2018), wetland mitigation banking (Robertson 2006), for-profit stream restoration (Lave 2012), and carbon offsets (Lansing 2012; Bumpus and Liverman 2008; Osborne 2013; Lockhart 2015), among others. Critically, much of this work centers the neoliberal state as market-maker and regulator, recognizing that many tradeable permit systems, for example, would not be able to function without heavy and ongoing state involvement. Creating these new environmental commodities and making them commensurable with other more established ones also requires an immense amount of labor. As Morgan Robertson (2012) has shown, marketizing an ecosystem service involves a complex process that implicates bureaucrats, scientists, capitalists, and others in a process of classifying, categorizing, unbundling of ecosystems into distinct environmental values or property rights, and then stacking—or packaging—those ecosystem functions into an appealing bundle.

Conservation is one arena where many of these new markets in ecosystem services and environmental commodities have arisen to try and make nature pay dividends. Jessica Dempsey (2016) identifies this move toward remaking conservation as an economically-rational policy-making decision as the imperative for nature to become "enterprising" in order to ensure its own survival. Similarly, Arsel and Büscher (2012) describe the same process as the construction of "Nature Inc.". There is an ample literature on what scholars have termed "neoliberal conservation" (Brockington and Duffy 2011; Büscher et al. 2012; Arsel and Büscher 2012). This work recognizes that as conservation is remade in the image of neoliberal policy imperatives, new subjectivities are formed, new partnerships and collaborations are forged with non-state and for-profit actors, and profitable activities like ecotourism come to take center stage. Protected areas often become "spectacularized" as a means of attracting attention, amplifying concern, and driving visitors (Igoe 2021).

Box 10.1 Studying the state: legal and financial research methods

There are a range of qualitative and quantitative methods used by political ecologists to study the monetization of nature, many of which are reviewed in the chapters in this volume, including those by Guthman and Butler (Chapter 13) and Huber (Chapter 12). Given the centrality of the state to the ongoing commodification and monetization of natures, including through its ability to create and enforce property rights, devolve its responsibilities or share them with the private sector, and create new markets, exchanges, and tradeable permit systems, I focus on legal and financial analysis as alternative methodologies which center the state, its processes, and effects.

In her work, socio-legal scholar Katharina Pistor underscores the importance of engaging with law in order to understand the functioning of capitalism; as she eloquently puts it, law is "the cloth from which capital is cut" (2019, 4). In recognition of this, political ecologists and environmental legal geographers have played a critical role in illuminating the value of engaging with legislation, regulation, and legal processes as a means of understanding the "how" of nature's monetization (Borgias 2018; Kay 2016; Perry and Gillespie 2019; Andrews and McCarthy 2014). As one notable recent example, John Casellas Connors and Chris Rea (2022) consider the US Pittman Robertson Act, a piece of legislation that earmarks an excise tax on firearms for environmental conservation and restoration. The authors argue that increased firearms sales in the US have meant that guns are playing an expanding role in financing land conservation, and as a result, are changing conservation aims and priorities as well. Another example comes from David Turton (2015), who, rather than looking at laws themselves, centers lawyers in his research on unconventional gas exploration in Australia. As a substitute for interviews or participant observation, Turton turns to publicly accessible recordings of lawyers participating in community forums to understand their role in both aiding and resisting a controversial emerging energy source.

Additionally, Amanda Kass (2020) offers an innovative approach to studying the financialization of the state with what she calls critical financial analysis. Drawing on her experience working in public administration, Kass compellingly argues for critical geographers to gain greater fluency in accounting and actuarial norms, balance sheets, and other financial data. As she puts it: "financial data provides rich information; however, its technocratic nature enables it to be used as a tool of power to limit who can weigh in on debates concerning how public funds should be spent and what constitutes prudent activities to those with financial expertise" (ibid., 107). As many scholars remind us, data is never value-neutral (see Chapter 8, this volume), thus, critical financial analysis provides a means of engaging deeply with discourses of state austerity and illuminating the politics embedded in what may seem like quotidian decision-making processes. While Kass' case study focuses on low-income housing in Chicago, her methods could be highly relevant for a wide range of political ecology research on political economy and governance.

From Commodity to Financial Asset

In recent decades, we have witnessed the financialization of economies across both the Global North and South, defined by Greta Krippner as "a pattern of accumulation in which profits accrue primarily through financial channels rather than through trade and commodity production" (2005, 174). This financialization process has led to a growing interest in revenue-generating assets (including land) and the increased power and presence of the financial, insurance, and real estate (FIRE) sectors in society, among other notable changes. Scholars interested in the political economy of the environment have taken an interest in how the financialization of nature has materially reconfigured environments to generate financial returns, and how finance's broadening and deepening engagement with the environment has altered socio-natural relations in a range of contexts.

Particularly in light of the 2008/2009 financial crisis, investors became interested in seeking out "real assets," like agricultural areas or timberland (Fairbairn 2021; Gunnoe and Gellert 2011). This great global "land grab" as it has been termed (Borras et al. 2011) has meant a significant transformation in who owns land, how they use it, and how this impacts neighboring communities, as well as workers (Li 2011). As Ryan Isakson (2017) has shown in a review of the literature on food and finance, not only are financial actors playing a key role through their increased ownership of farmland, but they are increasingly becoming involved in the supply chain, ranging from provisioning agricultural inputs, to food processing, to food retailing. This financialization of the entire agricultural supply chain has been shown to have negative impacts for both smaller-scale agricultural producers, as well as for laborers throughout the value chain.

While it requires substantial work to make land investable, for many financial actors, alternative asset classes like farmland provide a desirable means of diversifying their portfolios to include assets with multiple and flexible revenue streams (Borras et al. 2016). This allows for the interplay of financial profit-making and underlying and ongoing commodity production (Ouma 2020), or what scholars have termed "getting between M-C-M'" (Ouma et al. 2018, 500), in reference to Marx's understanding of the generation of surplus value. Beyond land, political ecologists have studied the broadening and deepening of the financialization of nature. Case studies on novel environment/finance entanglements range from financial investment in water and energy infrastructure (Loftus et al. 2019; Knuth 2018), decarbonization (Bridge et al. 2020; Bracking 2019), green bonds (Cousins and Hill 2021), and climate risk insurance (Knudson 2018; Johnson 2015). Assembling these novel asset types also requires a substantial amount of work to generate buy-in and acceptance, much of which is discursive and performative (Dempsey and Suarez 2016). With the growing interest in impact investing and in light of growing critiques of global agri-investment, Kish and Fairbairn, for example, show how critical the performance of "moral value" is to "the creation and maintenance of economic value" (2018, 583). And, while her case focuses on state—not private—investment, Clare Beer's concept of "resource spectacle" is also instructive for understanding how environmental investments and assets are made and reinforced through spectacular imagery and other forms of marketing (Beer 2023).

As more natures become financialized, scholars have raised questions about how our understandings of value, profit, and accumulation might shift. Brett Christophers (2018), as one example, raises critical questions about the applicability of Marxian value theory to understanding financial profit-making, suggesting that financial value production vis-à-vis nature hinges on the commodification of risk. For many researchers, what makes the financialization of the environment distinct, aside from the presence of new actors and new motives, is the centrality of rent extraction (Huber 2022). Recognizing that financial profits are by nature redistributive, several frameworks have been proposed by political ecologists to better contextualize the centrality of rent extraction to the financialization of nature. Andreucci et al.'s (2017) concept of "value grabbing," aims to place rent at the center of political ecological research, raising the point that the extraction of rents is analytically distinct from accumulation and that rent relations have their own particular intra-class conflicts. The value grabbing framework also compellingly distinguishes how the commodification and financialization of nature might differ by highlighting "that the central dynamic at play is the instituting of property rights that are not used exclusively or even mainly to *produce* new commodities, but rather are mobilized to *extract* value through rent relations" (ibid., 29). Similarly, drawing on Diane Elson's "value theory of labor," Purcell et al. (2020) propose a triadic conceptual schema that they call "rent-value-finance" and consider the financialization of water as an illustrative case to demonstrate the ways that rent enters into both the circulation and accumulation of financial value. In recent years, political ecologists have taken up these calls to engage more closely with rent theory. The literature also offers ample case studies showcasing the centrality of rent-seeking to the financialization of a range of resources and environments, including forests (Gunnoe 2014; Kay 2017), agricultural land (Fairbairn 2021), former coalfields (Schwartzman 2022), industrial fisheries (Campling and Havice 2014), international climate change agreements (Felli 2014), weather index insurance (Johnson 2022), mining policies (Emel and Huber 2008), and oil (Labban 2010).

Looking toward the future, scholarship on the financialization of nature has two major obstacles to contend with. First, as Christophers (2015) argues, the term financialization is so stretched and twisted that it has lost coherence. Thus, scholars may mean radically different things while using similar terms. Jessica Dempsey, recognizing this fact, suggests that at least some literature on the "financialization of nature" may actually be concerned with alternate processes, including economization, or "the extension of economic logics, practices, and calculations into new areas," and marketization, or "the creation of markets...where they did not exist before" (Dempsey 2017, 196). As finance continues to penetrate deeper into our societies and as the environment continues to become subject to investment, conceptual precision will become ever-more important in scholarship examining these processes. Second, researchers must recognize that the financialization of nature is a recursive process—natures are transformed to make them investable (Ouma 2020; Li 2014), and making investments in the environment work means rethinking priorities and fiduciary norms, often around metrics like corporate sustainability, carbon neutrality, or ESG (environment, social, governance). As Bregje van Veelen explains in a study on green finance for low-carbon agriculture, finance and environmental governance interact in plural ways, "new financial flows are not something 'done to' agriculture, but actively shaped and mediated by the 'agricultural field': the extraeconomic relations that shape and govern the agricultural sector" (2021, 136).

Conclusion

While the literature often considers the making of resources or property, neoliberal market-based environmental governance, and financialization as distinct processes, the actually-existing nature of timberland investment underscores the fact that nature's monetization is complex, distinct, and ever-changing. Many investor-owners capture revenue from the production of lumber and other forest products—a very traditional approach to making nature into a resource or commodity—alongside parceling off and selling property rights through hunting or recreational leasing, and participation in emerging environmental markets for carbon or wetland credits. Further still, the ever-changing nature of accumulation from the environment highlights the ways that nature/society relations, and physical environments, are constantly being remade to ensure they are amenable to ever-changing political economic conditions.

In response to these ever-changing conditions, there are many exciting new directions for political ecology scholarship on the monetization of nature. First, as scholars are becoming increasingly concerned with the question of just transitions, we see the question of energy landscapes and the making of new energy economies becoming increasingly critical. With the growing strategic importance of lithium for electric car batteries, as one example, we are seeing in real time the birth of a new high-value resource and the reconfiguration of nature/society relations in lithium hot spots (Riofrancos 2022; Hernandez and Newell 2022). Further still, the land-intensive nature of renewable energy technologies means the ongoing reconfiguration of property rights and extension of new enclosures, in ways that are likely to impact vulnerable rural communities around the globe (McCarthy 2015). Second, the emerging literature on infrastructure and storage highlight the critical importance of circulation—a relatively understudied aspect of Marx's oeuvre—to our understanding of how environments are monetized (Randle 2022; Cousins and Hill 2021). Finally, the growing interest in environmental repair by political ecologists (e.g., Huff and Brock 2023) offers another exciting new direction for research on the political economy of the environment. Harkening back to O'Connor's second contradiction of capitalism, the rise of a "growth economy of repair" raises key questions about how capitalism is able to profit from restoration as well as destruction of environments (ibid., 2).

For political ecologists, it can be dizzying to try and keep up with the ever-growing suite of markets and commodities being produced from the biophysical world. While we should not always take these new markets at face value, they warrant further study for what they reveal about nature/economy/state relations, and many scholars are doing important work on this front. Furthermore, while neoliberalism, for example, is archetypically understood as a retreat of the state, this chapter aims to underscore the fact that states play a vital role in this remaking of nature–society relations, and should be centered in our understandings of how broader and more novel natures come to circulate on markets, as resources, commodities, permits, and credits. Looking forward, political ecologists who are interested in nature's monetization should engage closely with the state, its tools, effects, and processes. Finally, it is also critical to remember that nature's monetization is always contextual, always partial, always incomplete, and ever-shifting. As Leigh Johnson (2015) reminds us through reference to the inconstant commodity status of camels for African pastoralist societies, there are also many environments and more-than-human communities that successfully exist in liminal spaces beyond the capitalist gaze (see also Chapter 18, this volume).

References

Andreucci, D., García-Lamarca, M., Wedekind, J., and Swyngedouw, E. (2017). "Value grabbing": A political ecology of rent. *Capitalism Nature Socialism*, *28*(3), 28–47.

Andrews, E., and McCarthy, J. (2014). Scale, shale, and the state: Political ecologies and legal geographies of shale gas development in Pennsylvania. *Journal of Environmental Studies and Sciences*, *4*, 7–16.

Arsel, M., and Büscher, B. (2012). Nature™ Inc.: Changes and continuities in neoliberal conservation and market-based environmental policy. *Development and Change*, *43*(1), 53–78.

Bakker, K. (2010). The limits of "neoliberal natures": Debating green neoliberalism. *Progress in Human Geography*, *34*(6), 715–735.

Bakker, K. J. (2004). *An Uncooperative Commodity: Privatizing Water in England and Wales*. Oxford University Press.

Bakker, K. (2005). Neoliberalizing nature? Market environmentalism in water supply in England and Wales. *Annals of the Association of American Geographers*, *95*(3), 542–565.

Banoub, D., Bridge, G., Bustos, B., Ertör, I., González-Hidalgo, M., and de los Reyes, J. A. (2021). Industrial dynamics on the commodity frontier: Managing time, space and form in mining, tree plantations and intensive aquaculture. *Environment and Planning E: Nature and Space*, *4*(4), 1533–1559.

Beer, C. M. (2023). "A cold, hard asset": Conservation resource spectacle in Chilean Patagonia. *Geoforum*, *143*, 103773.

Bellamy Foster, J. (1999). Marx's theory of metabolic rift: Classical foundations for environmental sociology. *American Journal of Sociology*, *105*(2), 366–405.

Bellamy Foster, J. (2000). *Marx's Ecology: Materialism and Nature*. NYU Press.

Bigger, P. (2018). Hybridity, possibility: Degrees of marketization in tradeable permit systems. *Environment and Planning A: Economy and Space*, *50*(3), 512–530.

Blaikie, P. and Brookfield, H. (1987). *Land Degradation and Society*. London: Methuen.

Borgias, S. L. (2018). "Subsidizing the State:" The political ecology and legal geography of social movements in Chilean water governance. *Geoforum*, *95*, 87–101.

Borras Jr, S. M., Hall, R., Scoones, I., White, B., and Wolford, W. (2011). Towards a better understanding of global land grabbing: An editorial introduction. *The Journal of Peasant Studies*, *38*(2), 209–216.

Borras Jr, S. M., Franco, J. C., Isakson, S. R., Levidow, L., and Vervest, P. (2016). The rise of flex crops and commodities: Implications for research. *The Journal of Peasant Studies*, *43*(1), 93–115.

Boyd, W., Prudham, W. S., and Schurman, R. A. (2001). Industrial dynamics and the problem of nature. *Society and Natural Resources*, *14*(7), 555–570.

Bracking, S. (2019). Financialisation, climate finance, and the calculative challenges of managing environmental change. *Antipode*, *51*(3), 709–729.

Bridge, G., Bulkeley, H., Langley, P., and van Veelen, B. (2020). Pluralizing and problematizing carbon finance. *Progress in Human Geography*, *44*(4), 724–742.

Brockington, D., and Duffy, R. (Eds.). (2011). *Capitalism and Conservation* (Vol. 45). John Wiley & Sons.

Bumpus, A. G., and Liverman, D. M. (2008). Accumulation by decarbonization and the governance of carbon offsets. *Economic Geography*, *84*(2), 127–155.

Büscher, B., Sullivan, S., Neves, K., Igoe, J., and Brockington, D. (2012). Towards a synthesized critique of neoliberal biodiversity conservation. *Capitalism Nature Socialism*, *23*(2), 4–30.

Campling, L., and Havice, E. (2014). The problem of property in industrial fisheries. *The Journal of Peasant Studies*, *41*(5), 707–727.

Castree, N. (2008a). Neoliberalising nature: The logics of deregulation and reregulation. *Environment and Planning A*, *40*(1), 131–152.

Castree, N. (2008b). Neoliberalising nature: Processes, effects, and evaluations. *Environment and Planning A*, *40*(1), 153–173.

Christophers, B. (2015). The limits to financialization. *Dialogues in Human Geography*, *5*(2), 183–200.

Christophers, B. (2018). Risking value theory in the political economy of finance and nature. *Progress in Human Geography*, *42*(3), 330–349.

Connors, J. P. C., and Rea, C. M. (2022). Violent entanglements. *Conservation and Society*, *20*(1), 24–35.

Cousins, J. J., and Hill, D. T. (2021). Green infrastructure, stormwater, and the financialization of municipal environmental governance. *Journal of Environmental Policy and Planning*, *23*(5), 581–598.

Dempsey, J. (2016). *Enterprising Nature: Economics, Markets, and Finance in Global Biodiversity Politics*. John Wiley & Sons.

Dempsey, J. (2017). The financialization of nature conservation?. *Money and Finance after the Crisis: Critical Thinking for Uncertain Times*, 191–216. John Wiley & Sons.

Dempsey, J., and Suarez, D. C. (2016). Arrested development? The promises and paradoxes of "selling nature to save it". *Annals of the American Association of Geographers*, *106*(3), 653–671.

Emel, J., and Huber, M. T. (2008). A risky business: Mining, rent and the neoliberalization of "risk". *Geoforum*, 39(3), 1393–1407.

Fairbairn, M. (2021). *Fields of Gold: Financing the Global Land Rush*. Cornell University Press.

Fannin, M. (2021). Human tissue economies: Making biological resources. In *The Routledge Handbook of Critical Resource Geography* (pp. 389–400). Routledge.

Felli, R. (2014). On climate rent. *Historical Materialism*, *22*(3–4), 251–280.

Gómez-Baggethun, E., De Groot, R., Lomas, P. L., and Montes, C. (2010). The history of ecosystem services in economic theory and practice: From early notions to markets and payment schemes. *Ecological Economics*, *69*(6), 1209–1218.

Gunnoe, A. (2014). The political economy of institutional landownership: Neorentier society and the financialization of land. *Rural Sociology*, 79(4), 478–504.

Gunnoe, A., and Gellert, P. K. (2011). Financialization, shareholder value, and the transformation of timberland ownership in the US. *Critical Sociology*, *37*(3), 265–284.

Hartwick, E. (1998). Geographies of consumption: A commodity-chain approach. *Environment and Planning D: Society and Space*, *16*(4), 423–437.

Harvey, D. (2003): *The New Imperialism*. Oxford: Oxford University Press.

Hernandez, D. S., and Newell, P. (2022). Oro blanco: Assembling extractivism in the lithium triangle. *The Journal of Peasant Studies*, *49*(5), 945–968.

Heynen, N., Kaika, M., and Swyngedouw, E. (Eds.). (2006). *In the Nature of Cities: Urban Political Ecology and the Politics of Urban Metabolism* (Vol. 3). Routledge.

Heynen, N., McCarthy, J., Prudham, S., and Robbins, P. (Eds.). (2007). *Neoliberal Environments: False Promises and Unnatural Consequences*. Routledge.

Huber, M. T. (2022). Resource geography III: Rentier natures and the renewal of class struggle. *Progress in Human Geography*, 46(4), 1095–1105.

Huff, A., and Brock, A. (2023). Introduction: Accumulation by restoration and political ecologies of repair. *Environment and Planning E: Nature and Space*, 6(4), 2113–2133.

Igoe, J. (2021). *The Nature of the Spectacle: On Images, Money, and Conserving Capitalism.* University of Arizona Press.

Isakson, S. R. (2017). Food and finance: The financial transformation of agro-food supply chains. In *New Directions in Agrarian Political Economy* (pp. 109–136). Routledge.

Johnson, L. (2015). Catastrophic fixes: Cyclical devaluation and accumulation through climate change impacts. *Environment and Planning A*, *47*(12), 2503–2521.

Johnson, L. (2022). Rents, experiments, and the perpetual presence of concessionary weather insurance. *Annals of the American Association of Geographers*, *112*(5), 1224–1242.

Kama, K. (2020). Resource-making controversies: Knowledge, anticipatory politics and economization of unconventional fossil fuels. *Progress in Human Geography*, *44*(2), 333–356.

Kass, A. (2020). Working with financial data as a critical geographer. *Geographical Review*, *110*(1–2), 104–116.

Kay, K. (2016). Breaking the bundle of rights: Conservation easements and the legal geographies of individuating nature. *Environment and Planning A: Economy and Space*, *48*(3), 504–522.

Kay, K. (2017). Rural rentierism and the financial enclosure of Maine's open lands tradition. *Annals of the American Association of Geographers*, *107*(6), 1407–1423.

Kay, K., and Kenney-Lazar, M. (2017). Value in capitalist natures: An emerging framework. *Dialogues in Human Geography*, *7*(3), 295–309.

Kish, Z., and Fairbairn, M. (2018). Investing for profit, investing for impact: Moral performances in agricultural investment projects. *Environment and Planning A: Economy and Space*, *50*(3), 569–588.

Knudson, C. (2018). One size does not fit all: Universal livelihood insurance in St. Lucia. *Geoforum*, 95, 78–86.

Knuth, S. (2018). "Breakthroughs" for a green economy? Financialization and clean energy transition. *Energy Research and Social Science*, *41*, 220–229.

Krippner, G. R. (2005). The financialization of the American economy. *Socio-economic Review*, *3*(2), 173–208.

Labban, M. (2010). Oil in parallax: Scarcity, markets, and the financialization of accumulation. *Geoforum*, *41*(4), 541–552.

Lansing, D. M. (2012). Performing carbon's materiality: The production of carbon offsets and the framing of exchange. *Environment and Planning A*, *44*(1), 204–220.

Lave, R. (2012). *Fields and Streams: Stream Restoration, Neoliberalism, and the Future of Environmental Science* (Vol. 12). University of Georgia Press.

Li, T. M. (2014). What is land? Assembling a resource for global investment. *Transactions of the Institute of British Geographers*, *39*(4), 589–602.

Li, T. M. (2011). Centering labor in the land grab debate. *The Journal of Peasant Studies*, *38*(2), 281–298.

Lockhart, A. (2015). Developing an offsetting programme: Tensions, dilemmas and difficulties in biodiversity market-making in England. *Environmental Conservation*, *42*(4), 335–344.

Loftus, A., March, H., and Purcell, T. F. (2019). The political economy of water infrastructure: An introduction to financialization. *Wiley Interdisciplinary Reviews: Water*, 6(1), e1326.

Lunstrum, E., and Massé, F. (2021). Conservation and the production of wildlife as resource. In *The Routledge Handbook of Critical Resource Geography* (pp. 358–368). Routledge.

Mansfield, B. (2004). Neoliberalism in the oceans: "rationalization," property rights, and the commons question. *Geoforum*, *35*(3), 313–326.

McAfee, K. (1999). Selling nature to save it? Biodiversity and green developmentalism. *Environment and Planning D: Society and Space*, *17*(2), 133–154.

McCarthy, J. (2004). Privatizing conditions of production: Trade agreements as neoliberal environmental governance. *Geoforum*, *35*(3), 327–341.

McCarthy, J. (2015). A socioecological fix to capitalist crisis and climate change? The possibilities and limits of renewable energy. *Environment and Planning A*, *47*(12), 2485–2502.

McCarthy, J., and Prudham, S. (2004). Neoliberal nature and the nature of neoliberalism. *Geoforum*, *35*(3), 275–283.

O'Connor, J. (1991). On the two contradictions of capitalism. *Capitalism Nature Socialism*, *2*(3), 107–109. doi: 10.1080/10455759109358463.

Osborne, T. (2013). Fixing carbon, losing ground: Payments for environmental services and land (in)security in Mexico. *Human Geography*, *6*(1), 119–133.

Ostrom, E. (1990). *Governing the Commons: The Evolution of Institutions for Collective Action.* Cambridge University Press.

Ouma, S. (2020). *Farming as Financial Asset: Global Finance and the Making of Institutional Landscapes* (p. 220). Agenda Publishing.

Ouma, S., Johnson, L., and Bigger, P. (2018). Rethinking the financialization of "nature". *Environment and Planning A: Economy and Space*, *50*(3), 500–511.

Peck, J., and Tickell, A. (2002). Neoliberalizing space. *Antipode* 34, 380–404.

Perry, N., and Gillespie, J. (2019). Restricting spatial lives? The gendered implications of conservation in Cambodia's protected wetlands. *Environment and Planning E: Nature and Space*, 2(1), 73–88.

Pistor, K. (2019). *The Code of Capital: How the Law Creates Wealth and Inequality*. Princeton University Press.

Polanyi, K. (2001 [1944]). *The Great Transformation: The Political and Economic Origins of Our Time*. Beacon Press.

Purcell, T. F., Loftus, A., and March, H. (2020). Value–rent–finance. *Progress in Human Geography*, *44*(3), 437–456.

Randle, S. (2022). Holding water for the city: Emergent geographies of storage and the urbanization of nature. *Environment and Planning E: Nature and Space*, 5(4), 2283–2306.

Riofrancos, T. (2022). The security–sustainability nexus: Lithium onshoring in the Global North. *Global Environmental Politics*, 1–22.

Robertson, M. M. (2006). The nature that capital can see: Science, state, and market in the commodification of ecosystem services. *Environment and Planning D: Society and Space*, 24(3), 367–387.

Robertson, M. (2012). Measurement and alienation: Making a world of ecosystem services. *Transactions of the Institute of British Geographers*, *37*(3), 386–401.

Robertson, M. M., and Wainwright, J. D. (2013). The value of nature to the state. *Annals of the Association of American Geographers*, *103*(4), 890–905.

Schwartzman, G. (2022). Climate rentierism after coal: Forests, carbon offsets, and post-coal politics in the Appalachian coalfields. *The Journal of Peasant Studies*, *49*(5), 924–944.

Smith, N. (2007). Nature as accumulation strategy. *Socialist Register*, *43*.

Smith, N. (2010). *Uneven Development: Nature, Capital, and the Production of Space*. University of Georgia Press.

St. Martin, K. (2001). Making space for community resource management in fisheries. *Annals of the Association of American Geographers*, *91*(1), 122–142.

Turner, M. D. (2017). Political ecology III: The commons and commoning. *Progress in Human Geography*, *41*(6), 795–802.

Turton, D. J. (2015). Lawyers in Australia's coal seam gas debate: A study of participation in recorded community forums. *The Extractive Industries and Society*, 2(4), 802–812.

van Veelen, B. (2021). Cash cows? Assembling low-carbon agriculture through green finance. *Geoforum*, *118*, 130–139.

Watts, M. J. (2013). *Silent Violence: Food, Famine, and Peasantry in Northern Nigeria* (Vol. 15). University of Georgia Press.

11 Protecting Nature

Political Ecologies of Conservation through the Lens of Peace Parks

Maano Ramutsindela

Introduction

The South African based *Getaway Magazine* has been instrumental in publicizing peace parks in Southern Africa for tourism purposes. In 1996, it published the map of peace parks in the region in red but subsequently changed the colour to green to emphasize the conservation ecotourism value of these parks. The political meaning of this change was an attempt to delink peace parks from the colonial project in which the imperialist Cecil John Rhodes painted Africa British red from the Cape to Cairo. The colour of peace parks maps has since remained green. These protected areas form part of the movement for rewilding different parts of the world (Corlett 2016; Cromsigt et al. 2018; Pettorelli et al. 2019) and are embraced by conservationists and non-governmental organizations as a progressive approach to nature conservation (Westing 1993; Busch 2008; Liu et al. 2020). They however raise broader questions of the colonial histories of conservation, conservation practices in global environmental crises, and the multiple proposals for the future of humanity and the planet. These questions are helpful for interrogating nature conservation beyond its simplistic goal of establishing, expanding, and managing protected areas.

Conservation has become a powerful tool for governing humans and nonhumans and continues to fundamentally (re)configure socioecological relations with huge implications for livelihoods, identities, international relations, and the (un)making of states (Escobar 1998; Dalby 2014; Death 2016; O'Lear 2018). All these are enabled by the mobilization of rationalities such as species extinction, the planetary emergency, state incapacity, problematic local communities/Indigenous peoples, and so on. These rationales not only assist in universalizing conservation thinking but also ground conservation practices that enjoy the support of non-governmental organizations, governments, and capitalists. The functions of conservation areas have expanded to include diplomatic relations, environmental governance, nation-building, and capitalist expansion (Ramutsindela 2004; West 2006; Okereke 2007; Death 2016; Kay and Kenney-Lazar 2017; Fletcher 2023). Protected areas are also sites of struggles over the control and use of natural resources and the consequent violence (Peluso and Watts 2001; Corson 2011; Ybarra 2012; Millner 2020).

A counternarrative to the violent histories of protected areas has emerged in the form of peace parks: protected areas established across the borders of two or more states often anchored on a national park. There are contested views on where the concept of a peace park originated, how it spread, and where it was first implemented (Ellis 1994; Chester

DOI: 10.4324/9781003165477-15

2006; Dorsey 1998; DaimlerChrysler 2001; Ali 2007). Such contestations centre around the terminology and its projected meanings, as well as official protocols associated with the establishment of peace parks. Geographically, peace parks can be established within and across the land and maritime borders of states. In this discussion I focus on cross-border protected areas variously referred to as transborder park, transboundary protected area (TPA), transnational park, peace park, and so on. The World Conservation Union (IUCN) estimated that TPAs increased from 59 in 1980 to 188 in 2005, and these continue to increase exponentially around the world (Zbicz and Green 1997; IUCN WCPA Transboundary Conservation Specialist Group 2023). A common feature of peace parks is their reliance on adjacent protected areas that either exist or must be created. Thus, the increase in the number of peace parks reflects an increase in that of protected areas and their geographic proximity.

Peace parks are a special designation of TPAs and are framed as new types of parks by expanding the conventional functions of a protected area (i.e., the protection of biodiversity) to include the promotion of peaceful co-existence among neighbouring nation states and fostering human interactions and development through ecotourism (Lejano 2006; Ali 2007; Quinn et al. 2012). They are presented as an avenue for advancing environmental diplomacy or a green approach to conflict resolution (Westing 1998; Roulin et al. 2017), despite their negative effects on intra- and inter-state relations (Ide 2020). Meanwhile, the cross-border nature of peace parks enables anarchist, scientific, romantic, managerial, and neoliberal discourses to thrive; all pursuing different environmental, economic, and political agendas (Wolmer 2003). The pursuit of various agendas through nature conservation is not unique to peace parks but is central to studies in political ecology. Conservation and resource control is considered one of the five big questions in political ecology, which draws attention to environmental regimes, livelihoods, and political action (Adams and Hutton 2007; Robbins 2019; Ponte et al. 2022). Conservation raises questions about constructing and policing ideas about space, humans, and nonhumans, and the consequent socioecological relations (Adams 2020). These questions are relevant to peace parks, and they broadly highlight many of the central and recurring themes in political ecology. In the following sections I use peace parks to foreground key themes shaping the political ecology of conservation research. These themes are: (Neo) colonialisms of protected areas, the role of borders and regions, the form and implications of environmental diplomacy, the relationality of conservation and development, and the shades of environmental philanthropy.

The (Neo)Colonialisms of Protected Areas

Conservationists and NGOs present the recent upsurge in the development of TPAs as an important and timely approach to the environmental crises. Yet, TPAs have a colonial history inseparable from the histories of protected areas more generally. Notwithstanding Dlamini's (2020) caution against universalizing the history of conservation, the colonial dimensions of protected areas can be identified in the former colonies and in the treatment of Indigenous populations in conservation spaces in highly industrialized countries. The colonialisms of protected areas is integral to the processes of extending extra-territorial political power by European countries, the subjugation of Indigenous populations, and the continuation of these processes through various mechanisms in the post-independence era (Memmi 1974; Said 1993; Mbembe 2001). Mignolo's (2011) concept of coloniality captures the multiple dimensions and mutations of colonialism under different contexts.

Two perspectives are helpful for understanding the (neo)colonialisms of protected areas. The first links protected areas to diverse (post)colonial processes. The establishment of protected areas was bound up with colonial processes in that they have been integral to the project of colonial conquest and plunder, they constituted geopolitical zones of friction between rival imperial actors, they institutionalized borders and regions of the empire, and formed part of Europe's civilizing mission (Gissibl et al. 2012; Ramutsindela 2020; Collins et al. 2021). They constitute "the colonial metanarrative of human hegemony over nature" (Coates 1998: 18). The hegemonic colonial concept of a national park has become the gold standard for the protection, use, and management of nature. The idea of a national park circulates through images of pristine nature empty of people, who ironically constitute a threat to the wilderness. The consequences of imposing such an idea on non-Western societies include the disruption of existing socioecological relations and livelihoods, the criminalization of local natural resource users through regulatory frameworks and management regimes, and the selective protectionist approach favouring certain animals, plants, and landscapes. In broad terms, protected areas are instruments of control. In post-independence, such control continues through the conditionalities of foreign aid, the militarization of conservation, and the transfer of control over natural resources from Southern states to foreign governments, private companies, local elites, and NGOs (Fairhead et al. 2012; Corson et al. 2013; Duffy et al. 2019).

The second perspective is that the exploitation of nonhumans and humans goes hand in hand (Adorno 2006; Collins et al. 2021). Nature is generally colonized through the imposition of totalizing Western views and ideologies of nature, the control and domination of nature by humans, and the transformation of landscapes into images that serve human interests. These processes are enabled by control over natural resources, including land. Land alienation has been central to the colonial project: it facilitated the establishment of settler colonies, the subjugation of Indigenous populations, and access to natural resources for industrialization. Land theft not only resulted in dual economies anchored on property regimes, but also transformed tillers of the land into cheap labour and tenants (Howitt et al. 1996; Stead and Altman 2019; Obeng-Odoom 2021). Though land dispossession inspired liberation struggles and the emergence of social movements, questions related to landownership and access to natural resources more broadly remain unresolved in most former colonies after independence (Moyo et al. 2012). Political ecologies of conservation should appreciate these histories to understand how protected areas mute debates on the coloniality of conservation and its enhancement by land/green grabbing, and the complicity of the state in the violence against its citizens (Fairhead et al. 2012; Ramutsindela et al. 2022).

The coloniality of conservation manifests in biocolonialism, in which multinational companies search for plants, animals, and human genes mostly in the Global South as raw material for "the multi-billion dollar industries of the twenty-first century biotechnology revolution" under the pretext that profits from patenting plants and genes would be shared with Indigenous and local communities (Ashcroft et al. 2013: 29). Such biopiracy is a form of colonial theft (Shiva 2016). The coloniality of conservation is further expressed through multiple refusals, including the state's refusal to return the ownership of land to local communities and Indigenous people that had been violently expropriated for protected areas without compensation.

Borders and Regions

The establishment of transborder protected areas and their use in regional integration projects open the possibilities for integrating analytical heuristic devices and concepts relevant to political ecology and political geography. The relationship between these two fields has been made explicit by reference to the political geography of the environment and the inclusion of political ecology in the revised aims and scope of the journal *Political Geography* (Benjaminsen et al. 2017). Geographic concepts such as regions and borders could be used to explore and express the intersection of the two fields, mainly because the concepts are interdisciplinary and flexible in their usage. Regions are understood as "social constructs based on social practice and discourse, and this is the real basis to evaluate their roles and functions" (Paasi et al. 2018: 3). The concept of a region has been used discursively, analytically, and materially to account for social, political, economic, and environmental relations in space. Its construction, performative nature, and relationality have been scrutinized in both political ecology and political geography.

Whereas perspectives in political ecology sought to understand regions in relations to natural resources, political geographers have deployed regions to reflect on, among other things, political economy, and regional security arrangements. For example, Regional Security Complex theorists focus on security as a defining element of region-formation with emphasis on the security of member states. In the realm of environmentalism, the concept of region has been deployed to understand bioregions and the appropriate structures and infrastructures to manage them. However, bioregionalism has also been deployed to advocate the return of social, political, and economic power to small-scale and self-governing regional communities, including secessionists and anarchists (Wolmer 2003; Evanoff 2018).

Research on new regionalism highlights that regions are not the product of states alone, as non-state actors and other external forces and processes are also involved in shaping the region at multiple levels. This scalar politics is linked to the governance of regions and could lead to shifting conceptions and functions of regions (Hettne and Söderbaum 2000; Hameiri 2013; Paasi et al. 2018). In the realm of the environment the new regionalism appears in the form of a supra-state entity with a spatial form of governance that involves multiple interest groups. Policy makers mobilize the entity towards regional integration and cross-border regionalization. To understand this process, we need to pay attention to the intersection of border politics and nature conservation, and the consequent political ecology of the border. Such a political ecology is "constitutive of border politics and human-environment dynamics, and it is driven by an environmental logic that largely hinges on certain endowed or presumed attributes of borders and borderlands" (Ramutsindela 2017: 107). There is a need to grasp how the border as one of the key concepts in (political) geography and widely used in social science is mobilized for conservation projects. The border continues to play an important role in shaping power relations, land use planning, and governance in transborder spaces, including TPAs (Dallimer and Strange 2015; Trillo-Santamaria and Pauel 2016), hence delimiting transborder areas requires attention to multiple factors (Medeiros 2020).

Border discourses and narratives are involved in the creation of cross-border conservation areas (Grichting and Zebich-Knos 2017). They frame the border as an impediment to the natural flow of the nonhuman and the management of ecological landscapes. In doing so, they denounced state borders long before the activists' political project of No Borders was publicized in a special issue of *Refuge: Canada's Journal on Refugees* in 2009.

The project focuses on the inequalities and injustices of the world system using refugees and migrants as reference points (Anderson et al. 2009). Nonetheless, conservation-driven border narratives essentialize the border as a fixed physical line in sharp contrast to conceptions of borders as spaces and processes, and a product of a set of practices (Johnson et al. 2011). Conservationists exploit the attributes of borderlands as marginal spaces with social, political, and economic dichotomies and variation in property regimes to establish cross-border protected areas with new sets of borders. Thus, imaginations of cross-border protected areas as borderless landscapes mask the new borders underpinning them.

Environmental Diplomacy

Diplomacy is a power-driven political activity facilitated by communication and influence on nation states. Its goal is to achieve agreements while protecting the interests of the parties involved (Berridge 2022). International relation scholars, political scientists, and legal studies have paid greater attention to diplomacy at the United Nations General Assembly to understand the evolution of resolutions, political posture, and negotiation strategies (Berridge and Jennings 1985; Comras 2010; Jones and Clark 2019). Global attention to environmental challenges has led to the emergence of environmental diplomacy. Such diplomacy captures the strategies for achieving commitments to environmental goals by self-interested states, encompasses dispute resolutions, and the prevention of conflict over environmental resources (Susskind 1994; Tolba 2008; Orsini 2020). The voluntary commitments by nation states are backed by science, they signify compromises on national interests (sometimes achieved through differential obligations) and are induced through selective incentives.

Environmental diplomacy also entails using the environment as a means for peacebuilding. A case in point is the attempt to resolve the historic political tension between North Korea and South Korea by turning the demilitarized zone separating the two countries into a peace park (Brady 2021; Kim 2022). It has been suggested that snow leopard peace parks should be created in central and south Asia to resolve the enduring conflict in that region (Maheshwari 2020). Similar attempts have been made to resolve the political tension between India and Pakistan by establishing the Siachen Peace Park in the Kashmir region (Wani et al. 2022). The proposal for the Russo-Japanese Peace Park in the Kuril Islands follows the same logic (Lambacher 2007). Environmental challenges are enrolled in the discourse of peace that links environmental degradation with conflict, though the association between the two has been questioned (Dalby 2014). Resolution of environmentally-induced conflict and attempts to save the planet are considered peace-building efforts that are worthy of recognition by the Nobel Peace Prize Committee as evidenced by the award of this Prize to the environmental activist Wangari Maathai in 2004. This way, the world's most prestigious prize services environmental diplomacy by rewarding past achievements as well as investing on the impact of the awardees (Lundestad 2019). These examples demonstrate that the linkages between the environment and peacebuilding bring together studies in international relations (IR), peace, geopolitics, and security.

The conventional view of environmental diplomacy focuses on inter-state interactions on environmental policy that gained traction at the Earth Summit in 1992 (Ali and Vladich 2016). It also influences assessments of environmental diplomacy (Susskind and Ali 2014; Li et al. 2020). The focus on environmentally mediated interactions between states is grounded on the belief that the most important environmental challenges are more global than regional or local and should be resolved at the global level (Susskind 1994; Tolba 2008). However, the global/local dichotomy is misleading as powerful states might conceal their national interests under the guise of global concerns to mobilize international support. The broadening of the concept of environmental diplomacy to include managing local conflicts over values, identities and distribution (Ali and Vladich 2016) has not fundamentally changed state centric approaches and is also limited to conflict resolution. A meaningful analysis of diplomacy in political ecologies of conservation should include paradiplomacy at sub-national level, and the various forms of environmental negotiations (formal and informal) that are not driven by conflict for at least three reasons. First, it is at the local level where the outcomes of global environmental agreements are felt. For example, the resettlement of local communities/Indigenous groups to achieve global biodiversity targets is a localized experience backed by national and international actors. Second, the sustainability of conservation cannot be achieved without the support of local inhabitants, and this is acknowledged through attempts to give local communities and Indigenous people a voice in international environmental summits/gatherings. Third, there have been negotiations over access to high value natural resources in communal lands. For example, the private sector persuades local communities to enter ecotourism ventures with the promise of shared profit. These examples point to the need to conceptualize, evaluate, and theorize environmental diplomacy beyond the IR contexts.

Box 11.1 Green violence

The responses to poaching and the controversial links between terrorism and the illegal wildlife trade has led to the rise in the militarization of conservation, which entails the deployment of the army to protected areas and the development and application of military style approaches and tactics to protect nature (Lunstrum 2014; Duffy et al. 2019). The militarization of conservation is not a new phenomenon but represents the intensification in the use of armed forces in conservation spaces. It also constitutes the material aspect of green violence. More broadly, green violence refers to "the deployment of violent instruments and tactics towards the protection of nature and ideas and aspirations related to nature conservation" (Büscher and Ramutsindela 2016: 10). Materially, green violence disrupts the livelihood of ordinary people living in the vicinity of protected areas, transforms professional rangers into paramilitary units, and incorporate local inhabitants into intelligence networks. Green violence turns protected areas into spaces of exception to turn a blind eye to human rights abuses, which include murder, unlawful arrests, and the use of new surveillance technologies that infringe on people's privacy (Sandbrook 2015). It is "characterised by dehumanising practices targeting non-western societies, especially marginalised cultural groups whose way of life do not chime with instituted protected areas and the ideologies of conservation" (Ramutsindela et al. 2022: 41). Green violence also takes a social form. Social violence denotes social orders are used at various levels to protect one group while exposing another to risks and vulnerabilities. Foreign governments, conservation lobby groups, and individuals (ab)use social power to protect nature at the expense of other people and institutions. Social platforms have been used to instigate, express, and mobilize groups to violence in conservation spaces.

Conservation and Development

Scholars have interrogated the relationship between conservation and development and how this has shifted under different conditions with varying consequences for humans and nonhumans (Martinez-Alier 2002; Adams 2008; Brockington and Wilkie 2015; Sloan et al. 2019). Development and its underpinning ideology of civilizing "the other" has been criticized for its adverse effects on the environment and on colonial subjects. For environmentalists, development projects constituted a threat to the protection of species and landscapes hence the negative outcomes of Western oriented forms of development spurred the rise of environmental and social movements. The unhealthy relations between developmentalism and environmentalism was mitigated by reframing the two as mutually beneficial and found expression in the concept of sustainable development, which gained traction since the 1970s. Conservationists are divided on the implications of sustainable development for protected areas. Some hold the view that conservation should focus on the protection of nature and should not be burdened by the task of improving the living conditions of people (Terborgh 1999). They see development projects as a waste of the much-needed resources and as falling outside the primary conservation mandate. This view underestimates the dynamic links between conservation and poverty (Adams et al. 2004).

Others hold the alternative view that sees worsening human conditions as a threat to conservation projects and argue that the success of protected areas is intertwined with human wellbeing (Folke 2006; Erbaugh et al. 2020). They consider ecotourism as a win-win solution for conservation and poverty alleviation through local economic development, and as an enterprise that would inspire community interests in protected areas and an avenue for forging different forms of partnerships necessary for depressed economies (Brandon 1996; Stronza et al. 2019). Ecotourism has expanded beyond protected areas to include geological features, hence geotourism and the resultant geoparks have developed as an approach to tourism (Dowling and Newsome 2006, 2018; Zouros 2016). Geotourism offers a platform to celebrate all aspects of nature and bridges the gap between earth sciences and tourism studies.

Attempts to cement the environment-development nexus through responsible tourism have been accompanied by the penetration of capitalism into protected areas. This has led to the neoliberalization of conservation as part of the broader process of neoliberalism anchored on market-based solutions to environmental problems (free market environmentalism), and guided by principles and practices of privatization, marketization, commodification, decentralization, deregulation, and reregulation (Harvey 1996; McAfee 1999; Igoe and Brockington 2007; Castree 2008; Fletcher 2023). As a form of TPA, peace parks are a conservation inspired business model that relies on frontier capitalism (Ramutsindela 2007; Büscher 2013), especially the tourism industry (Chiutsi and Saarinen 2017; Więckowski 2021). Proponents of peace parks argue that they increase revenue from tourism than from protected areas operating in isolation and that they offer the best business to shareholders (DaimlerChrysler 2001; Hanks 2003; Hanks and Myburgh 2015). They enable neoliberal conservation by denationalizing environmental assets to free them from state control and to create transnational spaces of accumulation through regulations in the form of treaties and memoranda of understanding.

The neoliberalization of peace parks fits into the general understanding of the ways in which the nonhuman world becomes subjected to market forces to facilitate the expansion of capitalism (Büscher 2013). Freehold land is a necessary condition for these processes because of its collateral effect and as a factor of production (in the Marxist sense). Private landowners and companies have capitalized on the establishment of peace parks to gain from land transactions and tourism ventures while conservation NGOs use them for fund-raising. Neoliberal conservation speaks to the intertwined logics of conservation and capitalism, namely capitalism's search for new markets in protected areas and conservationists' view of capitalism as a mechanism to finance conservation policies, strategies, and practices (Holmes 2012; Beer 2022). These logics reconcile the dialectical relationship between capitalism and nature conservation and forge the alliance between the two. As the section below will illustrate philanthropy facilitates and strengthens this alliance.

Environmental Altruism: Unmasking Neoliberal Ideology

The creation of the philanthropic movement, *The Giving Pledge*, by Warren Buffett, Melinda French Gates, and Bill Gates in 2010 commits philanthropists to give "the majority of their wealth to charitable causes either during their lifetimes or in their wills" (https://givingpledge.org). This movement makes the roles of philanthropists in the twenty-first century more visible than in the past by asking each member to write a letter explaining the reasons for making the pledge and by publishing the letters online. Though *The Giving Pledge* confirms philanthropy as a product of the generosity of the human spirit for the public good, it continues to raise questions about the power of philanthropy. The main critiques of philanthropy revolve around politics and economy. Politically, philanthropy is seen (in the Bourdieusian and Weberian sense) as a mechanism used by the bourgeoisie to deploy its accumulated capital for interclass struggles over status and cultural capital, and to align society with its interest (Ramutsindela et al. 2011). Powerful states mobilize philanthropy to advance their geopolitical goals. The case in point is the promotion of US foreign policy interests in Africa in the post-war era through philanthropic foundations like Ford, Rockefeller, and Carnegie (Barker 2008). The power of philanthropic organizations makes them what Eikenberry and Mirabella (2018) call a neoliberal voluntary state. It is a state made up of elite philanthropists and businesspeople rather than elected officials who "decide who receives social welfare and support and how it is received" (Eikenberry and Mirabella 2018: 43).

In the economic domain, critiques of philanthropy are captured by the concept of philanthrocapitalism, which refers to "the growing role for private sector actors in addressing the biggest social and environmental challenges facing the planet" (Bishop and Green 2015: 541). Philanthrocapitalism represents a shift from a religious, ethical or moral conscience to help those in need to a business strategy within a capitalist system (Ramutsindela et al. 2011; Liu and Baker 2016). Though philanthrocapitalism is as diverse as capitalism, it has been instrumental in promoting a form of democracy supportive to a free-market economy, mitigating the inequalities and social ills created by capitalism, and influencing policies (Eikenberry and Mirabella 2018; Beer 2022). It is also viewed as an "ideological framework that proposes its own diagnoses and prognoses" (Mediavilla and Garcia-Arias 2019: 857). The Social Impact Bond (SIB) illustrates the workings of such a framework. The SIB involves private investors raising money to fund a non-profit effort to tackle a problem which saves the government money. If successful, the government uses the savings to pay the private investor a financial return (Bishop and Green 2015). The involvement of philanthropists in protecting nature has led to the emergence of environmental philanthropy as a subfield of philanthropy focusing on preserving nature and promoting conservation-related activities at various scales while also shaping conservation outcomes and influencing environmental governance (Beer 2022; Betsill et al. 2022).

Conclusion

The concepts of borders, diplomacy, philanthropy, development, and (neo)colonialisms are bound up with peace parks but also with political ecologies of conservation more broadly. They are useful for forging and deepening the links between political ecology and political geography. In this chapter I have shown that peace parks provide a platform for expanding the scope of political ecology through the assemblage of concepts. This assemblage broadens analyses and critiques of socioecological systems and processes and draws attention to the relevance of social nature to other fields of inquiry. The political ecology of peace parks reveals forms of (neo)colonialisms that transform geographic spaces into conservation spaces aligned with the interests and aspirations of dominant powers and institutions. This process invokes border discourses to achieve two related objectives: to deconstruct national space and to construct a transnational space through the medium of conservation (Ramutsindela 2017). The emergent spatial configuration is a microregion, which appeals to regionalists, large-scale landscape projects, and entrepreneurs, including capitalists. This challenges any simplistic analysis of regions as either a subject of ecology or politics and invites us to consider regions in their problematic complexity (Walker 2016).

This chapter affirms that peace parks are a neocolonial artefact. Their colonialism is not only rooted in the colonial histories of protected areas that form core areas of transborder conservation but includes discriminatory conservation practices and outcomes that reflect, and are embedded in, inequalities and racism. Post-independence and the "new conservation" have sustained these practices hence the coloniality of peace parks is visible at the conceptual level as well as on the ground. Conceptually, transborder peace parks are a reinvention of the idea of a park with its own borders. On the ground, they exhibit forms of violence reminiscent of protected areas more generally. The nomenclature of parks for peace masks their material, social, and symbolic violence (Büscher and Ramutsindela 2016).

What sets peace parks apart from other protected areas is the intensity, types, and scale of diplomacy involved but also required. Negotiations by nation states to solve international environmental problems have resulted in various environmental agreements. Such negotiations are considered low politics in which commitments to an environmental cause are made through persuasion rather than the high politics of war. NGOs and conservation entrepreneurs broker negotiations for peace parks between states and between states and local inhabitants with the financial backing of philanthropists. Political ecological research should harness themes emerging from peace parks to enrich political ecologies of conservation and to broaden the scope of both political ecology and political geography and their intersectionality.

References

Adams, B., 2008. *Green development: Environment and sustainability in a developing world.* London: Routledge.

Adams, W.M., 2020. Geographies of conservation III: Nature's spaces. *Progress in Human Geography* 44(4): 789–801.

Adams, W.M. and Hutton, J., 2007. People, parks and poverty: Political ecology and biodiversity conservation. *Conservation and Society*, 5(2): 147–183.

Adams, W.M., Aveling, R., Brockington, D., Dickson, B., Elliott, J., Hutton, J., Roe, D., Vira, B. and Wolmer, W., 2004. Biodiversity conservation and the eradication of poverty. *Science*, 306(5699): 1146–1149.

Adorno, T.W., 2006. *History and freedom: Lectures 1964–1965*. Cambridge: Polity.
Ali, S.H. (ed). 2007. *Peace parks: Conservation and conflict resolution*. Cambridge, MA: MIT Press.
Ali, S.H. and Vladich, H.V., 2016. Environmental diplomacy. In Constantinou, S.M., Kerr, P. and Sharp, P. (eds). *The Sage handbook of environmental diplomacy*, pp. 601–616. London: Sage.
Anderson, B., Sharma, N. and Wright, C., 2009. Why no borders? *Refuge*, 26, p. 5.
Ashcroft, B., Griffiths, G. and Tiffin, H., 2013. *Post-colonial studies: The key concepts*. London: Routledge.
Barker, M., 2008. The liberal foundations of environmentalism: Revisiting the Rockefeller-Ford connection. *Capitalism Nature Socialism*, 19(2): 15–42.
Beer, C.M., 2022. Bankrolling biodiversity: The politics of philanthropic conservation finance in Chile. *Environment and Planning E: Nature and Space*, p. 25148486221108171.
Benjaminsen, T.A., Buhaug, H., McConnell, F., Sharp, J. and Steinberg, P.E., 2017. Political geography and the environment. *Political Geography*, 100(56): A1–A2.
Berridge, G.R., 2022. *Diplomacy: Theory and practice*. Cham: Springer Nature.
Berridge, G. and Jennings, A. eds., 1985. *Diplomacy at the UN*. London: Macmillan.
Betsill, M.M., Enrici, A., Le Cornu, E., and Gruby, R.L., 2022. Philanthropic foundations as agents of environmental governance: A research agenda. *Environmental Politics*, 31(4): 684–705.
Bishop, M. and Green, M., 2015. Philanthrocapitalism rising. *Society*, 52(6): 541–548.
Brady, L.M., 2021. From war zone to biosphere reserve: The Korean DMZ as a scientific landscape. *Notes and Records*, 75(2): 189–205.
Brandon, K., 1996. *Ecotourism and conservation: A review of key issues*, World Bank Group. United States of America. Retrieved from https://policycommons.net/artifacts/1460416/ecotourism-and-conservation/2101349/ on 26 February 2023.
Brockington, D. and Wilkie, D., 2015. Protected areas and poverty. *Philosophical Transactions of the Royal Society B: Biological Sciences*, 370(1681), p. 20140271.
Busch, J., 2008. Gains from configuration: The transboundary protected area as a conservation tool. *Ecological Economics*, 67(3): 394–404.
Büscher, B., 2013.*Transforming the frontier: Peace parks and the politics of neoliberal conservation in Southern Africa*. Durham NC: Duke University Press.
Büscher, B. and Ramutsindela, M. 2016. Green violence: Rhino poaching and the war to save southern Africa's peace parks. *African Affairs* 115(458): 1–22.
Castree, N., 2008. Neoliberalising nature: The logics of deregulation and reregulation. *Environment and Planning A*, 40(1): 131–152.
Chester, C.C., 2006. *Conservation across borders: Biodiversity in an independent world*. Washington: Island Press.
Chiutsi, S. and Saarinen, J., 2017. Local participation in transfrontier tourism: Case of Sengwe community in great Limpopo transfrontier conservation area, Zimbabwe. *Development Southern Africa*, 34(3): 260–275.
Coates, P. 1998. *Nature*. Oxford: Blackwell.
Collins, Y.A., Macguire-Rajpaul, V, Krauss, J.E., Asiyanbi, A., Jiménez, A., Bukhi Mabele, M., and Alexander-Owen, M. 2021. Plotting the coloniality of conservation. *Journal of Political Ecology*, 28(1). https://doi.org/10.2458/jpe.4683
Comras, V.D., 2010. *Flawed diplomacy: The United Nations and the war on terrorism*. Potomac Books.
Corlett, R.T., 2016. The role of rewilding in landscape design for conservation. *Current Landscape Ecology Reports*, 1: 127–133.
Corson, C., 2011. Territorialization, enclosure and neoliberalism: Non-state influence in struggles over Madagascar's forests. *Journal of Peasant Studies*, 38(4): 703–726.
Corson, C., MacDonald, K.I. and Neimark, B., 2013. Grabbing "green": Markets, environmental governance and the materialization of natural capital. *Human Geography*, 6(1): 1–15.
Cromsigt, J.P., Te Beest, M., Kerley, G.I., Landman, M., le Roux, E., and Smith, F.A., 2018. Trophic rewilding as a climate change mitigation strategy? *Philosophical Transactions of the Royal Society B: Biological Sciences*, 373(1761), p. 20170440.
DaimlerChrysler, 2001 *The parks of peace*. Stuttgart: DaimlerChrysler AG.
Dalby, S., 2014. Environmental geopolitics in the twenty-first century. *Alternatives*, 39: 3–16.
Dallimer, M. and Strange, N., 2015. Why socio-political borders and boundaries matter in conservation. *Trends in Ecology and Evolution*, 30(3): 132–139.

Death, C., 2016. *The green state in Africa*. New Haven: Yale University Press.
Dlamini, J.S., 2020. *Safari nation: A social history of the Kruger National Park*. Athens OH: Ohio University Press.
Dorsey, K., 1998. *The dawn of conservation diplomacy: US–Canadian wildlife protection treaties in the progressive era*. Seattle: University of Washington Press.
Dowling, R. and Newsome, D. eds., 2018. *Handbook of geotourism*. Edward Elgar.
Dowling, R.K. and Newsome, D. eds., 2006. *Geotourism*. London: Routledge.
Duffy, R., Massé, F., Smidt, E., Marijnen, E., Büscher, B., Verweijen, J., Ramutsindela, M., Simlai, T., Joanny, L., and Lunstrum, E., 2019. Why we must question the militarisation of conservation. *Biological Conservation*, 232: 66–73.
Eikenberry, A.M. and Mirabella, R.M., 2018. Extreme philanthropy: Philanthrocapitalism, effective altruism, and the discourse of neoliberalism. *Political Science and Politics*, 51(1): 43–47.
Ellis, S., 1994. Of elephants and men: Politics and nature conservation in South Africa. *Journal of Southern African Studies*, 20: 53–69.
Erbaugh, J.T., Pradhan, N., Adams, J., Oldekop, J.A., Agrawal, A., Brockington, D., Pritchard, R. and Chhatre, A., 2020. Global forest restoration and the importance of prioritizing local communities. *Nature Ecology and Evolution*, 4(11): 1472–1476.
Escobar, A., 1998. Whose knowledge, whose nature? Biodiversity, conservation, and the political ecology of social movements. *Journal of Political Ecology*, 5(1): 53–82.
Evanoff, R., 2018. Bioregionalism. In Castree, N., Hulme, M. and Proctor, J.D. *Companion to Environmental Studies*, pp. 13–16. London: Routledge.
Fairhead, J., Leach, M. and Scoones, I., 2012. Green grabbing: A new appropriation of nature? *Journal of Peasant Studies*, 39: 237–261.
Fletcher, R., 2023. *Failing forward: The rise and fall of neoliberal conservation*. University of California Press.
Folke, C., 2006. The economic perspective: Conservation against development versus conservation for development. *Conservation Biology*, 20(3): 686–688.
Gissibl, B., Höhler, S. and Kupper, P. eds., 2012. *Civilizing nature: National parks in global historical perspective*. New York: Berghan.
Grichting, A. and Zebich-Knos, M. eds., 2017. *The social ecology of border landscapes*. London: Anthem Press.
Hameiri, S., 2013. Theorising regions through changes in statehood: Rethinking the theory and method of comparative regionalism. *Review of International Studies*, 39(2): 313–335.
Hanks, J., 2003. Transfrontier Conservation Areas (TFCAs) in Southern Africa: Their role in conserving biodiversity, socioeconomic development and promoting a culture of peace. *Journal of Sustainable Forestry*, 17(1–2): 127–148.
Hanks, J. and Myburgh, W., 2015. The evolution and progression of transfrontier conservation areas in the Southern African Development Community. In van der Duim, R., Lamers, M. and van Wijk, J. eds., *Institutional arrangements for conservation, development and tourism in Eastern and Southern Africa: A dynamic perspective*, pp. 157–179. Dordrecht: Springer.
Harvey, D., 1996. *Justice, nature and the geography of difference*. Oxford: Blackwell.
Hettne, B. and Söderbaum, F., 2000. Theorising the rise of regionness. *New Political Economy*, 5(3): 457–472.
Holmes, G., 2012. Biodiversity for billionaires: Capitalism, conservation and the role of philanthropy in saving/selling nature. *Development and Change*, 43 (1): 185–203.
Howitt, R., J. Connell and P. Hirsh, eds., 1996. *Resources, nations and Indigenous Peoples*. Melbourne: Oxford University Press.
Ide, T., 2020. The dark side of environmental peacebuilding. *World Development*, 127: 104777.
Igoe, J. and Brockington, D., 2007. Neoliberal conservation. *Conservation and Society*, 5(4): 432–449.
IUCN WCPA Transboundary Conservation Specialist Group, 2023. Overview and description. Availablefrom:www.iucn.org/our-union/commissions/group/iucn-wcpa-transboundary-conservation-specialist-group# (accessed, 27 February 2023).
Johnson, C., Jones, R., Paasi, A., Amoore, L., Mountz, A., Salter, M., and Rumford, C., 2011. Interventions on rethinking 'the border' in border studies. *Political Geography*, 30(2), 61–69.
Jones, A. and Clark, J., 2019. Performance, emotions, and diplomacy in the United Nations assemblage in New York. *Annals of the American Association of Geographers*, 109(4): 1262–1278.

Kay, K. and Kenney-Lazar, M., 2017. Value in capitalist natures: An emerging framework. *Dialogues in Human Geography*, 7(3): 295–309.
Kim, E.J., 2022. *Making peace with nature: Ecological encounters along the Korean DMZ*. Durham NC: Duke University Press.
Lambacher, J., 2007. Nesting cranes: Envisioning a Russo–Japanese peace park in the Kuril Islands. In S.H. Ali ed., *Peace parks: Conservation and conflict resolution*. MIT Press.
Lejano, R.P., 2006. Theorizing peace parks: Two models of collective action. *Journal of Peace Research*, 43(5): 563–581.
Li, G., Zakari, A. and Tawiah, V., 2020. Does environmental diplomacy reduce CO_2 emissions? A panel group means analysis. *Science of The Total Environment*, 722: 137790.
Liu, H. and Baker, C., 2016. Ordinary aristocrats: The discursive construction of philanthropists as ethical leaders. *Journal of Business Ethics*, 133: 261–277.
Liu, J., Yong, D.L., Choi, C.Y. and Gibson, L., 2020. Transboundary frontiers: An emerging priority for biodiversity conservation. *Trends in Ecology and Evolution*, 35(8): 679–690.
Lundestad, G., 2019. *The world's most prestigious prize: The inside story of the Nobel Peace Prize*. Oxford: Oxford University Press.
Lunstrum, E., 2014. Green militarization: Anti-poaching efforts and the spatial contours of Kruger National Park. *Annals of the Association of American Geographers*, 104(4): 816–832.
McAfee, K., 1999. Selling nature to save it? Biodiversity and green developmentalism. *Environment and Planning D: Society and Space*, 17(2): 133–154.
Maheshwari, A., 2020. Ease conflict in Asia with snow leopard peace parks. *Science*, 367(6483): 1203–1203.
Martinez-Alier, J., 2002. *The environmentalism of the poor: A study of ecological conflicts and evaluation*. Cheltenham: Edward Elgar.
Mbembe, A., 2001. *On the postcolony*. Berkeley: University of California Press.
Mediavilla, J. and Garcia-Arias, J., 2019. Philanthrocapitalism as a neoliberal (development agenda) artefact: Philanthropic discourse and hegemony in (financing for) international development. *Globalizations*, 16(6): 857–875.
Medeiros, E., 2020. Delimiting cross-border areas for policy implementation: A multi-factor proposal. *European Planning Studies*, 28(1): 125–145.
Memmi, A., 1974. *The colonizer and the colonized*. London: Earthscan.
Mignolo, W., 2011. *The darker side of Western modernity: Global futures, decolonial options*. Durham NC: Duke University Press.
Millner, N., 2020. As the drone flies: Configuring a vertical politics of contestation within forest conservation. *Political Geography*, 80: 102163.
Moyo, S., Yeros, P. and Jha, P., 2012. Imperialism and primitive accumulation: Notes on the new scramble for Africa. *Agrarian South: Journal of Political Economy*, 1(2): 181–203.
Obeng-Odoom, F., 2021. *The commons in an age of uncertainty: Decolonizing nature, economy, and society*. Toronto: University of Toronto Press.
Okereke, C., 2007. *Global justice and neoliberal environmental governance: Ethics, sustainable development and international co-operation*. London: Routledge.
O'Lear, S., 2018. *Environmental geopolitics*. Lanham: Rowman & Littlefield.
Orsini, A., 2020. Environmental diplomacy. In Balzacq, T., Charillon, F., and Ramel, F. eds. *Global diplomacy: An introduction to theory and practice*, pp. 239–257. London: Palgrave Macmillan.
Paasi, A., Harrison, J., and Jones, M. 2018. New consolidated regional geographies. In Paasi, A., Harrison, J. and Jones, M. (eds) *Handbook on the geographies of regions and territory*. Cheltenham: Edward Elgar.
Peluso, N.L. and Watts, M. eds., 2001. *Violent environments*. Ithaca NY: Cornell University Press.
Pettorelli, N., Durant, S.M. and Du Toit, J., 2019. *Rewilding*. Cambridge: Cambridge University Press.
Ponte, S., Noe, C., and Brockington, D., 2022. *Contested sustainability: The political ecology of conservation and development in Tanzania*. Oxford: James Currey.
Quinn, M.S., Broberg, L., and Freimund, W., 2012. *Parks, peace, and partnership: Global initiatives in transboundary conservation* p. 576. University of Calgary Press.
Ramutsindela, M., 2004. *Parks and people in postcolonial societies: Experiences in Southern Africa*. Dordrecht: Kluwer.

Ramutsindela, M., 2007. *Transfrontier conservation in Africa: At the confluence of capital, politics and nature*. Wallingford: CABI.

Ramutsindela, M., 2017. Greening Africa's borderlands: The symbiotic politics of land and borders in peace parks. *Political Geography* 56: 106–113.

Ramutsindela, M., 2020. National parks and (neo)colonialisms. In M. Legun, J. Keller, M. Bell and M. Carolan eds. *The Cambridge handbook of environmental sociology Vol 1*, pp. 206–222. Cambridge: Cambridge University Press.

Ramutsindela, M., Matose, F. and Mushonga, T. eds., 2022. *The violence of conservation in Africa: State, militarization and alternatives*. Cheltenham: Edward Elgar.

Ramutsindela, M., Spierenburg, M. and Wels, H., 2011. *Sponsoring nature: Environmental philanthropy for conservation*. New York: Earthscan.

Robbins, P., 2019. *Political ecology: A critical introduction*. John Wiley & Sons.

Roulin, A., Rashid, M.A., Spiegel, B., Charter, M., Dreiss, A.N., and Leshem, Y., 2017. "Nature knows no boundaries": The role of nature conservation in peacebuilding. *Trends in Ecology and Evolution*, 32(5): 305–310.

Said, E., 1993. *Culture and imperialism*. London: Vintage.

Sandbrook, C., 2015. The social implications of using drones for biodiversity conservation. *Ambio*, 44(Suppl 4): 636–647.

Shiva, V., 2016. *Biopiracy: The plunder of nature and knowledge*. Berkeley: North Atlantic Books.

Sloan, S., Campbell, M.J., Alamgir, M., Engert, J., Ishida, F.Y., Senn, N., Huther, J. and Laurance, W.F., 2019. Hidden challenges for conservation and development along the Trans-Papuan economic corridor. *Environmental Science and Policy*, 92: 98–106.

Stead, V. and Altman, J., 2019. *Labour lines and colonial power: Indigenous and Pacific Islander labour mobility in Australia*. Canberra: Australian National University Press.

Stronza, A.L., Hunt, C.A., and Fitzgerald, L.A., 2019. Ecotourism for conservation? *Annual Review of Environment and Resources*, 44: 229–253.

Susskind L.E., 1994. *Environmental diplomacy: Negotiating effective environmental agreements*. New York NY: Oxford University Press.

Susskind, L.E. and Ali, S.H., 2014. *Environmental diplomacy: Negotiating more effective environmental agreements, 2nd edn*. New York NY: Oxford University Press.

Terborgh, J., 1999. *Requiem for nature*. Washington DC: Island Press.

Tolba, M.K., 2008. *Global environmental diplomacy: Negotiating environmental agreements for the world, 1973–1992*. Cambridge MA: MIT Press.

Trillo-Santamaria, J.M. and Pauel, V., 2016. Transboundary protected areas as ideal tools? Analyzing the Gerês-Xurés transboundary biosphere reserve. *Land Use Policy*, 52: 454–463.

Walker, P.A., 2016. On "Reconsidering Regional Political Ecologies" 13 years on. *Journal of Political Ecology*, 23(1), 123–125.

Wani, M.U.D., Dada, Z.A. and Shah, S.A., 2022. Building peace through tourism: The analysis of an ongoing Siachen Glacier dispute between India and Pakistan. *Asian Journal of Comparative Politics*, 7(4): 836–848.

West, P., 2006. *Conservation is our government now: The politics of ecology in Papua New Guinea*. Durham NC: Duke University Press.

Westing, A.H., 1993. Biodiversity and the challenge of national borders. *Environmental Conservation*, 20: 5–6.

Westing, A.H., 1998. Establishment and management of transfrontier reserves for conflict prevention and confidence building. *Environmental Conservation*, 25, 91–94.

Więckowski, M., 2021. How border tripoints offer opportunities for transboundary tourism development. *Tourism Geographies*, 1–24.

Wolmer, W., 2003. Transboundary conservation: The politics of ecological integrity in the Great Limpopo Transfrontier Park. *Journal of Southern African Studies*, 29(1): 261–278.

Ybarra, M., 2012. Taming the jungle, saving the Maya Forest: Sedimented counterinsurgency practices in contemporary Guatemalan conservation. *Journal of Peasant Studies*, 39(2): 479–502.

Zbicz, D.C. and Green, M.J., 1997. Status of the world's transfrontier protected areas. *Parks*, 7(3): 5–10.

Zouros, N., 2016. Global geoparks network and the new UNESCO Global Geoparks Programme. *Bulletin of the Geological Society of Greece*, 50(1): 284–292.

12 Degrading Nature

Production and the Hidden Ecology of Capital

Matthew T. Huber

Introduction: Inside the Hidden Ecology of Capital

The ammonia production facility hovers over the small riverside town with its giant white plumes of steam enveloping the surrounding air. This facility alone was responsible for the highest level of greenhouse gas emissions in the entire US chemical sector. The bulk of this is CO_2 from both burning natural gas and a process called "steam reforming" which produces hydrogen gas (H_2) from methane (CH_4). This hydrogen is combined with nitrogen freely available in the atmosphere to make ammonia (NH_3).

While on a tour of the facility, I asked the guide to show me where the CO_2 was vented. It was a thin and totally harmless looking vent pipe. The most significant form of ecological degradation coming from this facility is barely discernible. As such, the environmental justice activists in the region were unaware of the massive carbon footprint of the facility and rightly prioritized other facilities that pose more immediate and deadly toxic threats to local workers and residents.

Given its massive contribution to the crisis of global heating this exemplified a "political" ecology. But I wondered whether or not the managers of this gigantic carbon metabolism worried about their "carbon footprint" the same way carbon conscious consumers do. When I asked if they considered alternatives to the source of carbon emissions (natural gas), they excitedly answered yes. During the early-to-mid 2000s, natural gas prices spiked which forced many nitrogen plants to close in the United States. The managers proudly explained how efficiency measures taken at their plant allowed them to survive the crisis, but forced them to contemplate replacing natural gas with coal or pet coke (two of the dirtiest and most carbon intensive feedstocks one could imagine).

It then dawned on me that the people who controlled this "ecology" were not like you and me. They were not concerned with their carbon footprint. They were concerned with how to turn fossil fuels and air into fertilizer commodities sold for profit. Just as Marx ([1867] 1976: 280) argued the "secret of profit-making" required entering the "hidden abode of production" this too is the secret to understanding the *ecology of capital.* Although much of the ecological degradation in the world occurs outside the factory gates, the *source* of degradation is often private owners of production. Whether we are looking at mining (Arboleda 2020), farmland investors (Fairbairn 2020), or fishery conglomerates (Campling and Colás 2021), the root cause of much ecological degradation is often capital seeking a return (M-C-M') via a production process that turns nature into surplus value.[1]

This poses two problems for political ecology. First, our theories of the ecology of capitalism often don't focus on production, which is often assumed as a more "orthodox"

DOI: 10.4324/9781003165477-16

domain of traditional Marxist capital-labor relations. Second, it poses *methodological* questions of how we research production. It is of course a domain of private property surrounded by fences, security, and as Marx quipped, a sign, "No admittance except on business." While I was extremely lucky to gain admittance, this is not usually an option for political ecological researchers.

In this chapter, I trace what a political ecology of industrial production might look like. First, I revisit the "classical" political ecological approach to environmental degradation (Blaikie and Brookfield 1987) and suggest widening our view of who counts as a "land manager." Second, I argue an understanding of the ecology of capital must not only pay attention to the "concrete" processes Marx famously detailed between humanity and nature in the labor process, but also the *abstract* flow of value (M-C-M') defining the valorization process. Third, I review ecological theories of capital and their avoidance of production. Finally, I examine the methodological challenges for researching the "hidden abode" of production with concrete examples and challenges from my own nitrogen fertilizer research.

Before we get into the details, I want to review some key concepts in what follows. At its core, Marxism and its general analytical framework of historical materialism is concerned with the differences in how societies organize the *production* of material life and subsistence. Thus, for historical materialists production is an ecological concept or, as Ellen Meiksins Wood (1986: 188) observed, "production is essential to human existence." Yet, the bulk of Marx's critique of capitalism is how material production becomes hijacked by another abstract process of *valorization*, or the subjection of material production to the competitive exigencies of value and the imperative to produce more than you started with (i.e. profit or surplus value). Marx insisted we can only understand where this surplus comes from by examining this "hidden abode" of production—hidden because it is literally private property out of view of the public. For Marx, investigating this hidden abode reveals horrific worker exploitation. My contention here is that political ecologists can also learn about the secrets of ecological degradation by studying this hidden abode of production—and such investigation will also unveil horrific environmental exploitation. This kind of empirical work is also important theoretically. While we are accustomed to treating ecology as a "second contradiction of capitalism" (O'Connor 1998)—where capital undermines its own *conditions of production* external to the hidden abode—we need to theorize the ecology of capital as internal to the logic of production itself.

Land Degradation and Society Revisited

The classical political ecological approach to environmental degradation was laid out by Piers Blaikie and Harold Brookfield in *Land Degradation and Society* (1987). They argued traditional research approaches that examine rural environmental degradation like deforestation or soil erosion tend to blame the most marginalized peasants and other smallholders. Alternatively, Blaikie and Brookfield (1987, 27) suggested we must situate "land managers" in wider "chains of explanation" that highlight major national and global structural forces outside their control like commodity markets, state policies, and the international flows of capital. It is these political economic forces—not marginalized land practices—that must be at the center of any analysis of environmental degradation.

The wager of this chapter is we need not only focus on the marginalized rural communities near sites of degradation to understand the role of capital in ecological degradation.

Moreover, we can also pay attention to other agents—the managers of capital itself—who shape investments that ultimately hit the ground in destructive ways (Arboleda 2015). This class of investors are an altogether different but more powerful set of global ecological managers who can be studied alongside the "land managers" more proximate to degraded landscapes. Of course, Marx aimed for a critique that avoids blaming individual capitalists themselves focusing more on the "laws of motion" of capital itself. Do these laws include ecological attributes?

The Dual Character of Ecology Under Capital: Valorization and Metabolism

Before we could confront the political ecology of capital we must ask the basic question—what *is* capital? Marx had two answers. First, "capital is not a thing, it is a definite social relation of production pertaining to a particular historical social formation" (Marx [1894] 1981: 953). More specifically, the capital relation is a historically specific class relationship between capitalists and wage labor. But there is an even more basic definition of capital *as a process*. In Chapter 4 of *Capital* Volume 1, Marx ([1867] 1976: 247–257) lays out the "General Formula of Capital" as simply M-C-M'. Capitalists start the process with a sum of money (M), invest that money in commodities (C) (namely means of production and labor power) and hope to come out at the end with more money (M'). Capital is simply a process of money making more money, or as Marx described, "the ceaseless augmentation of value" (Marx [1867] 1976: 254).

In Volume 2, Marx ([1885] 1978) offers a more involved formula characterized by three circuits of capital: (1) the circuit of money-capital populated by banks, financial and other money capitalists, (2) the circuit of productive-capital where labor power and means of production are put to work producing commodities, (3) the circuit of commodity-capital where commodities are sold on the market. Figure 12.1 represents a schematic diagram of the three circuits.[2] From an ecological perspective, there is really only one moment in the process where capital touches down to earth to confront natural systems—the productive circuit. The money circuit is a world almost fully subsumed in the value form (money)—although its actors wield massive power over what actually happens to ecologies around the world. The commodity circuit includes the sale of material commodities—varieties of "processed natures"—but also might include various forms of "nature experiences" like eco-tourism (Ojeda 2012; Douglas 2014; Devine and Ojeda 2017), or even "green" commodities like carbon offsets (Bumpus and Liverman 2008; Osborne 2015; Bryant 2018). While production includes manufacturing, the knowledge economy, and many other supposedly humanized spaces, all forms of socionatural production take place through and with nature such as primary industries like agriculture

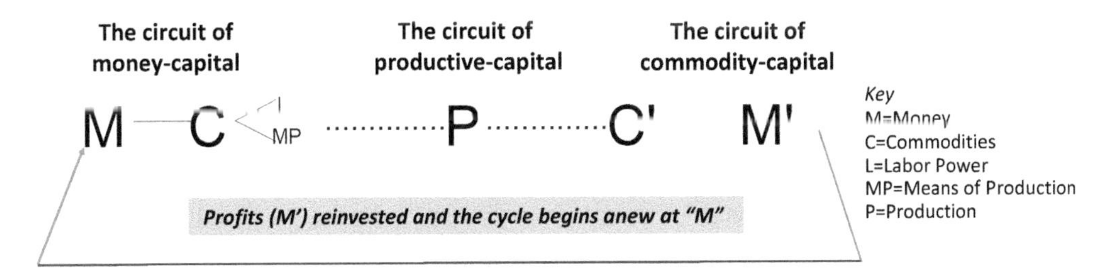

Figure 12.1 Marx's three circuits of capital.

(Kansaga et al. 2019; Vercillo 2022), mining (Marston and Perreault 2017; Arboleda 2020), timber (Ekers 2015; Kay 2017), fishing (Barbesgaard 2018; Campling and Colás 2021), and other extractive industries. In an editorial published two decades ago, Gavin Bridge and Andrew Jonas (2002: 761) proposed renewing a focus on such moments of production: "the direct appropriation of natural processes in each of these sectors (as raw material, waste sink, or as in agriculture the production process itself) renders the metabolism of nature and economy very visible." Nevertheless, an undue attention to primary production risks ignoring the massive ecological impacts of secondary industrial production. All forms of production contain ecological linkages and effects.

In sum, if ecology is ultimately an understanding of real, material *life processes*, then it is clear the fulcrum of the ecology of capital exists in what Marx called the "hidden abode of production." This is significant because Marx ([1867] 1976: 280) also insists we must focus on production to understand the "secret of profit-making" in the exploitation of living labor and the production of surplus value. He insisted much of classical political economy focused instead on the realm of market exchange. This is not simply relevant to nineteenth-century ideologues who idolize the realm of market exchange as a zone of freedom, choices, and equal property rights. Similarly today we are told all our ecological problems can be solved via the realm of market exchange and pricing ecosystem services or pollution—a topic of much political ecological research as of late (McAfee 1999; Heynen et al. 2007; Dempsey and Robertson 2012; Osborne 2015; Bryant 2018).

Much like Marx uncovered brutal labor exploitation, the hidden abode of production is also a source of *ecological* exploitation. It is quite clear production has the most ecological impacts—whether it be the industrial sector's massive share of carbon emissions (Huber 2022), or industrial capital in the primary industries clearcutting, monocropping, or mining the earth, and despoiling soil, air and water in the process. The hidden abode of production can include factories, farms, forests, and mines, but we can also include a software office building or a film production studio—any geographical place where capital puts workers and means of production *to work* producing commodities for sale. These spaces are always material and thus enmeshed in ecological networks.

In the deindustrialized Global North, one gets the impression it is only consumption and "ecological footprints" that matter in terms of ecological impact. But all these moments of immoral consumerism can be traced back to forms of capital seeking profit through the *amoral* logic of M-C-M'. Under capitalism, consumers, or more accurately workers, merely meet their needs with commodities, whereas capital seeks to turn money into more money with sheer indifference to the ecological consequences. So what can we say about the ecology of capital that hinges upon the "hidden abode" of production? It is tempting to fixate on right after Marx announces our entry to the hidden abode with his description of the "labour process" where, "Labour is, first of all, a process between man and nature..." (Marx [1867] 1976: 283).[3] Despite the rhetorical flair of this section—and its rich descriptions of the metabolic interchange between labor and nature—Marx's descriptions in this section are transhistorical. That is, they are meant to lay out some very basic points about how *all societies*—capitalist or not—are required to labor through nature to produce and reproduce life. This general idea on labor and nature is meant to set up a more historically specific examination of the labor process or production under capitalism. For capitalist production it is *not* the metabolic interchange of labor and nature that predominates (even if it can't be escaped), but rather the abstract movement of value or what he calls *the valorization process*. Just like the commodity has a dual character between use and exchange value, and labor a dual character between abstract and concrete, the production process too has a dual character between the labor process (metabolism) and the valorization process (M-C-M').

Marx is clear that the capitalist's primary focus is on valorization and they are largely *indifferent* to the material or concrete consequences of production. "Here we are no longer concerned with the quality, the character and the content of the labour, but merely with its quantity. And this simply requires to be calculated" (Marx [1867] 1976: 296). Thus, the valorization process is defined by what Derek Sayer (1987) called the *violence* of abstraction—subjecting the lived, concrete world to the rule of abstract value and profitmaking. As Kohei Saito (2017: 93) argues this does not mean capital can escape the materiality of the nature; quite the contrary it is capital's indifference that intensifies the ecological crisis itself. Returning to our fertilizer capital in the introduction, if natural gas prices imperil those profits, why not shift to coal or petcoke? The local and planetary effects are not of importance here. This—the subjection of material ecological production to the abstract logic of M-C-M'—is the heart of the ecology of capital.

Theorizing the Ecology of Capital

An ecological approach to production would rely on core Marxist theories of exploitation and the contradiction between the forces and relations of production. I have argued elsewhere (Huber 2022)—along with Hampton (2015)—that Marx's theory of exploitation based on relative surplus value is helpful in understanding how capital cheapens commodities through (largely) fossil-fuel based automation (with obviously massive ecological effects).

There are certain Marxist frameworks which do focus on production. Neil Smith's ([1984] 2008: 49–91) "production of nature" thesis situates nature and environmental crisis more broadly squarely within the dynamics of value and accumulation. However, despite the fact "production" is in the name of the thesis, there is little direct analysis of the dynamics of exploitation and surplus value in this elaboration. Smith pays much more attention to the importance of the rule of markets and geographical expansion under the "law of value" (see pp. 82–85). Another example is Boyd et al.'s (2001) articulation of the formal vs. real subsumption of nature under capital—a framework for understanding how capital shifts from simply extracting nature as it exists to actively producing the biological conditions of life itself (e.g., genetic engineering and other forms of "biocapital," see Rajan 2006). Yet, this formulation tends to assume Marx's more basic analysis of the formal vs. real subsumption of *labor* has little relevance to ecological analysis.

Meanwhile much Marxist ecological theory suggests we must look beyond production, or as Nancy Fraser (2014) has put it "behind Marx's hidden abode." This is despite the fact there exists a robust literature that examines the interplay of workers and ecology in the production of commodities (e.g., Hurley 1995; Bridge and Jonas 2002; Prudham 2005; Ekers 2015; Appel 2019; Werner 2022). James O'Connor's (1998) articulation of capital's *second* contradiction between capital and the "conditions of production" explicitly focused on ecological processes as "external conditions" to production. In this enumeration, O'Connor assumes the first contradiction—squarely focused on the relations and forces of production—is not ecological, or, at least, not useful to theorizing "new" environmental social movements. Thus, the first contradiction is one more useful to "old" socialist and labor movements who are assumed ill-equipped for the new ecological crisis.

Similarly, Nancy Fraser (2014) insists an ecological analysis of capitalism must go beyond production to interrogate the "background conditions" that allow capital to function in the first place (which also include care work of social reproduction and political investments in infrastructure). Like O'Connor—and her deep debt to his work is more substantially acknowledged later in Fraser (2021: 101)—she claims the "elders" of socialist or labor struggles are "burdened with blinders" and "have largely failed...to incorporate the insights of ecological thought systematically into their understanding of capitalism" (Fraser 2014: 56).

Additionally, Jason Moore's (2015a, 2015b) work is remarkably similar in its insistence that the ecology of capital lies *outside* of production and value relations—the traditional focus of Marxist economic analysis. For Moore, as with O'Connor and Fraser,[4] the focus should more properly focus on the "life-making" realms of care work and ecological life processes (see also, Mies 1986; Salleh 1997; Barca 2020). In fact, Moore sees the realms of production/exploitation—and the work of waged labor—as fundamentally separate from the "unpaid work/energy" of nature in the "web of life." Moore (2015b: 6) lays out this dualism clearly:

> I take paid work (capitalization) to be the domain of the capital-labor conflict over shares of value. This the question of exploitation. I take unpaid work to be a struggle over the forms and relations of capital to unmonetized social reproduction (as in "domestic labor") and to the "work of nature."[5]

To be clear, I don't want to imply all this work is not useful in examining "ecology" as something that is indeed "outside" production relations. Ultimately it is absolutely true that ecological processes and other "life-making" *does* take place outside the circuits of value and the hidden abode of production (e.g., childcare, soil microbes, water filtration, atmospheric cycles). More to the point, it is also clear that capital systematically destroys these "background conditions" upon which all life depends. Yet, it is also clear that the *root cause* of much of this degradation can often be traced to the traditional focus of Marxist analysis: exploitative production for profit. Thus, to understand the source of ecological degradation, we can make much more use of traditional Marxist analysis of surplus value, exploitation, and the relations/forces of production. In other words, while many assume *pace* Fraser, it is an elderly and outdated Marxism that must revise and update its theories to incorporate ecological thought, and we should consider that ecological thought could learn a lot from more traditional Marxist analysis.

Finally, there is one other framework for understanding the ecology of capitalism that *does* insist traditional Marxism offers an ecological critique of capitalism: the metabolic rift school (see Foster et al. 2010; Foster and Burkett 2016; Saito 2017; Foster 2020). This approach finds evidence of an ecological critique in Marx's own writings that capitalism is fundamentally based on a disruption of the flows of waste and nutrients through ecological systems. While Marx applied this to agriculture and soil science, it can also be applied to climate change and other cases where capitalism produces waste or pollution in excess of the capacities of natural systems to absorb them (see Clark and York 2005). There is much to learn from this approach, but it tends to assume that the real and true "ecological" Marx must be discovered deep in Volume 3 of *Capital* or in obscure unpublished notebooks (Saito 2017). While I would not deny the utility of these breakthrough insights on Marx's ideas on soil, ecology and metabolism, my point is there is much to be gained from an *ecological* analysis of more traditional Marxist categories. So how do we *study* the "hidden abode" of production?

Doing Research on Production: Barriers and Options

Marx's critique of political economy condemns bourgeois economics for only focusing on the supposed freedom of market exchange. Going beyond exchange requires going beyond the *surface* of economic life: "this noisy sphere, where everything takes place on the surface and in full view of everyone" (Marx [1867] 1976: 279). It also requires entering production as a zone of *class power* where capital rules. The rule of capital is made stark with Marx's depiction of a sign that says "No admittance except on business" (ibid: 280).

This is obviously a *methodological* problem for those political ecologists who wish to study production. Much political ecology studies "marginalized" populations and environmental social movements (McCarthy 2002). This indeed comes with its own ethical dilemmas in terms of research access and sharing of findings in ways that can aid wider struggles (Nager and Geiger 2007). Notwithstanding these challenges, such populations might often be quite willing to share their perspectives with ally scholar-activists. The challenge of "studying up" (Nader 1972; Robbins 2002) can present different kinds of problems: intransigence and lack of access. This refusal is backed by the legal power of property (over both "intellectual" and "material" aspects of production). Freedom of enterprise means capital is under zero obligation to speak to researchers. Below I examine several strategies for overcoming these barriers.

Entering the "Hidden Abode"

If you want to research the "hidden abode" of production, one methodological strategy is to gain access to sites of production. But this is not without its challenges. Readers might be surprised I gained access to a private corporate nitrogen fertilizer facility. The reason is both random and somewhat ridiculous luck. I simply drove up to a facility in a cluster of chemical facilities near Sarnia, Ontario (relatively close to where I live). I wore my only suit and tried to marshal my PhD credentials as simply a scholar interested in how the nitrogen production process works (of course, this is accurate). I was quite surprised when the manager agreed to let me in. Upon welcoming me, he bluntly declared, "Listen, normally we would never allow someone like you in here, but it just so happens my daughter goes to Syracuse University." It was through this contact that I got wider access to the company and other facilities in regions I was more interested (specifically the US Gulf Coast).

While this luck was just that, I do think a lesson is it is worth *trying* to gain access to production facilities. I can say it was extremely stressful driving up and making my pitch, but it did pay off. Even if you can't gain access to a facility there are obviously other ways to *learn* about what goes on inside. You could attempt to gain access to workers who work in the facility outside the property of the facility itself. I also found community leaders, environmental activists and local reporters were extremely fruitful sources of information. These sources can speak not only to what goes on inside production, but its wider impacts on workers, residents, and the environment outside.

Political-Industrial Ecology: Go with the Flow

You can learn a lot about what happens at a particular factory by studying general information on how industrial processes work in a given sector. Industrial ecology offers a sophisticated set of tools to trace the flows of materials, energy, and wastes through "industrial metabolism" research (Ayres 1989; Fischer-Kowalski 2003). For example, nitrogen fertilizer production relies on the process of chemical transformation known as "ammonia synthesis" which requires certain thermodynamic conditions of heat and pressure to occur (see Huber 2017a). In other words, it takes a significant amount of energy to simply generate the heat and pressure required. It also requires material "feedstocks" for the key elements of ammonia—nitrogen and hydrogen (NH_3). While decarbonization efforts project a shift to "green hydrogen" the vast majority of hydrogen production worldwide still relies on either natural gas or coal as the hydrogen feedstock (Tabuchi 2021). An industrial ecology approach would also point to waste byproducts of these processes. For ammonia synthesis reliant on steam reforming to create hydrogen via natural gas, the most eyebrow raising waste product is, unsurprisingly, carbon dioxide. As such, ammonia production accounts for about 1.8% of global carbon emissions (Royal Society 2020: 4); and these numbers do not account for the significant methane leakage as gas flows to and within a given facility (see Zhou et al. 2019).

It should be acknowledged that industrial ecology approaches fit squarely into the long tradition of "apolitical ecologies" (Robbins 2020). Simply tracing the material flows of the industrial metabolism tells us little about the social and political relations of those flows. Moreover, the fulcrum of the ecology of capital perspective laid out above is that capital is largely *indifferent* to the material nature of production and more concerned with the "immaterial but objective" flows of abstract value (Harvey 2010: 33).

Newell et al. (2017) advocate exactly this kind of "political-industrial ecology" which would use the tools of industrial ecology like material flows and life-cycle analysis toward more explicitly political analysis. For example, Deutz et al. (2017) show how high alkaline wastes from a variety of industrial processes are subject to political struggles over regulatory control and efforts to create a more "circular economy." My own analysis (Huber 2017b) argues for a class analysis rooted in the political power of those who own the means of production—and thus control and profit from the critical flows of material and energy inputs and wastes (see also Deutz 2014). My research on nitrogen capital reveals an industry who treats fossil fuel-based hydrogen production as a fixed 100-year-old technology that cannot be escaped even when clean alternatives do exist (like electrolysis of water). More importantly, the industry stridently argues against any national regulation of its prime waste product: carbon dioxide. When speaking with industry think tank policy experts they referred to such regulation as "carbon constraints" and argued such regulation is unfair unless applied to the entire planet.

Corporate Conferences: Where Capital Meets and Wants to Talk

If capital won't allow you into the hidden abode, you can seek out more public facing venues. Industrial conferences represent sites where capital's spokespeople meet and talk out in the open. These conferences not only feature rich presentations on industrial strategy, environmental regulatory challenges and market conditions, the participants also come *eager* to talk about their fields of expertise.

In November 2014 I attended the Fertilizer Outlook and Technology Conference in Savannah, Georgia. At the conference, I gained crucial insights about how the industry thinks about and manages its many ecological challenges. First, many presentations on sustainability or environmental stewardship made it entirely clear the industry sees their environmental problems—mostly focused on eutrophication and water pollution from fertilizer runoff[6]—as a problem of *farmer behavior* and not the industrial production of fertilizer itself. Since most water pollution—and a sizable amount of carbon emissions—revolve around farmer application of fertilizers, the industry consistently promoted a narrative that ecological problems could be solved by reforming farmer behavior. I saw several presentations on "precision agriculture" and the importance of the "Four Rs" of nutrient management—ensuring farmers use the right source, rate, time, and place of application.[7] These narratives seek to politically *displace* industrial production from the terrain of politics. I also was able to approach industry spokespeople for interviews. One interviewee stunningly revealed the contradiction of the industry's purported efforts to improve farmer efficiency. I asked her if such efforts to get farmers to *lower* demand contradicted the industry's goal to maximize production and profits. She admitted she hears this a lot from ammonia retailers: "Everybody says 'Why is the industry doing that if it is potentially reducing their ... sales?'" This is where the logic of capital trumps the ecological logic of efficiency and stewardship: all the fertilizer "wasted" through runoff and pollution means *more* sales and profits for those who own and control production. There is really no serious reason for the industry to dramatically reform farmer behavior.

Secondary Source Research: Listen to Capital

Finally, the most obvious way to learn about the hidden abode of production is to examine the industry's own voluminous written output like websites, annual reports, media statements, and press releases. Another rich source of data is trade journals where experts aligned with the industry publish reports and opinions. These trade journals can also give us a *historical archive* of industry strategy. I have been particularly interested in the massive expansion of fertilizer consumption in the postwar United States, and subsequently around the world under the banner of the "Green Revolution" (Patel 2013). When I read trade journals like *Agricultural Chemicals* and *Agricultural Ammonia News*,[8] I found an industry concerned about surplus capacity and eager to expand markets and consumption in the early years after World War II. An editorial in *Agricultural Chemicals* titled "Nitrogen: A Study of Productive Capacity" (Agricultural Chemicals 1954: 55) explained, "How long can we keep expanding output before anhydrous ammonia, and as a matter of fact nitrogen in all forms, will be coming out our ears?" Despite this alarm, the journal remained confident that "no oversupply should result if it is used by the American farmer at rates that are currently recommended as economically sound" (p. 149). These trade journals are also inundated with advertisements aimed toward farmers on the benefits of fertilizer. The industry used such narratives and marketing to actively produce the expanding market it needed to fuel the accumulation of nitrogen capital. This is information you actually wouldn't understand by simply entering the hidden abode, or tracing the material flows. The ecology of capital seeking M-C-M' operates in a world of markets and money with urgent needs to develop strategies to secure continued conditions of profitability.

Conclusion

When one thinks of the environmental crisis of fertilizer one imagines irresponsible farmers dumping excess ammonia onto crops and letting the greenhouse gases kill the climate and runoff destroy water systems. We simply don't think of the production facilities that I visited in my field work. Similarly, when one imagines the climate crisis the villains often appear as irresponsible *users* of energy like those driving an SUV or flying in a private jet. Both narratives erase the fundamental role of production and profit in making those moments of consumption possible. If political ecologists don't research the relationship between ecological degradation and production, who will? These spaces are hidden from view and consequently environmental politics too often remains fixated on the surface of everyday life.

Similarly, much policy and political ecological attention has for good reason fixated on the bizarre attempts to turn nature or ecosystem services into exchangeable commodities with prices in the realm of exchange (McAfee 1999; Dempsey and Robertson 2012; Buller 2022). While critiques of neoliberal market-based environmental policies will always have their place, we can also demonstrate the folly of "green capitalism" by investigating the hidden abode of production where capital's exploitation of nature is more nakedly brutal. We also have opportunities to extend "what counts as nature" (Haraway 1990: 8) in the realm of production to not only include primary extractive activities, but all forms of production.

Ultimately, a Marxist perspective is an ecological perspective concerned with how societies produce life itself. While political ecology has relied on ecological Marxist approaches that show how capitalism destroys the ecological conditions of life, we need more class precision. Capitalism is a system ruled by capital and capital is ruled by capitalists. As Utah Phillips memorably put it, "The Earth is not dying. It is being killed. And those that are killing it have names and addresses" (quoted in Klein 2000: 325). So understanding the system (capitalism) that is degrading the ecological basis of all life, we need more research on the capitalist centers of production driving these processes.

I tried to explain the theoretical justification for such a research program—as well as some methodological guideposts—but researching the ecology of capital cannot be prescribed ahead of time. Like most political ecological research, our efforts focus on understanding a particular ecological problem—deforestation or climate change. Tracing the social relations of capital (M-C-M') causing such devastation can go in multiple directions and will likely confront numerous obstacles along the way.

Notes

1 Marx would insist, however, that commodified labor power is the *source* of this surplus value (even if nature materially contributes significantly to the use values sold).
2 This differs from how Marx presents them in Volume 2, which I find more confusing. Also, I tend to refer to them in simpler terms as the money, productive and commodity circuits, respectively.
3 The annals of ecological Marxism are overstuffed with quotes from this ten pages of text (see, e.g., Benton 1989; Foster 2020; Swyngedouw and Heynen 2003).
4 Andreas Malm (2018: 190n34) points out that Moore fails to mention his debt to O'Connor in *Capitalism in the Web of Life* (although he had mentioned it previously).
5 One might protest that Marx's analysis of so-called "paid work" is of course premised on the *exploitation* of unpaid work harvested as surplus value from living labor.
6 Fertilizer induced toxic algae blooms had recently afflicted Lake Erie and much of the Toledo drinking water system (Wines 2014).
7 See https://nutrientstewardship.org/4rs
8 I must credit my tireless Research Assistants for actually collating and scanning the most relevant articles and advertisements—Jonathan Erickson and Carlo Sica. This was made possible by generous funding from the National Science Foundation (NSF Award no. 1437248).

References

Agricultural Chemicals. (1954) Nitrogen: A study of productive capacity. *Agricultural Chemicals* 9 (September): 55–57, 149.
Appel, H. (2019) *The Licit Life of Capitalism*. Durham, NC: Duke University Press.

Arboleda, M. (2020) *Planetary Mine: Territories of Extraction under Late Capitalism*. London: Verso.
Arboleda, M. (2015) Financialization, totality and planetary urbanization in the Chilean Andes. *Geoforum* 67: 4–13.
Ayres, R. (1989) Industrial metabolism. In, H.E. Sladovich, J. Ausubel, and J.H. Ausubel (eds.) *Technology and Environment*. Washington DC: National Academies Press, 23–49.
Barbesgaard, M. (2018) Blue growth: Savior or ocean grabbing? *Journal of Peasant Studies* 45(1): 130–149.
Barca, S. (2020) *Forces of Reproduction: Notes for a Counter-Hegemonic Anthropocene*. Cambridge: Cambridge University Press.
Benton, T. (1989) Marxism and natural limits: An ecological critique and reconstruction. *New Left Review* I/189: 51–86.
Blaikie, P. and H. Brookfield. (1987) *Land Degradation and Society*. London: Routledge.
Boyd, W., W.S. Prudham, and R.A. Schurman. (2001) Industrial dynamics and the problem of nature. *Society and Natural Resources* 14(7): 555–570.
Bridge, G. and A.E.G. Jonas (2002) Governing nature: The reregulation of resource access, production, and consumption. *Environment and Planning A* 34: 759–766.
Bryant, G. (2018) Nature as accumulation strategy? Finance, nature, and value in carbon markets. *Annals of the American Association of Geographers* 108(3): 605–619.
Buller, A. (2022) *The Value of a Whale*. Manchester, UK: Manchester University Press.
Bumpus, A. and D. Liverman (2008) Accumulation by decarbonization and the governance of carbon offsets. *Economic Geography* 84(2): 127–155.
Campling, L. and A. Colás (2021) *Capitalism and the Sea: The Maritime Factor in the Making of the Modern World*. London: Verso.
Clark, B. and R. York (2005) Carbon metabolism: Global capitalism, climate change and the biospheric rift. *Theory and Society* 34: 391–428.
Dempsey, J. and M. Robertson (2012) Ecosystem services: Tensions, impurities, and points of engagement within neoliberalism. *Progress in Human Geography* 36(6): 758–779.
Deutz, P. (2014) A class-based analysis of sustainable development: Developing a radical perspective on environmental justice. *Sustainable Development* 22(4): 243–252.
Deutz, P., H. Baxter, D. Gibbs, W.M. Mayes, and H.L. Gomes. (2017) Resource recovery and remediation of highly alkaline residues: A political-industrial ecology approach to building a circular economy. *Geoforum* 85: 336–344.
Devine, J. and D. Ojeda (2017) Violence and dispossession in tourism development: A critical geographical approach. *Journal of Sustainable Tourism* 25(5): 605–617.
Douglas, J.A. (2014) What's political ecology got to do with tourism? *Tourism Geographies* 16(1): 8–13.
Ekers, M. (2015) On the concreteness of labor and class in political ecology. In, T. Perreault, G. Bridge and J. McCarthy (eds.) *The Routledge Handbook of Political Ecology*. London: Routledge, 545–557.
Fairbairn, M. (2020) *Fields of Gold: Financing the Global Land Rush*. Ithaca, NY: Cornell University Press.
Fischer-Kowalski, M. (2003) On the history of industrial metabolism. In D. Bourg, S. Erkman and J. Chirac (eds.) *Perspectives in Industrial Ecology*. New York: Greenleaf, 35–45.
Foster, J.B. (2020). *The Return of Nature: Socialism and Ecology*. New York: Monthly Review Press.
Foster, J.B. and P. Burkett (2016) *Marx and the Earth: An Anti-Critique*. Chicago: Haymarket.
Foster, J.B., B. Clark and R. York (2010) *The Ecological Rift: Capitalism's War on the Earth*. New York: Monthly Review.
Fraser, N. (2014) Behind Marx's 'hidden abode': For an expanded conception of capitalism. *New Left Review* 86 (March–April): 55–72.
Fraser, N. (2021) A climate of capital: For a trans-environmental eco-socialism. *New Left Review* 127 (January–February): 94–127.
Hampton, P. (2015) *Workers and Trade Unions for Climate Solidarity: Tackling Climate Change in a Neoliberal World*. London: Routledge.
Haraway, D. (1990) Cyborgs at large: Interview with Donna Haraway (Interview by Constance Penley and Andrew Ross) *Social Text* 25/26: 8–23.
Harvey, D. 2010. *A Companion to Marx's Capital*. London: Verso.

Heynen, N., J. McCarthy, S. Prudham, and P. Robbins (eds.) (2007) *Neoliberal Environments False Promises and Unnatural Consequences*. London: Routledge.
Huber, M.T. (2017a) Chemical dialectics. *Geohumanities* 3(1): 165–166.
Huber, M.T. (2017b) Reinvigorating class in political ecology: Nitrogen capital and the means of degradation. *Geoforum* 85: 345–352.
Huber, M.T. (2022) *Climate Change as Class War: Building Socialism on a Warming Planet*. London: Verso.
Hurley, A. (1995) *Environmental Inequalities: Class, Race, and Industrial Pollution in Gary, Indiana, 1945–1980*. Chapel Hill, NC: University of North Carolina Press.
Kansaga, M.M., R. Antabe, Y. Sano, S. Mason-Renton and I. Luginaah (2019) A feminist political ecology of agricultural mechanization and evolving gendered on-farm labor dynamics in northern Ghana. *Gender, Technology and Development* 23(3): 207–233.
Kay, K. (2017) Rural rentierism and the financial enclosure of Maine's open lands tradition. *Annals of the American Association of Geographers* 107(6): 1407–1423.
Klein, N. (2000). *No Logo: Taking Aim at the Brand Bullies*. New York: Picador.
Malm, A. (2018) *The Progress of this Storm: Nature and Society in a Warming World*. London: Verso.
Marston, A. and T. Perreault (2017) Consent, coercion and *cooperativismo*: Mining cooperatives and resource regimes in Bolivia. *Environment and Planning A: Economy and Space* 49(2): 252–272.
Marx, K. ([1867] 1976) *Capital Volume 1*. London: Penguin.
Marx, K. ([1885] 1978) *Capital Volume 2*. London: Penguin.
Marx, K. ([1894] 1981) *Capital Volume 3*. London: Penguin.
McAfee, K. (1999) Selling nature to save it? Biodiversity and green developmentalism. *Environment and Planning D: Society and Space* 17(2): 133–154.
McCarthy, J. (2002) First World political ecology: Lessons from the Wise-use Movement. *Environment and Planning A* 34(7): 1281–1302.
Mies, M. (1986) *Patriarchy and Accumulation on a World Scale: Women and the International Division of Labor*. London: Zed Books.
Moore, J.W. (2015a) *Capitalism in the Web of Life: Ecology and the Accumulation of Capital*. London: Verso.
Moore, J.W. (2015b) Cheap food and bad climate: From surplus value to negative value in the capitalist world-ecology. *Critical Historical Studies* 2(1): 1–43.
Nader, L. (1972) Up the Anthropologist: Perspectives gained from studying up. In D. Hymes (ed.) *Reinventing Anthropology*. New York: Vintage Books, 284–311.
Nager, R. and S. Geiger (2007) Reflexivity, positionality and identity in feminist fieldwork revisited. In, A. Tickell, E. Sheppard, J. Peck and T. Barnes (eds.) *Politics and Practice in Economic Geography*. London: Sage, 267–278.
Newell, J.P., J.C. Cousins and J. Bakka. (2017) Political-industrial ecology: An introduction. *Geoforum* 85: 319–323.
O'Connor, J. (1998) *Natural Causes: Essays in Ecological Marxism*. London: Guilford.
Ojeda, D. (2012) Green pretexts: Ecotourism, neoliberal conservation and land grabbing in Tayrona National Natural Park, Colombia. *Journal of Peasant Studies* 39(2): 357–375.
Osborne, T. (2015) Tradeoffs in carbon commodification: A political ecology of common property forest governance. *Geoforum* 67: 64–77.
Patel, R. (2013) The long Green Revolution. *Journal of Peasant Studies* 40(1): 1–63.
Prudham, S. (2005) *Knock on Wood: Nature as Commodity in Douglas-Fir Country*. London: Routledge.
Rajan, K.S. (2006) *Biocapital: The Constitution of Postgenomic Life*. Durham, NC: Duke University Press.
Robbins, P. (2002) Obstacles to a First World political ecology? Looking near without looking up. *Environment and Planning A* 34(8): 1509–1513.
Robbins, P. (2020). *Political Ecology: A Critical Introduction*. 3rd edition. London: Wiley.
Royal Society (2020). *Ammonia: Zero-carbon Fertiliser, Fuel and Energy Store*. London: The Royal Society.
Saito, K. (2017) *Karl Marx's Ecosocialism: Capital, Nature, and the Unfinished Critique of Political Economy*. New York: Monthly Review Press.

Salleh, A. (1997) *Ecofeminism As Politics: Nature, Marx and the Postmodern*. London: Zed.
Sayer, D. (1987) *The Violence of Abstraction: The Analytic Foundations of Historical Materialism*. Oxford: Blackwell.
Smith, N. ([1984] 2008) *Uneven Development: Nature, Capital and the Production of Space*. Athens, GA: University of Georgia Press.
Swyngedouw, E. and N. Heynen (2003) Urban political ecology, justice and the politics of scale. *Antipode* 35(5): 898–918.
Tabuchi, H. (2021) For many, hydrogen is the fuel of the future. New research raises doubts. *New York Times*, August 13, A17.
Vercillo, S. (2022) A feminist political ecology of farm resource entitlements in Northern Ghana. *Gender, Place and Culture* 29(10): 1467–1496.
Werner, M. (2022) Geographies of production III: Global production in/through nature. *Progress in Human Geography* 46(1): 234–244.
Wines, M. (2014) Behind Toledo's water crisis, a long-troubled Lake Erie. *New York Times* August 5, A12.
Wood, E.M. (1986) *The Retreat from Class: A New "True" Socialism*. London: Verso.
Zhou, X., F.H. Passow, J. Rudek, J.C. von Fisher, S.P. Hamburg and J.D. Albertson. (2019) Estimation of methane emissions from the US ammonia fertilizer industry using a mobile sensing approach. *Elementa*, 7(1): 1–12.

13 Consuming Nature

From the Politics of Purchasing to the Politics of Ingestion

Julie Guthman and Michaelanne Butler

An unprecedented wave of protest followed the California Department of Pesticide Regulation's (DPR) April 2010 announcement of its intent to register methyl iodide as a replacement for the long used chemical compound, methyl bromide—an ozone-depleting substance that was destined for phase-out in compliance with the Montreal Protocol. Arguing that methyl iodide was even more acutely toxic and environmentally degrading than methyl bromide, an unusual array of activist organizations, including anti-pesticide, environmental, public health, farmworker, and "foodie" groups joined together to mount a major campaign to prevent methyl iodide from coming into use. They cited reports showing the chemical was a known neurotoxin and carcinogen, was associated with suppression of thyroid hormone synthesis, respiratory illness, and lung tumors, and a probable cause of miscarriages and birth defects. They made special efforts to communicate that those who would be subjected to the highest risk would be farmworkers, nearby residents, and any others within the vicinity of strawberry fields, especially because, unlike methyl bromide, methyl iodide would not dissipate into the upper atmosphere, staying close to the ground instead.

One of the major initiatives launched by activists against the approval of methyl iodide was an internet campaign encouraging the submission of public comments on the compound's registration. The initiative generated fifty-three thousand comments, with all but a handful opposing the chemical's registration. Researching this extraordinary response, Guthman coded the approximately 1,750 unique comments and found that many letter writers referred to actions they would take in the realm of consumption if the compound was approved. Some noted they would no longer *purchase* strawberries, in solidarity with farmworkers and concern over environmental toxicity and the unwanted absorption of chemicals through pesticide drift (Harrison 2011), while others said they would no longer *ingest* strawberries. This latter response could be read as an endeavor to circumvent embodied complicity with unjust agricultural practices but mostly reflected efforts to avoid personal exposure to toxicity, rooted in an inaccurate understanding that the fumigants would leave residue on the berries themselves (Guthman and Brown 2016).

Although in the case above refusing purchasing versus ingesting reflected very different impulses, consumption decisions more generally reflect the foremost manners in which publics have come to act in objection to a food system that is widely believed to be exploitative, inhumane, nutritionally bankrupt, and environmentally destructive. With its intimate, necessary, and everyday interactions, food is the primary sphere in which the politics of consumption are most routinely and recognizably enacted. Yet, food consumption is far from the only realm in which humans "consume nature." In the broadest sense, virtually

DOI: 10.4324/9781003165477-17

all human activity necessarily transforms and/or consumes nature, thereby leaving it in a different state from what it was (Moore 2015; Smith 2007). This kind of consumption can appear as the problem, as when humans leave a trail of waste or denudation in their wake (Falasca-Zamponi 2012). In other contexts, often involving food but also beyond, ethical or eco-friendly forms of human action in the form of green, or ethical, consumption are posited to be the solution (see Box 13.1). Through this lens, "doing political ecology" on green, or ethical consumption entails analyzing the political economies and cultural politics of the theories of change that drive efforts at ecological betterment through consumption.

This chapter looks to unpack the varying impulses, content, practices, and effects of a wide range of ethical consumption approaches. Some center around food but are applicable elsewhere; others are more specific and apply almost singularly in food contexts. As such, much of the work supporting this chapter is not traditionally associated with political ecology per se. Some draws from agrarian political economy which, as Galt (2013) has argued, has clear overlaps with first world political ecology, while other material draws from the amorphous field of food studies as well as social theory writ large, but maintains certain conceptual linkages with political ecology. Given this expansive source material, the highly selective use of citations regrettably falls far short of representing the vast scholarship that has contributed to the concerns discussed herein.

In constructing this chapter, we situate the multifarious impulses of green or ethical consumption along a continuum. In keeping with the opposing impulses evinced in the methyl iodide battle, at one end of the continuum are the political economic approaches to consumption that emphasize the capacity of consumers to leverage their purchasing power to undermine production practices that they deem as socially and environmentally detrimental and support those that are presumably better for people and the planet. At the other end are the ingestive approaches to consumption that emphasize embodiment

Box 13.1 Green consumption

As noted by Lorenzen (2014), green consumption can include the purchase of products and services that limit the use of energy and fossil fuels, water and raw materials (including packaging), and toxic chemicals in order to reduce waste and pollution. Well-known examples include hybrid or electric vehicles, solar panels, CFL or LED light bulbs, Energy Star appliances, and sustainably harvested forestry products. Green consumption also includes the purchase of second-hand goods, product sharing (p. 1064) or simply trying to consume less altogether. Green consumption is often facilitated by eco-labels, which certify that a product is more sustainably produced or harvested (Eden 2011). It is worth noting that green consumption is not without critique. Simon and Alagona (2013) perfectly capture the contradiction of green consumption, where buying is also using. Exploring the "leave no trace" ethic promulgated by an outdoor recreation industry, they point out the irony of encouraging the consumption of nature in the form of their products (think hiking boots) in order to consume nature in the form of enjoyment (hiking in a national park). While it is possible that hikers may not substantively alter nature through the act of hiking, the production of the boots may well have left a trace on the various ecologies from whence they were sourced.

and affect as gateways to change. The poles of this continuum roughly map onto Robbins's (2004) observations of two different traditions within political ecology of the "hatchet and the seed:" scholarship on the politics of purchasing has tended to be more critical, supporting opposition to current practices, while scholarship on the politics of ingestion has tended to be more sympathetic, supporting prefigurative practices, referring to those that attempt to model the world in which actors want to live. But since many of the practices discussed herein are overlapping and some even come full circle, we discuss them not as opposing approaches, but simply distinct ones.

Voting with Your Dollars

At the far end of the politics of purchasing, the meme "voting with your dollars" best captures the premise that consumers can favorably alter production practices by economically choking individual businesses and organizations whose practices they deem deleterious and supporting those they deem ameliorative (Lorenzen 2014). When pursued as a collectivity, withdrawing support is tantamount to a *boycott*, harkening back to the days when, as Johnston (2008) notes, "boycotts were used by unions, political activists, and individual consumers as a way to enact political preferences through anti-consumption behaviors" (p. 236). Examples of consumer power mobilized as a collectivity can be famously seen in the boycotts of grapes and lettuce on behalf of the United Farm Workers union in the late 1960s and early 1970s or of the multinational corporation, Nestlé, for its dubious marketing strategies of infant formula in low-income countries that ultimately undermined breast-feeding in the 1970s and 1980s (Sasson 2016). Remarkably, there are few other instances of boycotts in matters of food per se. Case in point, recent boycotts of palm oil were primarily motivated as a protest to the ecological impacts of oil palm cultivation on orangutan habitat (Fair 2021) rather than food production practices. And, indeed, some of the most robust contemporary instances of boycott-like strategies can be seen in the fossil fuel divestment movement (Ayling and Gunningham 2017). Perhaps the foment surrounding the companies that use and produce genetically-modified organisms (GMOs) would have made them a likely candidate for a boycott, but due to legal and practical difficulties of enacting a boycott, critics have instead tended to voice their protest through *buycotts* (Roff 2008).

Buycotting is a neologism that refers to purchasing practices that shift support towards producers whose approaches appear favorable or ethical (Neilson 2010), the most classic example being the growth of the organic foods industry since the early 1990s. Based on ideas that organically produced crops and livestock would employ less toxic, more soil friendly and more humane inputs and processes, buying organics became a way to contest a litany of concerns within food production, eventually including the use of GMOs (Guthman 2004a). This latter motivation was readily seen as consumers began to purchase organic milk over concerns about the potential effects of drinking milk from cows treated with recombinant bovine growth hormone (rBGH) (DuPuis 2000). Unlike boycotts, markets could be developed around buycotts and, moreover, impact might be effected solely through individual action, thus subsuming the need for the political organization required of boycotts. DuPuis has thus argued that although they do not emanate from organized campaigns, but rather from anxieties among middle- and upper-class consumers, these choices still count as political acts.

Effective buycotts, in a number of consumer markets, from food to timber, require mediating mechanisms to guarantee the veracity of alternative products. It is precisely

Box 13.2 Following the label

Hinging on consumer choice, effecting a buycott all but requires a label in order to communicate the key differences in how commodities have been produced. Efforts to establish mandatory labels for conventional products have often been met with fierce resistance from politically powerful incumbent industries who are reluctant to disclose practices that might be perceived as bad. While still generating opposition from agribusiness, voluntary labels promoting ethical food production practices have thus gained more traction as a means to identify and support good counterparts. Curiously, political ecologists have paid more attention to "following the thing" (see defetishizing the commodity) relative to following the label by analyzing the politics undergirding such labels. Yet, how these labels are operationalized matters tremendously in terms of who and what gets supported and how. Questions a political ecologist seeking to follow the label might ask include: Who decides what warrants a label and how do they do it? What standards do producers need to follow? How were those standards developed? What are the processes of verifying those standards were met? Who performs this verification and how are they remunerated? How does the label actually support producers? Do they get additional income? By what means? Formal or by market convention? (e.g., a promised price premium or an assumed one?) Does the label incentivize producers to create barriers to entry to eliminate price competition? For more on the importance of standard-setting in food governance see Hatanaka and Busch (2008).

this imperative, as well as evidence of fraudulent claims, that gave rise to the highly complex system of verification and certification behind the organic label that exists today (Guthman 2004a). Yet, as several scholars have shown, systems of standard-setting and certification are far from straight forward, and are not only subject to much contestation, but also add layers of costs and create significant barriers to entry for new producers (Guthman 2004a; Mutersbaugh 2002). As such, the arduous process of certification itself curtails the proliferation of the more ethical and sustainable production techniques that the labels are supposed to incentivize. Further, the added costs incurred in labeling are often passed down the supply chain, limiting access and making it difficult for low-income consumers to enact this kind of politics of consumption (Guthman 2007). So even though labels greatly differ in their standards, modes of verifications, and efficacy (see Box 13.2), this approach to ethical consumption overall is inherently limited.

Extending Solidarity

While enacting a politics of purchase through boycotts and buycotts emphasizes specific production practices, the closely related impulse behind *buying in solidarity* seeks to support specific producers. Intended to redistribute value to peasant producers, Fair Trade labeling serves as a prime example of this impulse (Mutersbaugh 2002; Raynolds 2002; Renard 2003). Sometimes this support is premised on the assumption that these smallholder producers utilize eco-friendly practices, such as growing coffee in the shade to preserve biodiverse habitats. Co-labeling fair trade alongside eco-labels such as organic, rainforest alliance, shade grown make this assumed relationship explicit (Bacon 2010).

Some scholars have emphasized that purchasing products with social labels such as Fair Trade goes beyond a material redistribution of dollars to promote an ethic of solidarity and reciprocity with farmers (or ranchers or timber producers) in the Global South (Clarke et al. 2007; Goodman 2004; Raynolds 2002; Renard 2003). For Goodman (2004, 896) the "political ecological imaginary" of Fair Trade "not only tells consumers how the commodity works, but ... demonstrates the progressive effects of their act of consumption on the particular community that grew what they are eating," thus encouraging moral reflexivity by consumers. Further, Barnett et al. emphasize that such campaigns of practicing ethics from afar present a "model of responsibility that connects individual and household consumption to broader mobilizations" such that "a narrow sense of individualized ethical responsibility is transformed into a practice of collective, political responsibility (Barnett et al. 2005, 42–43). In contrast, some suggest that such initiatives are more reflective of consumer desires to feel virtuous rather than true solidarity and thus question the efficacy of this approach (Dolan 2010; Shreck 2005; Wilson 2013).

Buying Local

Closer to home, *buying local* operationalizes a similar set of impulses and assumptions predicated on the idea that informed consumers can wield their purchasing power for good. Epitomized in the farm to table movement, this approach focuses on geographic proximity—the idea being that buying locally produced commodities or goods procured through "shortened supply chains" has manifold benefits. One is that, like Fair Trade, the enhanced trust engendered by "thickened" connections between producers and consumers (Bell and Valentine 1997; Crang 1996), presumably leads to more ecological production practices (Hinrichs 2003). Buying locally is also imagined to reduce the carbon footprint associated with transporting goods in the reduction of "food miles" it travels (McEntee 2010). These assumptions are evident elsewhere including in local water- and fiber-shed movements and decentralized energy production networks.

Despite the seemingly positive environmental effects of this approach, critiques are legion. Scholars are quick to point out that buying local is no panacea for environmental destruction. Laser focus on reduced energy in transportation, for instance, ignores the ever-present reality that farms themselves rely on fossil fuels and their byproducts (Mariola 2008). Further, critics note that "local" producers do not necessarily employ sustainable ecological methods—even the worst actors are local to some place and conventional commodity production is practiced somewhere (Hinrichs 2003). Finally, and perhaps the most damning criticism is that the concept conflates the geographical scale of production with its social, and hence ecological relations (Born and Purcell 2006; DuPuis and Goodman 2005). Put another way, proximity does not necessarily entail that producers share a particular set of concerns—social, ecological, or otherwise, making buying local somewhat of a chimera.

Defetishizing the Commodity

The oft-heard catchphrase of "knowing where your food comes from" that underpins the impulse to buy local also underpins another impulse, one that draws from the Marxian notion of *commodity fetishism*. Recognizing that knowledge itself is not enough, the theory of change behind defetishizing the commodity, which applies to many products beyond food, is that unmasking the exploitative conditions under which commodities are produced within capitalist systems will propel political action to change them. Core to this logic is an understanding that the very act of commodity exchange obscures or "fetishizes" the systems of production that brought a particular product into existence. Thus, some scholars have suggested that labels which disclose key materials and processes that producers might either utilize or avoid are a means of opening these ecological and social processes to scrutiny (Allen and Kovach 2000; Hartwick 1998). As Bryant and Goodman (2004) put it, rather than fetishizing, "these alternative commodities veritably shout to consumers about the socionatural relations under which they were produced." In a similar vein, but shorn of Marxian connotations, other scholars call for "biographies of production and distribution" (Cook 2004) or "following the thing" (Cook and Crang 1996) as means for Western consumers to make connections with the distant producers and traders who bring exotic goods to market.

Some, however, have noted that biographies of products from the Global South are often told through "carefully wrought images and texts" that do not always reveal the core problems surrounding the production of the given commodity (Bryant and Goodman 2004, 348). Consider products that are labelled as coming from the rainforest, which could suggest that the problem at hand is that humans have not sufficiently exploited enough rainforest products (Bryant and Goodman 2004). As such, other scholars have critiqued this approach, arguing that the labels or biographies that reveal can be their own fetish, leading consumers into feeling that knowledge is indeed enough while the labels themselves may not be as transparent as imagined (Freidberg 2003; Gunderson 2014; Guthman 2004b).

Not in My Body?

With the exception of organized boycotts, the impulses clustering toward the *politics of purchase* end of the continuum have been critiqued for operating at the level of the individual rather than the collective (cf. Barnett et al. 2005). Arguably, this impulse reaches it apotheosis in the *not-in-my-body* politics hinted at in the opening story, in which action appears even more individualistic and voluntary than not-in my-back-yard (NIMBY) politics (DuPuis 2000). Where some letter writers opposed to the use of methyl iodide threatened the withdrawal of their purchasing power, others promised to no longer ingest strawberries to avoid toxic exposure (Guthman and Brown 2016). Effecting what Szasz (2007) describes as an "inverted quarantine," this mirrors the dynamics enacted through support of organics and other labeling schemes by which empowered consumers who can afford to do so functionally buy their way out of product qualities they see as hazardous or unhealthful, with little solidarity with others. That said, the question of what to eat and its not-in-my-body sensibility has gained new valences in light of concern over animal agriculture in particular and is thus salient in the increased popularity of vegan and plant-based diets. Moving toward the *politics of ingestion* end of the spectrum, the quest for ethical consumption contra to ethical purchasing foregrounds the body itself as the site of both political contention and prefiguration, such that not in my body is the point.

Reflexive Consumption

Broadly, the notion of reflexive consumption responded to the need to theorize the consumer as a knowledgable, thoughtful and intentional agent of change in food systems rather than a dupe of industry—or its opposition (DuPuis 2000). In calling for reflexive localism, DuPuis and Goodman (2005) expanded the concept to connote consumption practices that reject a utopian politics of perfection and instead recognize the contradictions and complexity of food choices. Building on this approach, Probyn (2016) observes that reflexivity is crucial in the face of an incoherent politics of fully opting out of the negative implications of eating particular foods. Taking fish as an example, she highlights that these creatures may be consumed through unexpected avenues, like the meal commonly used to fertilize vegetable crops.

Serving as a particularly intriguing example of mindful ingestion, the nose-to-tail eating practices popularized by Fergus Henderson (2004) reflect a conscious decision to embrace the liveliness and edibility of the whole animal, even those parts rendered disposable through preference and factory farming practices (Weiss 2012). This aligns with Haraway's injunction, that there is "no way to eat and not to become one with other mortal beings to whom we are accountable, no way to pretend innocence and transcendence" (Haraway 2008, 295).

Enacting Visceral Politics

In contrast to the focus on cerebral deliberation and ethical debate captured in the notion of reflexive consumption, the idea of visceral politics is to foreground taste, affect, and tactile experience as guides to right and ethical eating. This approach implicitly sees the body as having an almost pre-discursive intuition that enables it to discern good from bad (Hayes-Conroy and Martin 2010). Proponents of Slow Food, for instance, foreground the gustatory pleasure of eating as an important tool for directing lay-eaters toward "right" food choices. Likewise, proponents of edible school yards imagine that by engaging the body directly, as a site for tactile education, they might "unlock" correct tastes (Hayes-Conroy and Hayes-Conroy 2013).

Scholars have noted, however, that certain enactment of food education, such as edible schoolyards, are based on presumptions that these interventions actually avoid the structural inequalities of race, class, and the like (Bruckner and Kowasch 2019; Hayes-Conroy and Hayes-Conroy 2013). They have noted that taste is highly subjective and influenced by social and structural factors, meaning that the right foods may not be recognizable by some objective standard (Mol 2009; Watson and Cooper 2021) Nonetheless, like reflexive consumption, the aim of visceral attunement is the restraint that emerges from good taste rather than perfect action (Mol 2009).

Box 13.3 On doing embodied research

Responding to a tacit vilification of the researcher's body as "lurk[ing] unseen, unruly, and uncontrollable in the shadows of the Great Halls of the academy" (Spry 2001, 720), food scholars in particular have foregrounded the visceral or "bodily realm where feelings, moods, and sensations are made manifest" (Hayes-Conroy 2010, 734) as both a site and "instrument" of research (Longhurst et al. 2008). Recognizing that ingesting food can be an incredibly affective experience, they have developed an approach that "takes the body seriously" and considers the biosocial dimensions of an individual or society's engagement with food (Hayes-Conroy and Martin 2010; Mudry et al. 2014; Sexton et al. 2017). Specific approaches might include: cooking and eating alongside subjects (Longhurst et al. 2008) or paying particular attention non-verbal interactions between people and would-be food through visual documentation and art installations (Roe 2006). Of course, similar approaches may be applied to experiences and embodied politics associated with other commodity-mediated human-environment interactions, such as using highly manicured lawnscapes, air purification systems, or bicycles and other self-powered modes of transport.

Refusal

A growing contingent of ethically oriented consumers see the restraint of reflexive and viscerally attuned eating as insufficient to address the problems endemic to food provisioning systems. Motivated by ethical, feminist, anarchist and post-colonial critiques—communities of practice have long foregrounded ingestive practices as a key location of political action through refusal—with the expectation that as prefigurative practices they can bring new worlds into being. As put by McGranahan (2016), refusal "can be generative and strategic, a deliberate move toward one thing, belief, practice, or community and away from another," making it the kind of prefigurative practice contained in Robbins's notion of the seed. In that way, the politics of refusal extends far beyond the realm of food and can be captured, say, in the de-growth movement that refuses to participate in capitalism's logic of necessary growth and the environmental havoc it precipitates (Leff 2021). Still, in the realm of food, nowhere is refusal more clearly visible than among vegan and vegetarians.

Contemporary vegetarian and vegan critiques build on an intellectual legacy that has historically framed carnivorous consumption as physically, intellectually, and spiritually defiling (Adams 2015; Gheihman 2021; Probyn 2000). Seeing such practices as complicit with oppression at a systems level, they are not only calling for an "alternative ending, veggie burgers instead of hamburgers" (Adams 2015, 79). Rather, by exercising personal choice as political action, practitioners situate ingestive refusal as an embodied resistance to domination in the present, including that of BIPOC bodies, as well as a bridge toward justice in the future (Gaard 2002; Harper 2016).

Far from homogeneous and too numerous to recapitulate here, critiques against these politics of refusal abound both within and outside ethical vegan and vegetarian camps. Ultimately, many amount to seeing such refusals as elitist, moralizing, over-zealous, and inaccessible (see Lee 2019 for summary). It is worth noting, however, that given the increased salience of climate change and public recognition of meat's "long shadow" as a major emitter of greenhouse gasses, proponents of "flexitarianism" and "reducetarianism" argue that moderate approaches, whereby eaters simply cut down on their consumption rather than abstaining from meat altogether, might have a broader appeal and thus result in a greater reduction of animal farming in absolute terms along with reductions in its associated environmental effects (Kateman 2017). Based on his examination of the pork industry and its obsession with efficiency, Blanchette (2020) makes a somewhat different point: that it is not so much the food that should be refused but the idea that everything must be subject to capitalist value.

Consuming Waste

The "freegan" movement operates on a similar politics of refusal and prefiguration. Contra "voting with your dollars," the portmanteau combining "free" and "vegan" captures a social movement that, as ethnographer Alex Barnard (2016, 25) describes, "attempts to directly, in the here and now, build a *new* society in the heart of the *old* one." As such, freegans seek to opt out of the capitalist production system entirely by refusing to purchase food or anything else (ErnstFriedman 2012). In this instance, however, the body's role as a site of resistance is much different from the embodied purity typified by many ethical vegans and vegetarians (Giraud 2021). Subsisting on what would otherwise be discarded, freegan practitioners open themselves to the risks of potential contamination (Tibbetts 2013), enrolling their very bodies as sites where the waste of capitalism is worked through. As a form of world making, however, there are inherent contradictions embedded in this approach. For instance, as freegans admit and critics point out, their eating practices are dependent on persistence of capitalism's tendencies to produce waste (Barnard 2016).

This dependency on continued waste is also the case for the decidedly less radical outcropping of waste-refusing politics found in the ugly food movement, which seeks to reduce supply side excess by redistributing off-spec foods toward willing eaters (Taber 2019). This iteration comes full circle to the politics of purchasing, while providing a striking counterpoint to the ironic consumption of nature discussed by Simon and Alagona (2013). Rather than using dollars to purchase commodities in order to enjoy nature "out there," the consumer purchases the waste of industrial agriculture production to alleviate the excess that is endemic to capitalist value formation (Gidwani and Reddy 2011).

The Tech Solution—the Antipolitics of Consumption

As discussed thus far, the politics of consumption has been exercised by concerned individuals and some collectivities in response to the current food system, notably via voting with dollars, insisting on defetishized commodities and their origins, or through attuned and intentional eating. Striking for its lack of engagement of consumers or, for that matter, politics, Silicon Valley's forays into food and agriculture, the topic of our current research, sheds light on a decidedly different approach. Claiming concern with the ecological problems of food production, especially climate change and animal welfare issues, many of the consumer-facing products being promulgated by the sector thus far have taken the form of what has been dubbed as *alternative proteins*, referring to plant-based simulacra of meats and dairy, insect-based ingredients, or cellular technologies that grow "meat" without the animal. Critically, inventors and promoters, many of whom are dyed-in-the-wool vegans, are motivated to produce products that attempt to mimic animal proteins because they do not believe consumers are willing to make sacrifices on behalf of nature and must be cajoled or even fooled into purchasing and consuming protein without animals (Biltekoff and Guthman in review; Jönsson et al. 2019; Sexton 2018).

This instantiates an *anti-politics (of consumption)* in which debate or deliberation about production practices is all but foreclosed. In a reversal of the enactive politics of consumption previously discussed, rather than requiring a degree of awareness among consumers of their role in either perpetuating or contesting current conditions, alternative protein innovators seek to make these choices purely a matter of taste, experience and price. Indeed, they imagine consumers as "lacking the knowledge, understanding and rationality required for meaningful engagement," thus "underscoring the mandate for actors in the sector to act on behalf of reluctant consumers" (Biltekoff and Guthman in review). Ironically, this approach relies on many sleights of hand, including the strategic obfuscation of the sources and processes that go into these products—in other words, that consume nature (Guthman and Biltekoff 2020).

Full Circle

The public comments turned out to have an ambiguous impact on the outcome of the registration decision regarding methyl iodide. DPR went on to approve the fumigant despite the thousands of comments and additional protest. Still, actual adoption of the chemical was slow. This can be partially attributed to the volume of negative comments, since they played a role in chilling the industry toward methyl iodide. However, it was a lawsuit brought against DPR that sealed the chemical's fate. The lawsuit complained that DPR did not follow its own statutes and procedures in registering the chemical. In the end, Arysta pulled methyl iodide from the US market, citing non-economic viability. In this light, the effectiveness of public comments in cementing methyl iodide's demise is unclear, but provides a useful window on how publics acting collectively can wield their consumption as a means of altering production practices.

Contrast that with the theory of change underpinning many of the approaches this chapter has described. They share a premise that action will scale up and displace bad practices, yet that has not been the case given specific political economies and availability. Many writing in the political ecology tradition have thus generally been critical of approaches that render consumption politics to the realm of individual choice, suggesting they reflect a constrained politics of the possible (Alkon 2014; Allen et al. 2003; Clay et al. 2020; Guthman 2008; Lavin 2009; Sexton et al. 2019). Clearly, the enactments of individual consumption choices, whether through purchasing or ingesting, are dependent on reflexive, "conscious" consumers, who are making particular connections between what to buy or eat and ecological outcomes. In particular, the politics of purchasing assumes a consumer who has wealth, information, and efficacy to redistribute (Bryant and Goodman 2004; Guthman 2008; Lorenzen 2014) whereas a politics of ingestion assumes the possibility of avoiding products and substances that bear any sort of "badness" and maintains the possibility of purity (Shotwell 2016). They are therefore inherently limited in their ability to effect broadly scaled, improved ecologies. Indeed, recognition of these limits is precisely what motivates Silicon Valley techies who believe their innovations can be brought to scale. Their take, however, eliminates choice altogether for publics wanting to transform the ecological (and social) relations of food (and other commodity) production, as if the entrepreneurs and their funders alone know best. A political ecologist might use the hatchet to critique the Silicon Valley approach as the ultimate in eco-managerial hubris while planting the seed of enacting politics at the collective level to create livable futures.

References

Adams, C. J. 2015. *The sexual politics of meat: A feminist-vegetarian critical theory*: Bloomsbury Publishing USA.
Alkon, A. H. 2014. Food justice and the challenge to neoliberalism. *Gastronomica: The Journal of Food and Culture* 14 (2): 27–40.

Allen, P., M. FitzSimmons, M. Goodman and K. Warner. 2003. Shifting plates in the agrifood landscape: The tectonics of alternative agrifood initiatives in California. *Journal of Rural Studies* 19 (1): 61–75.

Allen, P. and M. Kovach. 2000. The capitalist composition of organic: The potential of markets in fulfilling the promise of organic agriculture. *Agriculture and Human Values* 17 (3): 221–232.

Ayling, J. and N. Gunningham. 2017. Non-state governance and climate policy: The fossil fuel divestment movement. *Climate Policy* 17 (2): 131–149.

Bacon, C. M. 2010. Who decides what is fair in fair trade? The agri-environmental governance of standards, access, and price. *The Journal of Peasant Studies* 37 (1): 111–147.

Barnard, A. V. 2016. *Freegans: Diving into the wealth of food waste in America*: University of Minnesota Press.

Barnett, C., P. Cloke, N. Clarke and A. Malpass. 2005. Consuming ethics: Articulating the subjects and spaces of ethical consumption. *Antipode* 37 (1): 23–45.

Bell, D. and G. Valentine. 1997. *Consuming geographies: We are where we eat*. London: Routledge.

Biltekoff, C. and J. Guthman. 2023. Conscious, complacent, fearful: Agri-food tech's market-making public imaginaries. *Science as Culture*, 32(1): 58–82.

Blanchette, A. 2020. *Porkopolis: American animality, standardized life, and the factory farm*. Durham, NC: Duke University Press.

Born, B. and M. Purcell. 2006. Avoiding the local trap: Scale and food systems in planning research. *Journal of Planning Education and Research* 26(2): 195–207.

Bruckner, H. K. and M. Kowasch. 2019. Moralizing meat consumption: Bringing food and feeling into education for sustainable development. *Policy Futures in Education* 17 (7): 785–804.

Bryant, R. L. and M. K. Goodman. 2004. Consuming narratives: The political ecology of "alternative" consumption. *Transactions of the Institute of British Geographers* 29 (3): 344–366.

Clarke, N., C. Barnett, P. Cloke and A. Malpass. 2007. The political rationalities of fair-trade consumption in the United Kingdom. *Politics & Society* 35 (4): 583–607.

Clay, N., A. E. Sexton, T. Garnett and J. Lorimer. 2020. Palatable disruption: The politics of plant milk. *Agriculture and Human Values* 37 (4): 945–962.

Cook, I. 2004. Follow the thing: Papaya. *Antipode* 36 (4): 642–664.

Cook, I. and P. Crang. 1996. The world on a plate: Culinary culture, displacement and geographical knowledges. *Journal of Material Culture* 1 (2): 131–154.

Crang, P. 1996. Displacement, consumption and identity. *Environment and Planning A* 28 47–67.

Dolan, C. S. 2010. Virtual moralities: The mainstreaming of fairtrade in Kenyan tea fields. *Geoforum* 41 (1): 33–43.

DuPuis, E. M. 2000. Not in my body: RBGH and the rise of organic milk. *Agriculture and Human Values* 17 (3): 285–295.

DuPuis, E. M. and D. Goodman. 2005. Should we go "home" to eat? Towards a reflexive politics of localism. *Journal of Rural Studies* 21 (3): 359–371.

Eden, S. 2011. The politics of certification: Consumer knowledge, power, and global governance in ecolabeling. In *Global political ecology*, eds. Richard Peet, Paul Robbins and Michael Watts, 183–198. London: Routledge.

ErnstFriedman, K. 2012. Trash tours: Untying what freegans get out of the garbage. *Anthropology Now* 4 (3): 33–42.

Fair, H. 2021. Feeding extinction: Navigating the metonyms and misanthropy of palm oil boycotts. *Journal of Political Ecology* 28 (1).

Falasca-Zamponi, S. 2012. *Waste and consumption: Capitalism, the environment, and the life of things*. London: Routledge.

Freidberg, S. 2003. Cleaning up down south: Supermarkets, ethical trade, and African horticulture. *Journal of Social and Cultural Geography* 4 (1): 27–42.

Gaard, G. 2002. Vegetarian ecofeminism: A review essay. *Frontiers: A Journal of Women Studies* 23 (3): 117–146.

Galt, R. E. 2013. Placing food systems in First World political ecology: A review and research agenda. *Geography Compass* 7 (9): 637–658.

Gheihman, N. 2021. Veganism as a lifestyle movement. *Sociology Compass* 15 (5): e12877.

Gidwani, V. and R. N. Reddy. 2011. The afterlives of "waste": Notes from India for a minor history of capitalist surplus. *Antipode* 43 (5): 1625–1658.

Giraud, E. H. 2021. *Veganism: Politics, practice, and theory*. London: Bloomsbury Publishing.

Goodman, M. K. 2004. Reading fair trade: Political ecological imaginary and the moral economy of fair trade foods. *Political Geography* 23 (7): 891–915.
Gunderson, R. 2014. Problems with the defetishization thesis: Ethical consumerism, alternative food systems, and commodity fetishism. *Agriculture and Human Values* 31 (1): 109–117.
Guthman, J. 2004a. *Agrarian dreams: The paradox of organic farming in California*. Berkeley: University of California Press.
Guthman, J. 2004b. The "organic commodity" and other anomalies in the politics of consumption. In *Geographies of commodity chains*, eds. Alex Hughes and Suzanne Reimer. London: Routledge.
Guthman, J. 2007. The Polanyian way? Voluntary food labels as neoliberal governance. *Antipode* 39 (3): 456–478.
Guthman, J. 2008. Neoliberalism and the making of food politics in California. *Geoforum* 39 (3): 1171–1183.
Guthman, J. and C. Biltekoff. 2020. Magical disruption? Alternative protein and the promise of de-materialization. *Environment and Planning E: Nature and Space* 2514848620963125.
Guthman, J. and S. Brown. 2016. I will never eat another strawberry again: The biopolitics of consumer-citizenship in the fight against methyl iodide in California. *Agriculture and Human Values* 33 (3): 575–585.
Haraway, D. 2008. *When species meet*. Minneapolis: University of Minnesota Press.
Harper, A. B. 2016. Doing veganism differently: Racialized trauma and the personal journey towards vegan healing. In *Doing nutrition differently: Critical approaches to diet and dietary intervention*, eds. A. Hayes-Conroy and J. Hayes-Conroy, 151–168. London: Routledge.
Harrison, J. L. 2011. *Pesticide drift and the pursuit of environmental justice*. Cambridge, MA: MIT Press.
Hartwick, E. 1998. Geographies of consumption: A commodity-chain approach. *Environment and Planning D: Society and Space* 16: 423–437.
Hatanaka, M. and L. Busch. 2008. Third-party certification in the global agrifood system: An objective or socially mediated governance mechanism? *Sociologia Ruralis* 48 (1): 73–91.
Hayes-Conroy, A. 2010. Feeling slow food: Visceral fieldwork and empathetic research relations in the alternative food movement. *Geoforum* 41 (5): 734–742.
Hayes-Conroy, J. and A. Hayes-Conroy. 2013. Veggies and visceralities: A political ecology of food and feeling. *Emotion, Space and Society* 6: 81–90.
Hayes-Conroy, A. and D. G. Martin. 2010. Mobilising bodies: Visceral identification in the slow food movement. *Transactions of the Institute of British Geographers* 35 (2): 269–281.
Henderson, F. 2004. *The whole beast: Nose-to-tail eating*. New York: HarperCollins.
Hinrichs, C. G. 2003. The practice and politics of food system localization. *Journal of Rural Studies* 19 (1): 33–45.
Johnston, J. 2008. The citizen-consumer hybrid: Ideological tensions and the case of whole foods market. *Theory and Society* 37: 229–270.
Jönsson, E., T. Linné and A. McCrow-Young. 2019. Many meats and many milks? The ontological politics of a proposed post-animal revolution. *Science as Culture* 28 (1): 70–97.
Kateman, B. 2017. The reducetarian solution: How the surprisingly simple act of reducing the amount of meat in your diet can transform your health and the planet. In *The reducetarian solution: How the surprisingly simple act of reducing the amount of meat in your diet can transform your health and the planet*. New York: Penguin.
Lavin, C. 2009. Pollanated politics, or, the neoliberal's dilemma. *Politics and Culture* 2 (2).
Lee, A. 2019. The milkmaid's tale: Veganism, feminism and dystopian food futures. *Windsor Rev. Legal & Soc. Issues* 40: 27.
Leff, E. 2021. De-growth or deconstruction of the economy: Towards a sustainable world. In *Political ecology: Deconstructing capital and territorializing life*, ed. Enrique Leff, 209–219. Cham: Springer International Publishing.
Longhurst, R., E. Ho and L. Johnston. 2008. Using "the body" as an "instrument of research": Kimch'i and pavlova. *Area* 40 (2): 208–217.
Lorenzen, J. A. 2014. Green consumption and social change: Debates over responsibility, private action, and access. *Sociology Compass* 8 (8): 1063–1081.
Mariola, M. J. 2008. The local industrial complex? Questioning the link between local foods and energy use. *Agriculture and Human Values* 25 (2): 193–196.

McEntee, J. 2010. Contemporary and traditional localism: A conceptualisation of rural local food. *Local Environment* 15 (9–10): 785–803.
McGranahan, C. 2016. Theorizing refusal: An introduction. *Cultural Anthropology*. https://journal.culanth.org/index.php/ca/article/view/ca31.3.01/367
Mol, A. 2009. Good taste: The embodied normativity of the consumer-citizen. *Journal of Cultural Economy* 2 (3): 269–283.
Moore, J. W. 2015. *Capitalism in the web of life: Ecology and the accumulation of capital*: Verso Books.
Mudry, J., J. Hayes-Conroy, N. Chen and A. H. Kimura. 2014. Other ways of knowing food. *Gastronomica: The Journal of Food and Culture* 14 (3): 27–33.
Mutersbaugh, T. 2002. The number is the beast: A political economy of organic-coffee certification and producer unionism. *Environment and Planning A* 34 (7): 1165–1184.
Neilson, L. A. 2010. Boycott or buycott? Understanding political consumerism. *Journal of Consumer Behaviour* 9 (3): 214–227.
Probyn, E. 2000. *Carnal appetites: Foodsexidentities*. London: Routledge.
Probyn, E. 2016. *Eating the ocean*. Durham, NC: Duke University Press.
Raynolds, L. 2002. Consumer-producer links in fair trade coffee networks. *Sociologia Ruralis* 42 (4): 404–424.
Renard, M.-C. 2003. Fair trade: Quality, market and conventions. *Journal of Rural Studies* 19 (1): 87–96.
Robbins, P. 2004. *Political ecology*. Oxford: Blackwell.
Roe, E. J. 2006. Things becoming food and the embodied, material practices of an organic food consumer. *Sociologia Ruralis* 46 (2): 104–121.
Roff, R. J. 2008. No alternative? The politics and history of non-GMO certification. *Agriculture and Human Values* 26 (4): 351.
Sasson, T. 2016. Milking the third world? Humanitarianism, capitalism, and the moral economy of the Nestlé boycott: Milking the third world? *The American Historical Review* 121 (4): 1196–1224.
Sexton, A. E. 2018. Eating for the post-Anthropocene: Alternative proteins and the biopolitics of edibility. *Transactions of the Institute of British Geographers* 43 (4): 586–600.
Sexton, A. E., T. Garnett and J. Lorimer. 2019. Framing the future of food: The contested promises of alternative proteins. *Environment and Planning E: Nature and Space* 2 (1): 47–72.
Sexton, A. E., A. Hayes-Conroy, E. L. Sweet, M. Miele and J. Ash. 2017. Better than text? Critical reflections on the practices of visceral methodologies in human geography. *Geoforum* 82 200–201.
Shotwell, A. 2016. *Against purity: Living ethically in compromised times*. Minneapolis: University of Minnesota Press.
Shreck, A. 2005. Resistance, redistribution, and power in the Fair Trade banana initiative. *Agriculture and Human Values* 22 (1): 17–29.
Simon, G. L. and P. S. Alagona. 2013. Contradictions at the confluence of commerce, consumption and conservation; or, an REI shopper camps in the forest, does anyone notice? *Geoforum* 45 325–336.
Smith, N. 2007. Nature as accumulation strategy. *Socialist Register* 16: 1–36.
Spry, T. 2001. Performing autoethnography: An embodied methodological praxis. *Qualitative Inquiry* 7 (6): 706–732.
Szasz, A. 2007. *Shopping our way to safety: How we changed from protecting the environment to protecting ourselves*. Minneapolis: University of Minnesota Press.
Taber, S. 2019. Eating ugly food won't save the world: Farms aren't tossing perfect produce, you are. *Washington Post* March 8.
Tibbetts, J. 2013. Freegans risk the hazards of dumpster diving. *Canadian Medical Association Journal* 185 (7): E281.
Watson, D. L. B. and C. M. Cooper. 2021. Visceral geographic insight through a "source to senses" approach to food flavour. *Progress in Human Geography* 45 (1): 111–135.
Weiss, B. 2012. Configuring the authentic value of real food: Farm-to-fork, snout-to-tail, and local food movements. *American Ethnologist* 39 (3): 614–626.
Wilson, B. 2013. Delivering the goods: Fair trade, solidarity, and the moral economy of the coffee contract in Nicaragua. *Human Organization* 72 (3): 177–187.

Part IV

Political Ecologies of Identities, Difference, and Justice

14 Engaging Nature

Public Political Ecology for Transformative Climate Justice

Joel E. Correia and Tracey Osborne

Engaging the Natures of Climate Injustice (an Introduction)

Four Kichwa men stand in the mud in front of a damaged building and behind Mirian Cisneros, then president of Sarayaku Indigenous community in Ecuador's Amazon. Mirian addresses the camera to call for solidarity and draw attention to a devastating flood (Conaie Comunicación 2020): "we have very sad news because many families now have lost their homes, we have lost our plants and crops, we have lost on a large scale much of the work our families have done." As Mirian speaks, the video cuts to images that depict the effects of the floods: fields wiped clean by the rushing waters, mud and debris strewn throughout homes, twisted metal roofs and remnants of destroyed buildings, and damaged infrastructure. "Therefore" he says, "the Kichwa people of Sarayaku declare that we are in a state of emergency."

The March 17, 2020 flood devastated the community's material infrastructure just as the Covid-19 pandemic began and laid bare the lack of support from provincial and state authorities. Sarayaku representatives denounced that neither group provided a meaningful response to the 380 families that comprise the community one year following the flooding. Commemorating the second anniversary of the Rio Bobonaza flood, Sarayaku released a written statement via the community's website: "There is something that we must keep in mind, we are already feeling the effects of and are victims to climate change. This is the constant struggle of our people" (Sarayaku El Pueblo del Medio Día 2022).

The Ecuadorian Amazon is a site of rich biocultural diversity plagued by vast hydrocarbon reserves (Lu, Valdivia, and da Silva 2017). Petroleum extraction has driven development further into the forest and onto Indigenous lands since the 1940s. The legacy of major transnational oil corporations, particularly deforestation and pollution left in their wake, has generated direct environmental injustices for many Indigenous communities who have reported water contamination, increased cancer rates, and birth defects (Cepek 2018; Sawyer 2022). Ecuadorian oil extraction has far-reaching effects articulated to California's environmental justice communities and global climate change more broadly due to the combined impacts of deforestation and burning fossil fuels on carbon emissions. Ecuadorian oil represents the majority imported by the US state of California (California Energy Commission 2021), where refineries in Oakland have contaminated generations of marginalized communities with toxic pollution through the lasting legacies of redlining (Lee 2021). The effects of climate change across the Amazon Basin are locally distinct, though precipitation events in Ecuador are changing in their timing and intensity with acute impacts on riverine communities like Sarayaku (Funatsu et al. 2019).

The Rio Bobonaza flood in Sarayaku is a conjuncture where intersecting political economic, social, and ecological processes shed light on the urgency of the climate crisis and

DOI: 10.4324/9781003165477-19

importance of transformative climate justice. As political ecologists committed to engaged scholarship, we (i.e., both authors) have been witness to the intersectional impacts of climate change and already-existing forms of injustice that affect many of the communities we work with. Indeed, we have both worked in different capacities with members of the Sarayaku People on climate change mitigation and the rights of environmental defenders. The Rio Bobonaza flood and immediate effects of COVID are two moments in the Sarayaku history, yet they do not define Sarayaku.[1] Moreover, we are aware that political ecology scholarship can run the risk of reproducing harm by recounting painful events (Moulton et al. 2021). Our intent is not to exploit suffering for theory building but to identify pathways toward abolition ecologies (Heynen and Ybarra 2021) through public scholarship that rethinks status quo approaches to the politics of nature, particularly climate change.

The stakes of climate change and rapid action to mitigate its effects cannot be clearer. The Intergovernmental Panel for Climate Change (IPCC) released the Sixth Assessment Report in April 2022. Like the series of recent reports, this one was sobering and sounded the alarm more clearly than ever. Hoesung Lee, the IPCC Chair, said, "We are at a crossroads. The decisions we make now can secure a livable future. We have the tools and the know-how required to limit warming." But we have to act now and in new ways, because the climate crisis is here. Compounding the effects of climate crisis, or perhaps indicative of its broad reach, humans are witnessing the sixth mass extinction (Büscher 2021) while mass consumption exacerbates social injustice (Scheidel et al. 2018), plastic pollution is everywhere (Liboiron 2021), and authoritarian populism is on the rise globally (McCarthy 2019).

The tools to confront the challenges of the climate crisis exist. In this chapter, we argue that political ecology is one such tool but we call specific attention to scholarship that engages nature through a public political ecology (PPE) approach—one that is theoretically informed, advances a critical convergence approach, and conducted "through a community of praxis" (Osborne 2017, 844). More specifically, we provide an overview of PPE while showing how the approach can support transformative climate justice by engaging the politics of nature with attention to power, inequality, and social action. Newell et al. (2021, 2) argue that "transformative climate justice" expands the notion of climate justice beyond normative framings by foregrounding praxis, pluriversality, and decolonization. A transformative approach does not abandon questions of procedural, distributional, and representative justice (on different forms of justice see, e.g., Sze and London 2008; Walker 2009; Pellow 2017; Álvarez and Coolsaet 2018) but rather centers diverse understandings of justice as defined and practiced by myriad actors—not just academics and policymakers in the Global North. Engaging nature in an era of global climate crisis requires a praxis that leverages a suite of tools to effect systems change: from participatory action research to policy analysis.

This chapter briefly discusses some of the broad theoretical currents from which engaged political ecology scholarship emerges by focusing on feminist political ecology, decolonial research, and critical pedagogy. Next, we discuss key elements of doing public political ecology and emphasize that an ethical imperative drives such work to confront injustice. Instead of exhaustively surveying existing literature, we highlight the importance of relationality, decolonized methodologies, and participatory action research by providing examples of ongoing initiatives that demonstrate engaged political ecology in practice. The chapter closes with reflections on the future of PPE amidst the urgency of the global climate crisis.

For Public Political Ecologies of Justice

Political ecology provides a framework to critically assess complex social-ecological systems while centering unequal power relations, politics, and justice through trans-scalar analyses. Piers Blaikie, one of the field's founders, asks a provocative question, "what is political ecology for?" (2016, 26 our emphasis). In answering this question, Blaikie argues for engagement outside of the university, urging political ecologists to make their work relevant to different publics. PPE is *for* justice—climate, environmental, and otherwise. But this raises important questions. Who are the publics of this political ecology? How do political ecologists engage in ethical and meaningful research collaborations with those publics? These are questions we (i.e., both authors) have been debating together and in broader conversations for several years, in the Public Political Ecology Lab that Tracey Osborne hosted at University of Arizona (2011–2019), through a series of multi-session panels we organized at American Association of Geographers (2016 and 2017) and Dimensions of Political Ecology (2018) conferences, and also through critical reflection during our respective work. It seems that every political ecologist conducting public research has a distinct notion about what constitutes the "public" of their work. Rather than posit a precise definition, we highlight ethical, engaged research and critical scholarship that informs our work and through which we identify some of the publics of political ecology.

Environmental conflicts, power, and politics that animate political ecology research are diverse, though related through an enduring concern for (in)justice. With roots in Marxist analyses of structural factors that drive inequality in human-environment relations (Nietschmann 1979) rendered visible through systems analyses (Blaikie and Brookfield 2015), the focus on justice has always charged political ecology with a particular politics of the possible (Escobar 2020). Scholars shown this through distinct approaches that evoke liberation ecologies (Peet and Watts 2004), the enduring hope encapsulated in the seeds sewn following critique (Robbins 2012), and the demands for abolition such that new ecologies of justice might be forged (Heynen and Ybarra 2021; Ranganathan and Bratman 2021). Political ecology is simultaneously a research practice (Paulson et al. 2003) as it is a community (Robbins 2012; Osborne 2017).

Public political ecology (PPE) builds from a rich tradition of action research in political ecology that centers justice (Robbins 2012; Blaikie 2016; Sultana 2022; da Silva and Correia 2022; *inter alia*) and critically evaluates the politics of nature, from its varied meanings to struggles over how different actors seek to engage the so-called "natural" world (Escobar 1999; Paulson, Gezon, and Watts 2003; Heynen, Kaika, and Swyngedouw 2006; Mollett 2021; Osborne and Shapiro-Garza 2018; *inter alia*). PPE goes further by advancing a trans-disciplinary convergence approach guided by a commitment to decolonial methods, praxis, and transformative justice. Gready and Robins's (2014, 340) definition of transformative justice is helpful; they conceive it as "change that emphasizes local agency and resources, the prioritization of process rather than preconceived outcomes and the challenging of unequal and intersecting power relationships and structures of exclusion at both the local and the global level." Blaikie and Brookfield's (2015) germinal framing of the "chains of explanation" make clear the longstanding influence of systems thinking and multi-scalar analysis in political ecology. Much political ecology scholarship has done this work by focusing on case studies (Helmcke 2022). PPE shifts the scale of analysis from case studies to networked action because transformative climate justice requires engaged research across sites with a diverse suite of actors, including local communities, grass-roots organizations, civil society, state decision-makers, academic institutions, and international market actors.

Engaged, justice-focused scholarship purports an a priori normative ethic whereby research will be conducted in ways that deepen respect, cause no harm, and that the outcomes will advance more just ends for frontline communities. Despite such aspirations, "participatory" and engaged research comes with peril (Coombes et al. 2014), drawing attention to the politics of public research, particularly regarding data sovereignty and funding, as Bryan and Wood (2015) and Wainwright (2013) have shown. Merely employing participatory methods does little to alter existing power inequities, oppressive social relations, and the coloniality of geographic research (Radcliffe 2022). Bearing these pitfalls in mind, political ecology is rich with examples of engaged critical scholarship that advances nuanced analyses of injustice while laying the theoretical foundations for more ethical, engaged research practice (Rice et al. 2015; Ranganatha 2015; Bledsoe 2019; de Wit et al. 2021; *inter alia*). Thus, we hold to the transformative potential of engaged research while constantly grappling with the uneven power relations inherent to research *with*—not on or about—different actors and communities who comprise "the public" of this political ecology.

Public political ecology thus never takes for granted the responsibility for ethical research practice and a commitment to praxis informed by insights from feminist geographers, anti-colonial research methods, and the critical pedagogy of Paulo Freire (1993 [1970]) and bell hooks (1994). Both Freire and hooks draw attention to the importance of just educational practice whereby knowledge is shared mutually through processes of relational co-production that decenters expertise. Their insights inform PPE in the field and in the classroom. If education is the practice of freedom, as Freire (1993 [1970]) suggested and "teaching practices are a site of resistance" to the status quo and path to change as hooks (1994, 21) envisioned, then PPE envisions engaged research as a practice of freedom and tool for resistance (see also Brown and Strega 2015). It is through the lens of these vital insights from feminist and decolonial scholars that PPE takes its commitment to ethical practice both in the classroom and in "the field" where we work alongside and with collaborators for transformative climate justice.

Feminist geographers have shaped political ecologists' concern for justice and ethical, engaged research in important ways. Early feminist political ecology showed that power, access, and knowledge about the environment is never merely a question of class relations or inequality but that environmental change is always gendered (Schroeder 1993; Rocheleau et al. 1996). Political ecology analysis has benefited from feminist insights that have reshaped research practice through critical praxis (Elmhirst 2011; Sultana 2021). Particularly necessary are intersectional analyses of race and gender in environmental conflict (Mollett and Faria 2013; Mollett 2021), rethinking the politics of embodiment through exposure to environmental harms (Sultana 2011; Truelove 2011; Johnson et al. 2021), questioning "who" is the subject of political ecology research (Hawkins et al. 2011; Loftus 2020), and the power relations that shape whose knowledge counts in the production of environmental science and policy (Rocheleau 2008; Collard, Dempsey, and Sundberg 2015; Zanotti and Suiseeya 2020; Elias, Joshi, and Meinzen-Dick 2021). Feminist approaches also foreground care (Jarosz 2011; Vaz-Jones 2018; Harcourt and Bauhardt 2019) to center analyses on ethics, both personal and communal, where care in political ecology is "ethical practice" (Jarosz 2004).

As ethical practice, PPE is committed to decolonization. Coloniality drives racialization, gender violence, extractivism, and environmental conflicts (Quijano 2000) that many political ecologists investigate. PPEists advance decolonial praxis yet recognize that "inclusive" or "participatory" research does not alter colonial relations of appropriation and oppression (Alimonda 2019). We seek to enact an "other geography" that resists white supremacist, heteropatriarchal normativity (Oswin 2020) while confronting disciplinary geography's coloniality (Davis and Todd 2017; Daigle and Ramírez 2018), and the fraught use of decolonial discourse in academia that often disassociates the term from the political and material work it refers to (Liboiron 2021). Pivoting toward anti- and decolonial approaches enacts thought and practice against empire and the oppressions of colonial power (Moore and Joudah 2022) without co-opting the language of decolonization as an empty signifier. As Tuck and Yang make clear, "[d]ecolonization offers a different perspective to human and civil rights-based approaches to justice, an unsettling one, rather than a complementary one. Decolonization is not an 'and'. *It is an elsewhere*" (2012, 36 our emphasis). PPE does not seek the forms of commensurability that Tuck and Yang warn of but recognizes incommensurability as a condition of solidarity research that always holds ethical relations at the fore of such work. Here, we draw specific lessons and inspiration from Linda Tuhiwai Smith's enduring and powerful work *Decolonizing Methodologies* that warns of the intimate relations between empire, coloniality, and research while making clear that reciprocity and relationality are necessary to unsettle extractive logics that permeate much academic knowledge production.

Doing Engaged Political Ecology

How does one do "engaged" political ecology? What makes the approach we espouse "public" in relation to other ways of doing political ecology? These are common questions and important to consider. The short answer is that there is not one single path or prescription to share. Aside from the theoretical insights that we surveyed above and the community of praxis that exists among political ecologists and their collaborators, we also take seriously the promise that land-grant universities should serve the public through relevant research, outreach, and teaching. Indeed, we have each benefited from opportunities at land-grant universities, from our respective education to our current roles as faculty members at such institutions. However, the land grant model is clearly not without its problems, principle among them is a history of violence and land theft that must be acknowledged and rectified by supporting social justice, accessibility, and meaningful public engagement with historically marginalized communities (Goldstein, Paprocki, and Osborne 2019). Awareness of those histories underscores our imperative to leverage the power and resources of the university to support social-ecological justice through critical research, outreach, and education. That is not to say that PPE is expressly tied to land-grant universities, but that our approach to PPE is informed by a commitment to advance research that makes an impact beyond the walls of the university. We highlight three concepts helpful to all PPE and engaged political ecology research: relationality, participatory research, and decolonized methodologies.

PPE centers *relationality*. Relationality can be understood in many ways. Doreen Massey's (1991) analysis has shaped a current in political ecology scholarship that examines place through a relational frame (Bebbington and Batterbury 2001; Escobar 2001; Nelson and Seager 2005). Other analyses highlight relationality by examining how environmental politics are enmeshed with political economies rooted in place (Rocheleau and Roth 2007; Cantor et al. 2018). Indigenous and feminist political ecologists provide another way of understanding relationality, that of being in relation to and with one another, place, and other-than-human natures (Coombes, Johnson, and Howitt 2012; Sundberg 2014; Middleton 2015; Zanotti et al. 2020; Mollett 2021). Here, we are also thinking about relationality beyond hierarchical scale, attentive to non-human actors, and based on long-term collaborations (Blaser 2010; Larsen and Johnson 2017).

When done well, PPE fosters ethical relationality through a research process that attends to the insights and concerns of all involved parties. This is not easy to do. The time-intensive process views the role of academic research as *responding to* public interest and designed in collaboration with public partners (see, e.g., Osborne et al. 2021; Correia 2022), not solely for the benefit of academic researchers. Inherent to this approach is a rejection of parachute research intended to extract information for a specific output. Demands to publish or perish, to produce outputs for funders, and respond to diverse and sometimes divergent interests are not easily resolved. Indeed, these factors often undermine engaged research based on collaborative partnerships. But there are pathways toward this work and several existing examples that provide inspiration (see Table 14.1).

Table 14.1 Some key (but not all!) methodological considerations that inform PPE and engaged political ecology research

Participatory action research (PAR)	At its best, PAR is an approach based on prior consultation and collaboration with partners whose insights inform the research process from design and data collection to outputs. Effective PAR responds to public interests and requires time to build trust, establish responsibilities and ensure accountability. PAR should be the antonym to extractive research by empowering partners and ensuring meaningful outcomes. See also Austin (2004), Osborne (2017), Correia (2022).
Relationality	A complex concept too nuanced to fully detail here, political ecologists have used notions of relationality to evaluate trans-scalar processes, to critique the idea of hierarchical scale altogether, and to center other than "Western/modern" ontologies. We draw attention to relationality because it impels a research ethic based on care and long-term commitment (i.e., relations) with partners and place. See also Rocheleau and Roth (2007), Whyte (2020), Mollett (2021), and Liboiron (2021).
Praxis	Praxis is the synthesis of theory and practice in ways in which ideas become a material force for social change. See Gramsci (2012 [1932–1935]). "While Gramsci's work is compatible with other critical scholars of praxis—for example Paulo Freire and bell hooks who focus on liberatory pedagogy for people marginalized along lines of race, class and gender (Freire 1993 [1970]; hooks 2014)—it also provides an understanding of the *conjuncture*, the ways in which political economic and cultural forces come together in critical moments for social transformation" (Osborne 2017, 850).
Decolonialized methodologies	Decolonizing research avoids extractive practices that reinforce unequal power relations established under imperialism and colonialism. It centers the knowledge and worldviews of Indigenous peoples and marginalized communities with an aim toward transformative social justice (Tuhiwai Smith 2012). According to Middleton (2015, 564), Indigenous frames "become the primary ordering system within which intersecting ecological, political and economic factors are understood (alongside issues of scale, differentiated populations and time)." See also Tuck and Yang (2012); De Leeuw and Hunt (2018), Liboiron (2021).

(*Continued*)

Table 14.1 (Continued)

Convergence research	Beyond interdisciplinarity, convergence research intentionally draws together actors from different sectors through action research to confront the complexity of social-ecological challenges that many political ecologists study. It is an "approach to knowledge production and action that involves diverse teams working together in novel ways—transcending disciplinary and organizational boundaries—to address vexing social, economic, environmental, and technical challenges in an effort to ... promote collective well-being" (Peek et al. 2020).

Engaged political ecology should be about *partnerships*, not projects (see, e.g., Austin 2004). The former fosters mutual responsibility and more equal relations, while the latter is outcome-driven and often dictated by the pressure to publish rather than meet community interests. Indeed, much of this work employs *participatory action research* (PAR) to achieve its goals, whereby the research process is as important as any specific outcome (Radonic and Kelly-Richards 2015; Osborne 2017; Gustafson 2021). We want to emphasize that publicly-engaged research and long-term collaborations do not absolve unequal power—be that to access resources, to communicate findings, and other facets of community-based academic research (Lopez 2020). Power inequities always exist. Yet, committing to long-term partnerships based on collaboration and critical reflexivity invites a more horizontal and reciprocal research process that shifts power to non-academic collaborators (Zanotti et al. 2020).

PAR is not without its limits. But we have found that truly participatory approaches based on partnerships help ensure that research is not theft (Robbins 2006) but a process that empowers. Here we return to Osborne's (2017, 845) aspiration: "At its best, PPE might serve as a community of praxis where stakeholders in- and outside the academy work to develop an Earth Stewardship that integrates environmental sustainability and social justice and makes careful yet powerful use of innovations from public geographies such as participatory action research and mapping, service learning and social media." There is no template for building relations with public partners, ensuring meaningful PAR processes, or how to "apply" a relational approach that we can share. However, there are many critical environment scholars and activists collaborating through long-term commitments that provide helpful guidance. We have noted a few exciting PPE initiatives in Table 14.2. that demonstrate a variety of approaches, scales, and pathways for

Table 14.2 Public political ecology in action. Here we share several initiatives that embody the community of praxis and engaged research central to PPE. Although the people who have organized and are involved in these projects might not define them as PPE, we share them because each provides inspiration and insights about moving such work forward in different ways and across scales

Initiative	*Brief description*	*Website*
University of California Center for Climate Justice	An initiative that aims to "leverage and harness the power of the university to support, strengthen, and build an emergent climate justice ecosystem and social movement that solves the climate crisis through science, systems-thinking, and social-ecological justice."	https://centerclimatejustice.universityofcalifornia.edu

(*Continued*)

Table 14.2 (Continued)

Initiative	*Brief description*	*Website*
Decolonizing Water	"Our goal is to create a prototype of an Indigenous-led community-based water monitoring initiative that is rooted in Indigenous laws, and is a practical expression of Indigenous water governance"	https://decolonizingwater.ca
Environmental Justice Atlas	"The environmental justice atlas documents and catalogues social conflict around environmental issues ... It also attempts to serve as a virtual space for those working on EJ issues to get information, find other groups working on related issues, and increase the visibility of environmental conflicts."	www.ejatlas.org
Climate Alliance Mapping Project	"The Climate Alliance Mapping Project (CAMP) builds interactive climate justice story maps that bring together scientific data and digital stories produced by affected communities to educate the public, connect local communities with global climate justice networks, and inform policy decisions."	https://climatealliancemap.org
Civic Laboratory for Environmental Action Research (CLEAR)	"CLEAR's ways of doing things, from environmental monitoring of plastic pollution to how we run lab meetings, are based on values of humility, accountability, and anti-colonial research relations. We specialize in community-based and citizen science monitoring of plastic pollution, particularly in wild food webs, and the creation and use of anti-colonial research methodologies."	https://civiclaboratory.nl
Critical Ecology Lab	"Our mission is to create novel processes and spaces for communities of people with scientific and generational knowledge to destabilize oppressive systems and fight back against escalating social and planetary disaster. We aim to create impact-driven research that shifts public dialogue and civic action towards greater justice and, as a result, more stable and functioning ecological systems."	www.criticalecologylab.org
Governance and Infrastructure in the Amazon (GIA)	"The GIA project seeks to create, strengthen and expand a Community of Practice and Learning on the use of tools and strategies by conservation and development practitioners from NGOs, community organizations, government and academia. The GIA theory of change is that by bringing practitioners together to share experiences, reflect and dialogue, they will collaboratively learn and adapt, improve their use of tools and strategies to improve social-environmental governance."	https://giamazon.org

imagining how a PPE approach can be used to support transformational climate justice and attend to other critical issues that mark this conjuncture of environmental crisis.

Why Public Political Ecology and Why Now?

Simply, because the stakes are so high.

The chapter opens with the Rio Bobonaza flood in Sarayaku Ecuador to underscore the urgency of the climate crisis and the ways in which it is deeply entwined with a range of social, racial and political issues and injustices. Because of this, we suggest that political ecologists must do more than study nature-society relations. We must critically engage the politics of contemporary social-ecological crises through convergence research that advances decolonial praxis. The US National Science Foundation has charged scholars with tackling "grand problems" through research that transcends scale, sector, and discipline. The complexity of problems like the climate crisis requires collaboration, creativity, and collective action. It is necessary to document and evaluate how localized forms of environmental injustice manifest. Political ecology case studies are vital for this. Yet climate justice requires work beyond the local and must contend with other sites of power, decision making, and action. This is beyond a call for relevancy (Blaikie 2010). It is an ethical impetus to use political ecology research and the power of the university to effect broader social transformation outside of the academy needed now more than ever.

This is a historic conjuncture on a global scale. Scientific consensus makes it resoundingly clear that sweeping action across multiple scales and sectors must happen before 2030 if truly catastrophic climate change is to be avoided (Intergovernmental Panel on Climate Change 2022). This is not an unfounded alarmist argument but a robust response to the overwhelming peer-reviewed scientific data on climate change—from the physical drivers and processes to adaptation and mitigation. Experiences of communities we work with in Paraguay (Correia 2019, 2022) and Ecuador make clear the stakes of the climate crisis. Still, those stakes are visible in the annual cycle of devastating wildfires across the US west, the extreme heatwaves that threaten human life and destroy crops across the Indian subcontinent, the massive flooding events that have recently inundated cities across Europe, and sea-level rise encroaching on sites from Miami, Florida to low-lying island nations. Climate change is here. Yet we have the tools to make the changes for a livable future that minimizes climate injustice. Engaged political ecology inspired by critical hope is one of those tools, as the examples in Table 14.2 and much of the research we cite here show.

We also point to the recent IPCC reports and related advocacy as another example of PPE. The IPCC process takes convergence research to another scale, involving hundreds of scientists, thousands of studies, and countless hours of collaborative work and thought across disciplines. Contributing authors and scientists are committed to communicating their findings broadly to publics worldwide through pedagogies that break down barriers between academia and beyond. There is undeniably a need for more Indigenous and frontline community involvement in the IPCC process, but we are encouraged that a significant body of critical social science research led by and involving Indigenous peoples informs the findings. The threat of catastrophic climate change has never been more evident, nor is the pathway to address the problem.

Doing PPE and engaged political ecology requires hope alongside action research co-created with frontline communities navigating environmental change (see, e.g., Correia 2022), across networks of large academic institutions (see, e.g., UC Center for Climate Justice, https://centerclimatejustice.universityofcalifornia.edu), to community, NGO, academic, and state collaborations across large regions (see, e.g., the Governance and Infrastructure in the Amazon, https://giamazon.org). Herein lies a vital characteristic that distinguishes *public* political ecology from other approaches—there is an explicit commitment to Gramsci's praxis (2012 [1932–1935]) "wherein ideas become a material force for emancipatory social change" (Osborne 2017, 850) that is always rooted in a community of action. We aim to support transformative climate justice, though that is not the only path for PPE. What matters is using the tools to enact critically informed change. If praxis is "action and reflection on the world to change it," as bell hooks (1994, 14) argued, then political ecologists must be present at this conjuncture to use their tools to engage the politics of nature in the name of supporting transformative justice.

Note

1 Learn more about Sarayaku initiatives and history as written by community members at the following websites: https://report.territoriesoflife.org/territories/sarayaku-ecuador/ and https://kawsaksacha.org/

References

Alimonda, Hectór. 2019. The coloniality of nature: An approach to Latin American political ecology. *Alternautas*, 6(1): 102–142.

Álvarez, Lina and Coolsaet, Brendan. 2018. Decolonizing environmental justice studies: A Latin American perspective. *Capitalism Nature Socialism* 31(2): 50–69.

Austin, Diane. 2004. Partnerships, not projects! Improving the environment through collaborative research and action. *Human Organization* 63(4): 419–430.

Bebbington, Anthony J. and Batterbury, Simon P.J. 2001. Transnational livelihoods and landscapes: Political ecologies of globalization. *Cultural Geographies* 8(4): 369–380.

Blaikie, Piers. 2010. Should some political ecology be useful? The inaugural lecture for the Cultural and Political Ecology Specialty Group, annual meeting of the Association of American Geographers, April 2010. *Geoforum* 43(2): 231–239.

Blaikie, Piers. 2016. Towards an engaged political ecology. In *Alternative development: Unravelling marginalization, voicing change*. Catherine Brun, Piers Blaikie, and Michael Jones eds. pp. 25–38. London: Routledge.

Blaikie, P. and Brookfield, H. 2015. *Land degradation and society*. Routledge.

Blaser, Mario. 2010. *Storytelling globalization from the Chaco and beyond*. Durham: Duke University Press.

Bledsoe, Adam. 2019. Afro-Brazilian resistance to extractivism in the Bay of Aratu. *Annals of the American Association of Geographers* 109(2): 492–501.

Brown, Leslie and Strega, Susan. 2015. *Research as resistance: Revisiting critical, Indigenous, and anti-oppressive approaches*, 2nd edition. Toronto: Canadian Scholars' Press.

Bryan, J. and Wood, D. 2015. *Weaponizing maps: Indigenous peoples and counterinsurgency in the Americas*. Guilford Publications.

Büscher, Bram. 2021. Political ecologies of extinction: From endpoint to inflection-point. *Journal of Political Ecology* 28: 696–704.

California Energy Commission. 2021. *Foreign sources of crude oil imports to California 2020*. State of California Energy Commission. Accessed August 15, 2022 www.energy.ca.gov/data-reports/energy-almanac/californias-petroleum-market/foreign-sources-crude-oil-imports

Cantor, Alida, Stoddar, Elisabeth, Rocheleau, Dianne, Brewer, Jennifer F., Roth, Robin, Birkenholtz, Trevor, Foo, Katherine, Nirmal, Padini. 2018. Putting rooted networks into practice. *ACME An International Journal for Critical Geographers* 17(4): 958–987.

Cepek, Michael. 2018. *Life in oil: Survival in the petroleum fields of Amazonia*. Austin: University of Texas Press.

Climate Alliance Mapping Project. 2022. Climate justice story maps. Accessed April 10, 2022. https://climatealliancemap.org

Collard, Rosemary-Claire, Dempsey, Jessica, and Sundberg, Juanita. 2015. A manifesto for abundant futures. *Annals of the American Association of Geographers* 105(2): 322–330.
Conaie Comunicación. 2020. Inundación Sarayaku. 21 March. Accessed March 25, 2022. www.facebook.com/conaie.org/videos/inundaci%C3%B3n-sarayaku/229142805137629/
Coombes, Brad, Johnson, Jay T, and Howitt, Richard. 2012. Indigenous geographies I: Mere resource conflicts? The complexities in Indigenous land and environmental claims. *Progress in Human Geography* 36(6): 810–821.
Coombes, Brad, Johnson, Jay T, and Howitt, Richard. 2014. Indigenous geographies III: Methodological innovation and the unsettling of participatory research. *Progress in Human Geography* 38(6): 845–854.
Correia, Joel E. 2019. Unsettling territory: Indigenous mobilizations, the territorial turn, and the limits of land rights in the Paraguay-Brazil borderlands. *Journal of Latin American Geography* 18(1): 11–37.
Correia, Joel E. 2022. Between flood and drought: Environmental racism, settler waterscapes, and Indigenous water justice in South America's Chaco. *Annals of the American Association of Geographers* https://doi.org/10.1080/24694452.2022.2040351
Daigle, Michelle, and Ramírez, M., 2018. Decolonial geographies. In *Keywords in radical geography: Antipode at 50*, T. Jazeel, et al. ed., 78–84. London: Wiley-Blackwell.
da Silva, M.S.R. and Correia, J. 2022. A political ecology of jurisdictional REDD+: Investigating social-environmentalism, climate change mitigation, and environmental (in) justice in the Brazilian Amazon. *Journal of Political Ecology*, 29(1): 123–142.
Davis, Heather and Todd, Zoe. 2017. On the importance of a date, or decolonizing the Anthropocene. *Acme: An International Journal for Critical Geographies* 16(4): 761–780.
Decolonizing Water. 2022. "Our approach." Accessed April 28, 2022. https://decolonizingwater.ca/our-approach/
De Leeuw, Sarah and Hunt, Sarah. 2018. Unsettling decolonizing geographies. *Geography Compass* 12 (7): p. e12376.
De Wit, Maywa Montenegro, Shattuck, Annie, Iles, Alastair, Graddy-Lovelace, Garrett, Roman-Alcalá, Antonio, and Chappell, M. Jahi. 2021. Operating principles for collective scholar-activism: Early insights from the agroecology research action-collective. *Journal of Agriculture, Food Systems, and Community Development* 10(2): 319–337.
Elias, Marléne, Joshi, Deepa, and Meinzen-Dick, Ruth. 2021. Restoration for whom, by whom? A feminist political ecology of restoration. *Ecological Restoration* 39(1–2): 3–15.
Elmhirst, Rebecca. 2011. Introducing new feminist political ecologies. *Geoforum* 42(2): 129–132.
Escobar, Arturo. 1999. After nature: Steps to antiessentialist political ecology. *Current Anthropology* 40(1): 1–30.
Escobar, Arturo. 2001. Culture sits in places: Reflections on globalism and subaltern strategies of localization. *Political Geography* 20(2): 139–174.
Escobar, Arturo. 2020. *Pluriversal politics: The real and the possible*. Duke University Press.
Freire, Paulo. 1993 [1970]. *The pedagogy of the oppressed*. Trans. Myra Bergman Ramos. New York: Continuum International Publishing.
Funatsu, Beatriz M., Dubreuil, Vincent, Racapé, Amandine, Debortoli, Nathan S., Nasuti, Stéphanie, and Le Tourneau, François-Michel. 2019. Perceptions of climate and climate change by Amazonian communities. *Global Environmental Change* 57: 101923. https://doi.org/10.1016/j.gloenvcha.2019.05.007
Goldstein, Jenny E., Paprocki, Kasia, and Osborne, Tracey. 2019. A manifesto for a progressive land-grant mission in an authoritarian populist era. *Annals of the American Association of Geographers* 109(2): 673–684.
Gramsci, Antonio. 2012 [1932–1935]. *Selections from the prison notebooks of Antonio Gramsci*. Hoare, Q. and G. Nowell-Smith eds., New York: International Publishers.
Gready, Paul and Robins, Simon. 2014. From transitional to transformative justice: A new agenda for practice. *The International Journal of Transitional Justice* 8: 339–361.
Gustafson, Seth. 2021. Children breathe their own air. *Area* 53: 106–113.
Harcourt, W. and Bauhardt, C. 2019. *Feminist political ecology and the economics of care*. Routledge.

Hawkins, Roberta, Ojeda, Diana, Asher, Kiran, Baptiste, Brigitte, Harris, Leila, Mollett, Sharlene, Nightingale, Andrea, Rocheleau, Dianne, Seager, Joni, and Sultana, Farhana. 2011. A discussion. *Environment and Planning D: Society and Space* 29(2): 237–253.
Helmcke, Cornelia. 2022. Ten recommendations for political ecology case research. *Journal of Political Ecology* 29(1): 266–276.
Heynen, Nik, Kaika, Maria, and Swyngedouw, Erik. 2006. *In the nature of cities: Urban political ecology and the politics of urban metabolism*. London: Routledge.
Heynen, Nik and Ybarra, Megan. 2021. On abolition ecologies and making "freedom as a place". *Antipode* 53(1): 21–35.
hooks, bell. 1994. *Teaching to transgress*. New York: Routledge.
Intergovernmental Panel on Climate Change. 2022. Climate change 2022: Impacts, adaptation and vulnerability. Working Group II Sixth Assessment Report. United Nations Environmental Program. Accessed April 25, 2022 www.ipcc.ch/report/ar6/wg2/
Jarosz, Lucy. 2004. Political ecology as ethical practice. *Political Geography* 23(7): 917–927.
Jarosz, Lucy. 2011. Nourishing women: Toward a feminist political ecology of community supported agriculture in the United States. *Gender, Place and Culture* 18(3): 307–326.
Johnson, Adrienne, Zalik, Anna, Mollett, Sharlene, Sultana, Farhana, Havice, Elizabeth, Osborne, Tracey, Valdivia, Gabriela, Lu, Flora. 2021. Extraction, entanglements, and (im)materialities: Reflections on the methods and methodologies of natural resource industries fieldwork. *Environment and Planning E: Nature and Space* 4(2): 383–428.
Larsen, Soren C. and Johnson, Jay T. 2017. *Being in place: Indigenous coexistence in a more than human world*. Minneapolis: University of Minnesota Press.
Lee, Charles. 2021. Confronting disproportionate impacts and systemic racism in environmental policy. *Environmental Law Reporter* 51: 10207.
Liboiron, Max. 2021. *Pollution is colonialism*. Durham: Duke University Press.
Loftus, Alex. 2020. Political ecology III: Who are "the people"? *Progress in Human Geography* 44(5): 981–990.
Lopez, Christina W. 2020. Community geography as a model for improving efforts of environmental stewardship. *Geography Compass* 14: e12485.
Lu, Flora, Valdivia, Gabriela, Silva, Nestor. 2017. *Oil, revolution, and indigenous citizenship in Ecuadorian Amazonia*. New York: Palgrave.
Massey, Doreen. 1991. A global sense of place. *Marxism Today* (38): 24–29.
McCarthy, James. 2019. Authoritarianism, populism, and the environment: Comparative experiences, insights, and perspectives. *Annals of the American Association of Geographers* 109(2): 301–313.
Middleton, Beth Rose. 2015. Jahát Jatítotódom: Toward an indigenous political ecology. In *The International Handbook of Political Ecology*. Raymond L. Bryant ed. Cheltenham: Edward Elgar. pp. 561–576.
Mollett, Sharlene. 2021. Hemispheric, relational, and intersectional political ecologies of race: Centering land-body entanglements in the Americas. *Antipode* 53: 810–830.
Mollett, Sharlene and Faria, Caroline. 2013. Messing with gender in feminist political ecology. *Geoforum* 45: 116–125.
Moore, Adam and Joudah, Nour. 2022. The significance of W. E. B. Du Bois's decolonial geopolitics. *Annals of the American Association of Geographers* https://doi.org/10.1080/24694452.2022.2035667
Moulton, Alex A., Velednitsky, Stepha, Harris, Dylan M., Cook, Courtney B., and Wheeler, Britany L. 2021. On and beyond traumatic fallout: Unsettling political ecology in practice and scholarship. *Journal of Political Ecology* 28(1): 677–695.
Nelson, Lise and Seager, Joni. 2005. *A companion to feminist geography*. Malden: Blackwell.
Newell, Peter, Srivastava, Shilpi, Naess, Lars Otto, Torres Contreras, Gerardo A., and Price, Roz. 2021. Toward transformative climate justice: An emerging research agenda. *WIRES Climate Change* 12(6): e733. https://doi.org/10.1002/woc.733
Nietschmann, Bernard. 1979. *Between land and water: The subsistence ecology of the Miskito Indians, eastern Nicaragua*. London: Seminar Press.
Osborne, Tracey. 2017. Public political ecology: A community of praxis for earth stewardship. *Journal of Political Ecology* 24(1): 843–860.

Osborne, Tracey and Shapiro-Garza, Elizabeth. 2018. Embedding carbon markets: Complicating commodification of ecosystem services in Mexico's forests. *Annals of the American Association of Geographers* 108(1): 88–105.

Osborne, Tracey, Brock, Samara, Chazdon, Robin, Chomra, Susan, Garen, Eva, Gutierrez, Victoria, Lave, Rebecca, Lefevre, Manon, Sundberg, Juanita. 2021. The political ecology playbook for ecosystem restoration: Principles for effective, equitable, and transformative landscapes. *Global Environmental Change* 70: 102320. https://doi.org/10.1016/j.gloenvcha.2021.102320

Oswin, Natalie. 2020. An other geography. *Dialogues in Human Geography* 10(1): 9–18.

Paulson, Susan, Gezon, Lisa L., and Watts, Michael. 2003. Locating the political in political ecology: An introduction. *Human Organization* 62(3): 205–217.

Peek, Lori, Tobin, Jennifer, Adams, Rachel M., We, Haorui, and Mathews, Mason Clay. 2020. A framework for convergence research in the hazards and disaster field: The natural hazards engineering research infrastructure CONVERGE facility. *Frontiers in Built Environment* 6: 110. https://doi.org/10.3389/fbuil.2020.00110

Peet, Richard and Watts, Michael. 2004. *Liberation Ecologies*, 2nd Edition. New York: Routledge.

Pellow, David. 2017. *What is critical environmental justice studies?* London: Polity.

Quijano, Aníbal. 2000. Coloniality of power and Eurocentrism in Latin America. *International Sociology* 15(2): 215–232.

Radcliffe, Sarah. 2022. *Decolonizing geography: An introduction*. London: Wiley.

Radonic, Lucero and Kelly-Richards, Sarah. 2015. Pipes and praxis: A methodological contribution to the urban political ecology of water. *Journal of Political Ecology* 22: 357–465.

Ranganathan, Malini. 2015. Storm drains as assemblages: The political ecology of flood risk in post-colonial Bangalore. *Antipode* 47(5): 1300–1320.

Ranganathan, M. and Bratman, E., 2021. From urban resilience to abolitionist climate justice in Washington, DC. *Antipode* 53(1): 115–137.

Rice, Jennifer L., Burke, Brian J., and Heynen, Nik. 2015. Knowing climate change, embodying climate praxis: Experiential knowledge in Southern Appalachia. *American Association of Geographers* 105(2): 253–262.

Robbins, Paul. 2006. III, Research is theft: Environmental inquiry in a postcolonial world. In *Approaches to Human Geography*. Aitken, Stuart and Valentine, Gil eds. London: Sage. pp. 311–324.

Robbins, Paul. 2012. *Political ecology: A critical introduction*, 1st Edition. New York: Wiley.

Rocheleau, Dianne. 2008. Political ecology in the key of policy: From chains of explanation to webs of relation. *Geoforum* 39(2): 716–727.

Rocheleau, Dianne and Roth, Robin. 2007. Rooted networks, relational webs and powers of connection: Rethinking human and political ecologies. *Geoforum* 38(3): 433–437.

Rocheleau, Dianne, Thomas-Slayter, Barbara, Wangari, Esther. 1996. *Feminist political ecology*. London: Taylor & Francis Group.

Sawyer, Suzana. 2022. *The small matter of suing Chevron*. Durham: Duke University Press.

Sarayaku El Pueblo del Medio Día. 2022. Sarayaku después de la inundación. Accessed March 25, 2022. https://sarayaku.org/sarayaku-despues-de-la-inundacion/

Scheidel, Arnim, Temper, Leah, Demaria, Federico, and Martínez-Alier, Joan. 2018. Ecological distribution conflicts as forces for sustainability: An overview and conceptual framework. *Sustainability Science* 13: 585–598.

Schroeder, Richard A. 1993. Shady practice: Gender and the political ecology of resource stabilization in Gambian Garden/Orchards. *Economic Geography* 69(4): 349–365.

Sultana, Farhana. 2011. Suffering for water, suffering from water: Emotional geographies of resource access, control and conflict. *Geoforum* 42(2): 163–172.

Sultana, Farhana. 2021. Political ecology 1: From the margins to the center. *Progress in Human Geography* 45(1): 156–165.

Sultana, Farhana. 2022. Critical climate justice. *Royal Geographic Society* 188: 118–124.

Sundberg, Juanita. 2014. Decolonizing posthumanist geographies. *Cultural Geographies* 21(1): 33–47.

Sze, Julie and London, Jonathan K. 2008. Environmental justice at the crossroads. *Sociology Compass* 2(4): 1331–1354.

Truelove, Yaffa. 2011. (Re)conceptualizing water inequality in Delhi, India through a feminist political ecology framework. *Geoforum* 42(2): 143–152.
Tuck, Eve and Yang, K. Wayne. 2012. Decolonization is not a metaphor. *Decolonization: Indigeneity, Education and Society* 1(1): 1–40.
Tuhiwai Smith, Linda. 2012. *Decolonizing methodologies: Research and Indigenous peoples*. 2nd edition. London: Zed.
Vaz-Jones, Laura. 2018. Struggles over land, livelihood, and future possibilities: Reframing displacement through feminist political ecology. *Journal of Women in Culture and Society* 43(3): 711–735.
Wainwright, Joel. 2013. *Geopiracy: Oaxaca, militant empiricism, and geographical thought*. New York: Palgrave Pivot.
Walker, Gordon. 2009. Beyond distribution and proximity: Exploring the multiple spatialities of environmental justice. *Antipode* 41(1): 614–636.
Whyte, Kyle. 2020. Too late for indigenous climate justice: Ecological and relational tipping points. *WIREs Climate Change* 11(1): e603.
Zanotti, Laura and Suiseeya, Marion. 2020. Doing feminist collaborative event ethnography. *Journal of Political Ecology* 27(1): 961–987.
Zanotti, Laura, Carother, Courtney, Apok, Charlene, Huang, Sarah, Coleman, Jesse, Ambrozek, Charlotte. 2020. Political ecology and decolonial research: Co-production with the Iñupiat in Utqiagvik. *Journal of Political Ecology* 27: 43–66.

15 Gendering Nature

From Ecofeminism to Feminist Political Ecology

Deepti Chatti

In the summer of 2023, as I waited for a prenatal massage appointment to begin at a therapy center in Palo Alto, California, my eyes fell on the stack of reading material provided for the reading pleasure of those waiting. The latest *National Geographic* magazine was available. "8 BILLION," it said in bold text overlaid on the image of the earth. Underneath the image, in red text, was the tagline, "THE POPULATION PARADOX." I had followed the news about the world's population numbers exceeding eight billion people, and it did not surprise me that National Geographic had chosen this topic as their cover story. For most feminist scholars and activists, the personal and the political have always been inseparable and entangled. I had considered very carefully and critically for several years how a decision to bring another human into the world to join my multispecies kin would align with my intellectual, social, and political goals in the realm of transnational feminist, environmental, and climate justice. Geographers, development organizations, and people concerned about environmental degradation of various professional persuasions have long been interested in how many people exist in the world. Often debated as the term "population control," and proposed as a solution to "overpopulation," several environmentalists have concerned themselves with the reproductive choices of others, often low income women in the Global South (Sasser 2018); for an influential and highly critiqued text in this vein, see Paul Ehrlich's *The Population Bomb*, 1968.) Feminist scholars and activists interested in gender–environment relationships and reproductive justice have long pushed back against "population control," as the concept is mired in eugenics, coercive interventions against bodies deemed unsuitable and inferior to reproduce, informed by imperialist and colonial ideologies, tied in problematic ways to xenophobic and racist politics, and last but not least, totally ineffectual at the purported claims of "saving the environment" as its policies are always targeted towards poor people of color in the Global South who have a much smaller impact on the environment through their consumptive patterns compared with anyone who lives in the global North (Sasser 2018; Sayre 2008; Hartmann 2016; Hendrixson et al. 2020; Bhatia et al. 2020). Further, focusing on individual consumption serves as a distraction from the systemic causes of climate change (Haraway 2015; Moore 2017; Johnson et al. 2022) and turns attention away from the radical reorganization needed in the ways we account for value, applaud endless growth, make food, organize our cities, produce energy, and devalue non-human nature. While anxieties about overpopulation never truly went away in development circles, they have resurfaced with renewed vigor and visibility in the Anthropocene (Johnson et al. 2022), with varied motivations (Hartmann and Barajas-Román 2011) (Figure 15.1).

DOI: 10.4324/9781003165477-20

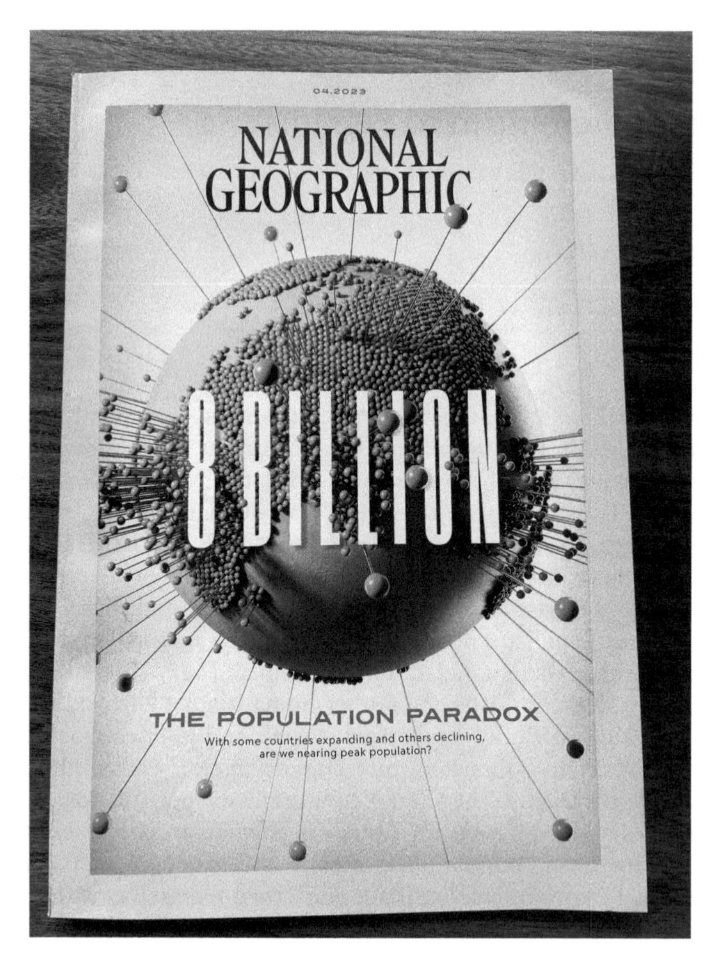

Figure 15.1 Cover of the April 2023 issue of *National Geographic*.

As Jade Sasser writes in her review of *Making Kin Not Population* (Clarke and Haraway 2018), an edited volume bringing together feminist Science and Technology Studies (STS) scholars to revisit the population question in the context of climate change from a radical, feminist, justice-oriented and anticolonial lens:

> Focusing on population-as-human-numbers lifts demographic trends out of the social contexts of racism, structural violence enacted against the poor, deeply entrenched poverty, grinding inequalities that lead to different rates of birth, sickness, and death, and the everyday rapacious violence done to the planet via capitalist systems of resource extraction.
>
> (Strathern et al. 2019: 162)

Some of the contributors of the edited volume align more closely with Sasser on the issue of population, focusing instead on the concept of reproductive justice as a condition for collective thriving (see Michelle Murphy's contribution), while others insist on the possibility of "a radical, nonhumanist demography embedded in multi-kinded/multi-species justice and care" (see Donna Haraway's response to Strathern and Sasser's review, ibid.: 170).

I use this example to illustrate two key points relevant to the chapter that follows. First, feminist scholars have long intervened in environmental narratives bringing in critical perspectives to more dominant discourses on environmental degradation and the gender/nature topic more broadly; and second, far from speaking in unison, feminist scholars have sometimes passionately disagreed with each other on environmental topics, and not shied away from debating each other, writing with, and collaborating across differences of opinion (see Subramaniam 2018 on how this debate models respectful dialogue and serves as an antidote to the erosion of public debate and discourse). These varied feminist positions have been collectively edifying to political ecologists, as a nuanced way to think through differences and sit with discomfort, especially with people and collectives with whom one has some shared social and political goals, and also to highlight the heterogeneity of ways of thinking about gender–environment relations.

The interconnections between gender injustice and the degradation of the environment has received continued interest from scholars and activists over several decades. This includes the connections between the concepts of gender and nature, how society's interactions with environments are gendered, how environmental impacts and hazards are gendered, how knowledge about the environment is gendered, and the interlinkages between (and consequences of) the feminization of nature and the naturalization of gender. As the reader can imagine, this is vast intellectual ground to cover and this chapter cannot include all of the above topics in rich detail. My goal is to provide the reader with an understanding of the connections between the concepts "Gender" and "Nature," to introduce allied concepts and literatures for further reading, and to provide some helpful conceptual tools for the contemporary moment of environmental and social change. Reflecting the feminist underpinnings of much of this work, rather than an unequivocal canon, or a single intellectual history of the field, there are plural voices emerging from varied epistemic traditions including political ecology, environmental anthropology, feminist science studies, environmental humanities, and development studies. The plural narratives have sometimes come together, amplifying their theoretical contributions, and at other times have spoken in tension with each other. In this chapter, I trace some of these intellectual trajectories for the reader, and also highlight the tensions between varied intellectual positions. Where appropriate, I provide suggestions for further reading.

Tracing the History of Thinking about Gender and Nature Together

The feminist anthropologist Sherry Ortner was one of the first scholars to describe the connection between gender and nature. In her widely cited article "Is female to male as nature is to culture?" Ortner sought to explain the "universality of female subordination" across different cultures. Ortner asks: "What could there be in the generalized structure and conditions of existence, common to every culture, that would lead every culture to place a lower value upon women?"(Ortner 1972: 71). Ortner theorizes that women's subordination across multiple different societies flows from the widespread association of women with the conceptual category of "nature"; even if not directly identified as being part of nature, as having closer affinity with, and as being "less transcendental of nature than men" (ibid.: 73). Ortner draws on another influential feminist text to make her arguments, Simone de Beauvoir's *The Second Sex* (1953 [1949]). As a feminist anthropologist and critical scholar, Ortner is clear that the very categories of nature and culture are a product of culture, that this "closeness" of women to nature is not "real" but that there are physical, social, and psychological reasons why it seems that way related to women's physiology, biology, and widespread roles in social reproduction.

Ortner's conclusion was to emphasize the need to change social institutions and cultural assumptions such that "both men and women can and must be equally involved in projects of creativity and transcendence" (Ortner 1972: 86). Another influential early text in thinking about the historical connections between gender and nature, especially within North American and European academia, was Carolyn Merchant's *Death of Nature* (1980). By describing the conceptual shift from thinking of nature as an organism to thinking of nature as a machine, Merchant, a feminist environmental historian, traces the structural linkages between the domination of women and the domination of nature made possible by the scientific revolution in Europe. In sixteenth-century Europe, Merchant argues that the dominant understanding of life, society, and the world was *organismic*, emphasizing interconnections between beings, and a belief that there was a vital force animating the environment. As Merchant writes, "Central to the organic theory was the identification of nature, especially the earth, with a nurturing mother: a kindly beneficent female who provided for the needs of mankind in an ordered, planned, universe" (Merchant 1980: 2). Simultaneously, there existed another view of nature as wild, uncontrollable, and chaotic—the source of storms, droughts, and other disasters. Both these views associated "nature" with "female," which in subsequent centuries led to a belief that nature required subjugation and control by machinery, industrialization, and commerce, which were seen in the domain of "male" pursuits. Thus arose the concept of "ecofeminism," which brought together the domination of women and the degradation of nature into theoretical conversation together. Within the frameworks of Western philosophy, the dyads nature/culture, female/male, mind/body could be seen as mapped onto each other with one being inferior to the other.

Ortner's article and Merchant's *Death of Nature* came on the heels of several precipitous happenings. The 1960s saw the rise of environmental movements around the world. In the global North, there was a disenchantment with the rampant embrace of industrialization which had led to high levels of pollution, and the threat of nuclear war loomed large due to geopolitical conflicts involving the United States and the Soviet Union. Rachel Carson's *Silent Spring* was published in 1962 (Carson 2002 [1962]). The United Nations held its first environmental conference in Stockholm in 1972. Mainstream environmental movements of the time focused on the threats posed to human and ecological health from industrialization and war, but were predominantly led by cis white Euro-American men, and as such, focused on limited concerns grounded in anthropocentrism. As Sherilyn MacGregor reminds us in her introduction to the *Routledge Handbook of Gender and Environment* (2017), feminists were disenchanted with the narrow focus of environmental movements at the time, and the lack of progressive gender politics within mainstream environmental movements in the global North (to say nothing of the lack of diversity along other axes such as race, class, nationality, immigration status, sexuality, disability, and so on), and frustrated with the anthropocentric underpinnings of much environmental thought (Plumwood 2004). These frustrations, along with the conceptual need to theorize the degradation of nature alongside gender-based injustice led feminists interested in environmental questions to organize separately under the banner of "ecofeminism." This strand of ecofeminism emerged in the particular context of social and ecological movements of North America, Europe, and Australia at that time.

Another parallel strand of ecofeminism emerged from the Global South which critiqued the capitalist frameworks of logics that simultaneously devalued gender and nature. Writing in the introduction to *Ecofeminism* (1993: 3), Maria Mies and Vandana Shiva describe the capitalist patriarchal world system which "emerged, is built upon, and maintains itself through the colonization of women, of 'foreign' peoples and their lands; and of nature, which it is gradually destroying." Mies and Shiva focus on the "closeness" of women and nature. Shiva was building on her earlier writing in *Staying Alive: Women, Ecology, and Survival in India* (1989) in which she argued that rural Indigenous women were the rightful (and best) caretakers of nature since they are the original givers of life, and understand the everyday struggles of conserving and maintaining life and being in relation with their environments. Drawing on environmental resistance movements like the Chipko movement in Uttarakhand in India, and on the philosophical concept of *prakriti* (which means Nature, powered by a feminine living force) from precolonial Hinduism, Shiva argued that colonialism, capitalism, patriarchy, science, modernity, and mainstream development destroy life and nature, and we need to eschew them in their current forms or else face widespread ecological destruction (Shiva 1989). Ariel Salleh, in her foreword to Maria Mies and Vandana Shiva's *Ecofeminism*, notes that while mainstream liberal feminism seek equal opportunities for women within the existing framework of capitalism, ecofeminists seek to highlight the simultaneous undervaluing of nature and women as two sides to the same coin within existing capitalist patriarchal socio-economic structures (Salleh 1993). Building on her work on ecofeminism, Vandana Shiva has become one of the most prominent critics of the way global trade agreements prioritize the profits of agricultural companies over the wellbeing of farmers whose livelihoods and survival depend on being able to grow food and sell it.

Ecofeminists from the global North and South have been critiqued within broader feminist scholarly circles for being essentialist, for uncritically naturalizing, homogenizing, and accepting the category of "women," for centering gender and ignoring other axes of identity that women have, being too focused on the "closeness" of women and nature, and for prioritizing the biological processes of reproduction over the social processes of devaluing reproductive labor (see Box 15.1). There continues to be much debate about whether feminists should embrace or shun the "closeness" with nature (Agarwal 2019), whether ecofeminism is essentialist, and if so, what the consequences are for scholarship and praxis within environmental and feminist circles. These debates are foundational for feminist political ecology research today.

Box 15.1 Is ecofeminism grounded in essentialism?

A key point on which feminist scholars disagree is whether the de-valuing of women as a category of persons and the undervaluing of labor that is gendered female in hetero-patriarchal-capitalist society is due to the "natural" (used here as synonymous with "biological") role that women play in reproductive labor, or due to the socially/culturally/politically constructed role of women as caregivers in a socio-economic structure that devalues carework (or indeed any other domestic labor) because it is gendered female. The former would argue that women's biological role in reproduction ties them to nature in specific inalienable and "real" ways, where the latter would argue that the primary role played by women in reproduction, and the lesser value placed on "reproductive labor" of all kinds (as opposed to "productive" labor) is socially constructed, and a symptom of patriarchy rather than an indication of a "natural" or "real" affinity that anyone may have for reproductive labor or care-work. Some ecofeminists, especially on the more essentialist end of the spectrum, draw upon spiritual connections between women and nature, emphasizing the socio-cultural-spiritual connections between life-giving properties of nature and women. Others have focused on women as a category of subjects within international projects of development and environmental conservation as being both victims of environmental destruction and caretakers of local environments. Yet others, from a post-structuralist lens, have focused on how both the categories of gender and nature are socially constructed and problematized binary modes of thinking that create false dichotomies like male/female, nature/culture, mind/body, humans/animals, and so on. Some of these scholars on the more constructivist end of the spectrum have called for an avoidance of using "women" as a subject category entirely, calling instead for a focus on how gender operates as an axis of power. Others have critiqued the focus on gender in isolation from other axes of identity like race, class, tribe, nationality, sexuality, dis/ability, and so on. For further reading on these debates, and for more historical grounding on the philosophical underpinnings of the feminisation of nature, see Agarwal (2019), Plumwood (2002 [1993], 2004), New (1996), MacGregor (2017), Seager (1993), Sandilands (1999), Ortner (1972), Lorentzen and Eaton (2002), Rao (2012), King (1995), MacCormack and Strathern (1980), Leach (2007), King and Plant (1989).

Women, Gender, Environment, Development

Another body of scholarship that has deeply influenced feminist political ecology is literature on gender, environment, and development from the varied disciplines of geography, anthropology, and development studies. In the 1980s, 1990s, and 2000s, closely tied to the ecofeminist victim/caretaker conception of women (especially women from the Global South whose subsistence and survival were tied in direct ways to engaging with their environments and using natural resources) were global policy debates about environment and development. Under the banners of "women in development" (WID), "women and development," and "women, environment, and development" (WED), "gender and development" (GAD), many development organizations and policy advocates foregrounded the material ways in which poor and Indigenous women from the Global South interacted with their environments and used natural resources to perform their caretaking and provisioning roles in society as women. Since they interacted with their environments and used natural resources so closely for their livelihoods and survival, the argument went, women had the most at stake in conserving and protecting them. Furthermore, policy advocates argued that women are most vulnerable to environmental degradation, and therefore simultaneously most interested in conserving and protecting their environments from harm, and could be persuaded to participate in movements to conserve and protect nature. These ideas persist in popular and policy discourses today. Thus, "women" as a category of development subjects became targets for environmental conservation projects. As a way to insert the concerns of women into mainstream development agendas (which until then often tended to ignore women altogether), gender advocates often simplified and homogenized the vast experiences and interests of varied people separated by several other axes of identity (such as class, caste, race, tribe, ethnicity, sexuality, livelihood, nationality, dis/ability) under an umbrella category of "women." For further reading on gender and development, see Harcourt (2016).

Discomfited with an overly simplified, homogenized, and unrealistic feminine subject, many feminist scholars critiqued development narratives that failed to disaggregate the category of "women," and assumed a static set of interests, behaviors, ecological knowledges, and motivations for all women regardless of their other identities and the social, cultural, economic, political, ecological, and geographic contexts that they inhabited. Prominent among these voices was the feminist economist Bina Agarwal, who proposed an alternative to ecofeminism called "feminist environmentalism" (Agarwal 2019). Agarwal argues that women experience gender-specific harms relating to the environment not simply because of spiritual, ideological or historical "closeness" with nature (as argued by many ecofeminists) but due to the material ways in which women interact with their environments. Agarwal called for a material grounding of feminist analyses of environmental struggles (in addition to the symbolic and historic analyses that ecofeminists were already doing), grappling with who has property, power, and control over natural resources, and paying close attention to daily livelihood activities of women differentiated by class, race, caste, tribe, and so on. Agarwal's article was published a few years after Chandra Mohanty's *Under Western Eyes* (1988), which drew attention to the ways in which Third World women are often homogenized discursively and politically by Western feminists into a "singular monolithic subject." Thus, the late 1980s and early 1990s led to increased recognition (although not enough) among development studies scholars, practitioners, and feminist scholars and organizers that much more nuance was needed to understand the lived experiences and knowledge of varied women around the world in their interactions with nature. There was a concerted call to move away from the simplistic victim/caretaker model initially proposed by ecofeminists. Similarly, feminist scholars in anthropology, geography, and development studies have made critical interventions in their disciplines to trouble the idea of hegemonic transnational development discourses that were opposed by "local women" to protect their environments and ways of life. For example, Rebecca Klenk (2004) complicated such simplistic narratives of resistance by analyzing the polysemic enthusiasm for development expressed by some of her research interlocutors in Kumaon, India. Klenk (2004) shows us that transforming gendered subjectivities is a central component of development projects, and that the subjects of this intended transformation bring their own critical perspectives, skills, strategies, and contexts to bear on the process of development, rather than being simple victims of development schemes, or victims of poverty, or victims of patriarchy.

For a deeper dive into the critical literature on gender, development, environment, and conservation to illustrate the parallel intellectual trends in the allied disciplines of geography, anthropology, and development studies, see Parreñas (2018), Govindrajan (2019, 2021), Sultana (2011, 2014), Ramamurthy (2003), Gururani (2002), Berry (2003, 2008), and Birkenholtz (2023).

Feminists Doing Political Ecology: Centering and Decentering Gender, and Creating Environmental Knowledges

In the 1990s, a parallel strand of scholarship studying gender and nature emerged from political ecologists who wanted to expand the tools of critical geography to analyze gendered power dynamics in relation to natural environments as a way to build more just and ecologically sustainable worlds. An influential early text was *Feminist Political Ecology: Global Issues and Local Experiences* (2013 [1996]), a polyvocal volume edited by Dianne Rocheleau, Barbara Thomas-Slayter, and Esther Wangari who sought to bring together the "separate but increasingly convergent critiques of sustainable development" from political ecology (in geography) and feminist scholarship in other disciplines and intellectual traditions. Feminist political ecologists study the multiple ways in which access to natural resources is mediated by gender and power (Rocheleau et al. 2013 [1996]) and how social identities are forged through people's interactions with nature and everyday material practices relating to the environments they inhabit (Sundberg 2016). Furthermore, influenced by feminist science studies scholars, feminist political ecologists study the ways in which environmental knowledges are gendered. Building on a longer intellectual trajectory of feminist geographers studying space, place, and gender (Massey 1994; McDowell 1999; McDowell and Sharp 1997), Feminist Political Ecology (FPE) studies how spaces and places are gendered, and draws attention to places understudied due to them being gendered female (for example see Maria Elisa Christie's work on *kitchenspaces*—Christie 2006). FPE analyzes the material interactions with environments through gendered embodied practices related to livelihoods and survival, and the environmental knowledges and expertise developed as a result (Gururani 2002; Gagné 2019; Simon et al. 2022). FPE draws attention to multiple scales of analysis ranging from the body, the home, to the city, region, and the nation state, emphasizing the connections between these scales (Truelove 2011; Guthman 2012).

Rocheleau's earlier work (1995) drew on poststructuralist feminist theory to suggest a politics of affinity which is dynamic and contextual rather than a politics of identity—that is to say, not simply "women studying women" (p. 159). In doing so, her work called attention to the politics of knowledge production ("who counts, who is counted, and in what context") rather than solely a politics of inclusion. While conceptualizations of FPE have been intersectional from their initial theorizations, analyzing gender alongside other axes of identity (Rocheleau et al. 2013 [1996]), in subsequent application and practice many FPE studies have privileged gender over other forms of difference in their scholarship and praxis.

This has led to feminist political ecologists of color from the global North and South pushing feminist political ecology as a field to be more intersectional in its analysis of power, calling attention to "the privileging of gender over other forms of difference in FPE" (Mollett 2017: 146). In particular, Mollett and Faria (2013) critique the absence of explicit discussions of race within most FPE scholarship. Drawing on poststructuralist scholars like Frantz Fanon, Homi Bhabha, Arturo Escobar, Anne McClintock, and James Ferguson, Mollett and Faria (2013) argue that this absence is especially glaring since many FPE scholars work in the Global South critiquing projects of international development, discourses around which flow directly from colonial processes of racialization. Mollett and Faria (2013: 117) propose three possible reasons for the absence of race: first, a "political wariness" with stressing differences among women, whiteness within academia, and the difficulties of theorizing "a messier notion of gender" which is more intersectional and accounts for racial differences and a politics of place. Calling for a *postcolonial intersectional analysis* in FPE, Mollett and Faria (2013) argue that FPE is well positioned to "examine racialized processes in the making of gendered subjectivities in the Global South" (p. 118) and that patriarchy and racism are "mutually imbued" in shaping human-environment relationships (p. 117). Explaining what they mean by postcolonial intersectionality in FPE, Mollett and Faria write:

> Postcolonial intersectionality acknowledges the way patriarchy and racialized processes are consistently bound in a postcolonial genealogy that embeds race and gender ideologies within nation- building and international development processes.

> This concept reflects the way women and men are always marked by difference whether or not they fit nicely in colonial racial categorizations, as cultural difference is also racialized ... postcolonial intersectionalities in FPE would help better differentiate among women in the same way feminism was forced to confront its historical engagement with "imperialist origins" ... against the construction of a "third world woman" and prioritizes a grounded and spatially informed understanding of patriarchy constituted in and through racial power.
>
> (Mollett and Faria 2013: 120)

Important contributions to political ecology on gender/nature relations have also been made by Feminist Science and Technology Studies (STS) scholars. Feminist STS scholars have drawn attention to key questions about knowledge production, epistemology, objectivity, and science, building on influential works within history, philosophy, and anthropology of science and technology carried out by feminist scholars even before an interdisciplinary field of study called "feminist STS" existed (see for example Evelyn Keller's (1985) *Reflections on Gender and Science*, Susan Bordo's (1986) *Cartesian Masculinization of Thought*, and Sharon Traweek's (1988) *Beamtimes and Lifetimes*). Influential early FSTS thinkers like Sandra Harding and Donna Haraway developed concepts like *partial objectivities* (Harding 1986) and *situated knowledges* (Haraway 1988) to draw attention to the ways in which *all* knowledge is partial and made by embodied knowers influenced by their environments, social identities, cultures, and positionalities, not just the knowledge claims of marginalized groups. The claims of neutral objectivity or the "view from nowhere" (Haraway 1988) was a power move that was only accorded to dominant social groups (usually wealthy white European men). These claims of neutrality and objectivity were dangerous as they led to certain embodied identities being accorded more power as knowledge producers than others. Several excellent edited volumes are available for the reader to familiarize themselves with the key intersections between political ecology, STS, and feminist theory. See for example Maralee Mayberry, Banu Subramaniam, and Lisa Weasel's volume *Feminist Science Studies: A New Generation* (2001), Sandra Harding's reader on *Postcolonial Science and Technology Studies* (2011) and Mara Goldman, Paul Nadasdy, and Matthew Turner's edited volume on *Knowing Nature: Conversations at the Intersection of Political Ecology and Science Studies* (2019).

Exciting contemporary scholarship in FPE builds on scholarship and activism from the last few decades, together with decolonial and anticolonial feminisms, to analyze and critique contemporary sustainable development, and environmental conservation paradigms (Ojeda et al. 2022). This new and rich body of work combines theory and praxis from Black, Indigenous, and Global South feminisms to make explicit the links between the varied oppressions (historic and contemporary) caused by colonialism and the destruction of lives (human and non-human) and living worlds captured under the umbrella of "nature." Building on the concept of "rooted networks" (Rocheleau and Roth 2007; Rocheleau and Nirmal 2015) that analyze human-ecological relationships embedded not only in place but also in histories and political economies that we co-create and inhabit, FPE scholars and activists today are "making and defending plural territories" (Ojeda et al. 2022: 158), "centering plural knowledges and worlds" (p. 159), striving towards "feminist, anticolonial multispecies justice" (p. 160), and forging alliances across differences for justice.

Emerging from this rich and varied intellectual foundation, feminist political ecology today encompasses a wide range of scholars working from different positionalities, epistemologies, and geographic locations. In the introduction to a recently published volume (Harcourt et al. 2023) Ana Agostino and colleagues sketch out the contours of contemporary FPE (Agostino et al. 2023). Embracing an open-ended approach in the edited volume, the authors describe current FPE scholarship as "politically meaningful knowledge" building on both empirical research and feminist theory, "shaped by every-day and embodied lives within damaged, dynamic, and contested environments, as well as by hope in collaborative ventures on the margins of academic practices" (ibid.: 2). The volume describes the interrogation of knowledge politics that many contemporary FPE scholars engage in across different themes—international climate action through the COP process, how to conduct research in non-extractive ways, feminist ideas of care practices which are different and contested across generations of scholars and across places, understanding of health in relation to environments, decoloniality, the population question, situated entanglements with water, how to cope with the new normals of pandemics, and climate crisis, layered on top of longstanding intersectional injustices.

Box 15.2 From "women" to "gender" to "intersectionality"

Scholarship and activism within gender and environment circles has changed from centering a subject constituted as *women*, to an analytic of *gender* in power relations, to embracing *intersectionality* where different and interlocking axes of identity and difference are analyzed (Sundberg 2016). While ecofeminists called for understanding the conceptual, philosophical, and spiritual underpinnings of the shared oppressions of "women" and "nature," feminist development studies scholars and practitioners called for a more material grounding of the connections between the vulnerabilities and exploitation of women and nature, especially foregrounding different class, race, nationality and other aspects of women's positionalities. Feminist political ecologists called for understanding gendered power relations and how they affect relationships to the environment, and influenced by feminists of color from the global North and South, intersectional FPE aims to think about gender in relation to other axes of identity and difference. The analytical and political subject of analysis has moved from being "women" to "gender" to "intersectionality"; thus, tracing the scholarly journey over the last several decades, we see the influences of feminist social theory on gender/nature conceptualizations.

Box 15.3 Feminist approaches, methodologies, and epistemologies in political ecological research

Since this edited volume is focused on *doing* political ecology, it is appropriate to ask the question: Is there a feminist methodology for doing political ecological research? Are there a specific set of tools that one can or should use? How do/should researchers bring a feminist lens and ethos to doing political ecology? What would such research look like? I would argue that rather than a select set of methods for data collection, analysis or writing, or even a limited set of theoretical frameworks, feminist political ecology is about orientating our research towards more just and liberatory goals. The contours of what that looks like depends on the context of the research being carried out. There are multiple generative feminist scholars and texts that inform my thinking in this section (Abu-Lughod 1990; Grewal and Kaplan 1994; Rocheleau et al. 2013 [1996]; Sprague 2016; Hesse-Biber et al. 1999; Wolf 1996; Visweswaran 1994). While none of these texts are necessarily prescriptive in their theory or methods, they inform my own work and offer directions for thinking about feminist approaches, methodologies, and epistemologies. I briefly highlight three aspects: praxis, reflexivity, interdisciplinarity.

(a) **Praxis**: From its inception, Feminist Political Ecology (FPE) has been attuned to praxis. Rocheleau et al. (2013 [1996]) drew on both social theory *and* social movements to theorize FPE. An orientation towards praxis and social movements suffuses much feminist work in academia, even when it advances theory (see for example Grewal and Kaplan 1994). Indeed, most feminist scholars would argue that the very binary between theory and praxis is problematic (see also Baksh-Soodeen and Harcourt 2015). (See Chapter 14 of this volume for further discussion of critical praxis in political ecology.)

(b) **Reflexivity**: Being self-reflexive about power asymmetries in research collaborations, writing, citational practices, mentoring, conference attendance, fieldwork, activism, and allyship comprise a feminist orientation to scholarship and praxis. While reflexivity has always been important when theorizing contemporary subjectivities, it is particularly salient as we confront "global" climate change. Poststructural scholars have shown us that the global/local binary replicates the center/periphery model of erroneous thinking that further shores up ethnocentric ways of thinking that flows of ideas, knowledge, technology, expertise, resources, capital, theories are unidirectional (flowing from the global North to South only). Instead, eschewing the romanticization of the "local," feminists have theorized that global and local are permeable constructs, and *every* location is hybrid and translocal (see also Hall 1997; Gupta and Ferguson 1997). Reflexivity demands accountability for the ways in which "one's privileges are linked to another woman's oppression or exploitation" (Grewal and Kaplan 1994: 19). Collaborations across difference and asymmetries are unavoidable, indeed I would say they are necessary to build transnational solidarities, but they must proceed with an understanding and acknowledgement of the deep historic and contemporary differences across various locations and positionalities of the individuals and institutions involved.

(c) **Interdisciplinarity**: Feminist research is interdisciplinary. For many decades, disciplinary boundaries, home departments, and established canons excluded women, feminists, and analysis of gender (Hesse-Biber et al. 1999). Thus, feminist scholars occupied locations at the margins of disciplines, in interdisciplinary feminist studies journals, and in women's and gender studies departments. Rising from that history, feminist political ecology draws upon multiple disciplines, and diverse ways of knowing and producing knowledge.

Conclusion and Future Directions: Towards Emancipatory Ecologies and More Just Worlds

The theoretical, methodological, and praxis-oriented influence of FPE is evident in much of political ecology, critical geography and cognate fields today. However, a lot of this work does not self-identify as "feminist political ecology." Despite being a thriving intellectual field making multiple significant theoretical, epistemological, methodological contributions with deep links to praxis, feminist political ecology remains at "the margins" of geography (Sultana 2021), with many of its contributions absorbed into mainstream critical geography and allied disciplines without acknowledgement to the intellectual debt many political ecologists owe to the work done in FPE (Elmhirst 2011, 2015). On one hand this speaks to the importance of the many interventions of FPE to critical geography, but on the other hand it also speaks to the epistemic power dynamics at work within the creation of the "canon" of political ecology. More recently, scholars have proposed feminist ecologies, and decolonial and anticolonial ecological feminisms (Ojeda et al. 2022) as theoretical concepts that build on ecofeminism and feminist political ecology, and provide theoretical tools for the current moment of multiple ecological crises towards building more emancipatory ecologies and towards feminist multispecies justice. Feminist political ecologists continue to advance new knowledge, dialogue across difference, and engage in theory and praxis in this contemporary moment of global environmental, health, economic and political crises (see for example Harcourt et al. 2023).

Some of the challenges for contemporary scholarship and activism inspired by feminist political ecologists include how to conceptualize multi-species ecojustice (and consider plants, animals, and all other earthly beings as "kin," following Donna Haraway) in this contemporary moment of climate crisis where we still have not made kin of all humans, nor reconciled racial, gender, and economic justice (see also David Pellow's *What Is Critical Environmental Justice?* 2017). Simultaneously, there are other important and emergent questions still to be tackled: How do we preserve the conditions for collective thriving across lines of difference? How do we think through sometimes incommensurable ways of knowing and multiple ontologies (see also Saxena et al. 2018) to collaborate on building more just worlds? And how do we bring our tools of critical analysis to do the careful and attentive work required to address the urgency of the climate crisis? Feminist methodologies, scholarship and orientation to praxis offer us helpful foundations towards building emancipatory ecologies and more just worlds.

References

Abu-Lughod, Lila. Can there be a feminist ethnography? *Women and Performance: A Journal of Feminist Theory* 5, no. 1 (1990): 7–27.

Agarwal, B. The gender and environment debate: Lessons from India. In *Population and environment* (pp. 87–124). Routledge, 2019.

Agostino, Ana, Rebecca Elmhirst, Marlene Gómez, Wendy Harcourt, and Panagiota Kotsila. Sketching out the contours. In *Contours of feminist political ecology*, pp. 1–18. Cham: Springer International Publishing, 2023.

Baksh-Soodeen, Rawwida, and Wendy Harcourt, eds. *The Oxford handbook of transnational feminist movements*. Oxford University Press, 2015.

Berry, Kim. Lakshmi and the scientific housewife: A transnational account of Indian women's development and production of an Indian modernity. *Economic and Political Weekly* (2003): 1055–1068.

Berry, Kim. Good women, bad women and the dynamics of oppression and resistance in Kangra, India. *Humboldt Journal of Social Relations* (2008): 4–38.

Bhatia, Rajani, Jade S. Sasser, Diana Ojeda, Anne Hendrixson, Sarojini Nadimpally, and Ellen E. Foley. A feminist exploration of "populationism": Engaging contemporary forms of population control. *Gender, Place and Culture* 27, no. 3 (2020): 333–350.

Birkenholtz, Trevor. Infrastructuring drip irrigation: The gendered assembly of farmers, laborers and state subsidy programs. *Environment and Planning E: Nature and Space* 6, no. 1 (2023): 132–152.

Bordo, Susan. The Cartesian masculinization of thought. *Signs: Journal of Women in Culture and Society* 11, no. 3 (1986): 439–456.

Carson, Rachel. *Silent spring*. Houghton Mifflin Harcourt, 2002 [1962].

Christie, Maria Elisa. Kitchenspace: Gendered territory in central Mexico. *Gender, Place and Culture* 13, no. 6 (2006): 653–661.

Clarke, Adele, and Donna Jeanne Haraway. *Making kin not population*. Prickly Paradigm Press, 2018.

De Beauvoir, Simone. *The second sex*. New York, 1953 [1949]. Originally published in French in 1949.

Ehrlich, P. *The population bomb*. Sierra Club/Ballantine Books, 1968.

Elmhirst, R. Introducing new feminist political ecologies. *Geoforum*, 42, no. 2 (2011): 129–132.

Elmhirst, Rebecca. Feminist political ecology. *The Routledge handbook of gender and development* (2015): 519–530.

Gagné, Karine. *Caring for glaciers: Land, animals, and humanity in the Himalayas*. University of Washington Press, 2019.

Goldman, Mara J., Paul Nadasdy, and Matthew D. Turner, eds. *Knowing nature: Conversations at the intersection of political ecology and science studies*. University of Chicago Press, 2019.

Govindrajan, Radhika. *Animal intimacies: Interspecies relatedness in India's central Himalayas*. University of Chicago Press, 2019.

Govindrajan, Radhika. Labors of love: On the political economies and ethics of bovine politics in Himalayan India. *Cultural Anthropology* 36, no. 2 (2021): 193–221.

Grewal, Inderpal, and Caren Kaplan, eds. *Scattered hegemonies: Postmodernity and transnational feminist practices*. University of Minnesota Press, 1994.

Gupta, Akhil, and James Ferguson, eds. *Anthropological locations: Boundaries and grounds of a field science*. University of California Press, 1997.

Gururani, Shubhra. Forests of pleasure and pain: Gendered practices of labor and livelihood in the forests of the Kumaon Himalayas, India. *Gender, Place and Culture: A Journal of Feminist Geography* 9, no. 3 (2002): 229–243.

Guthman, Julie. Opening up the black box of the body in geographical obesity research: Toward a critical political ecology of fat. *Annals of the Association of American Geographers* 102, no. 5 (2012): 951–957.

Hall, Stuart. The local and the global: Globalization and ethnicity. *Cultural Politics* 11 (1997): 173–187.

Haraway, Donna. Anthropocene, capitalocene, plantationocene, chthulucene: Making kin. *Environmental Humanities* 6, no. 1 (2015): 159–165.

Haraway, Donna. Situated knowledges: The science question in feminism and the privilege of partial perspective. *Feminist Studies*. Vol 14. No. 3. (1988).

Harcourt, Wendy, ed. *The Palgrave handbook of gender and development: Critical engagements in feminist theory and practice*. Springer, 2016.

Harcourt, Wendy, Ana Agostino, Rebecca Elmhirst, Marlene Gómez, and Panagiota Kotsila. *Contours of feminist political ecology*. Springer Nature, 2023.

Harding, Sandra. *The science question in feminism*. Cornell University Press, 1986.

Harding, Sandra, ed. *The postcolonial science and technology studies reader*. Duke University Press, 2011.

Hartmann, Betsy, and Elizabeth Barajas-Román. The population bomb is back–with a global warming twist. *The Women, Gender and Development Reader* (2011): 327.

Hartmann, Betsy, Anne Hendrixson, and Jade Sasser. "Population, gender equality and sustainable development." In Leach, M. (Ed.). (2016). *Gender Equality and Sustainable Development* (1st ed.), pp. 56–81. Routledge. https://doi.org/10.4324/9781315686455

Hawkins, Roberta, Diana Ojeda, Kiran Asher, Brigitte Baptiste, Leila Harris, Sharlene Mollett, Andrea Nightingale, Dianne Rocheleau, Joni Seager, and Farhana Sultana. A discussion. *Environment and Planning D: Society and Space* 29, no. 2 (2011): 237–253.

Hendrixson, Anne, Diana Ojeda, Jade S. Sasser, Sarojini Nadimpally, Ellen E. Foley, and Rajani Bhatia. Confronting populationism: Feminist challenges to population control in an era of climate change. *Gender, Place and Culture* 27, no. 3 (2020): 307–315.

Hesse-Biber, Sharlene Nagy, Christina K. Gilmartin, and Robin Lydenberg. *Feminist approaches to theory and methodology: An interdisciplinary reader*. Oxford University Press, 1999.

Johnson, Amy, Chris Hebdon, Paul Burow, Deepti Chatti, and Michael Dove. Anthropocene. In *Oxford research encyclopedia of anthropology* (2022).

Keller, Evelyn Fox. *Reflections on gender and science*. Yale University Press, 1985.

King, Ynestra, and Judith Plant. The ecology of feminism and the feminism of ecology. *Environmentalism: Critical Concepts* 2 (1989): 18–28.

King, Ynestra. Engendering a peaceful planet: Ecology, economy, and ecofeminism in contemporary context. *Women's Studies Quarterly* 23, no. 3/4 (1995): 15–21.

Klenk, Rebecca M. "Who is the developed woman?": Women as a category of development discourse, Kumaon, India. *Development and Change* 35, no. 1 (2004): 57–78.

Leach, Melissa. Earth mother myths and other ecofeminist fables: How a strategic notion rose and fell. *Development and Change* 38, no. 1 (2007): 67–85.

Lorentzen, Lois Ann, and Heather Eaton. Ecofeminism: An overview. In *The forum on religion and ecology at Yale*. 2002.

MacCormack, Carol, and Marilyn Strathern, eds. *Nature, culture and gender*. Cambridge University Press, 1980.

MacGregor, Sherilyn. Gender and environment: An introduction. In *Routledge handbook of gender and environment*, pp. 1–24. Routledge, 2017.

Massey, Doreen. *Space, place and gender*. Minneapolis: University of Minnesota, 1994.

Mayberry, Maralee, Banu Subramaniam, and Lisa H. Weasel, eds. *Feminist science studies: A new generation*. Routledge, 2001.

McDowell, Linda, and Joanne P. Sharp. *Space, gender, knowledge: Feminist readings*. Arnold, 1997.

McDowell, Linda. *Gender, identity and place: Understanding feminist geographies*. University of Minnesota Press, 1999.

Mehta, Lyla. Dianne Rocheleau: The feminist political ecology legacy and beyond. In *The Palgrave handbook of gender and development: Critical engagements in feminist theory and practice*, pp. 262–275. Palgrave Macmillan UK, 2016.

Merchant, Carolyn. *The death of nature: Women, ecology, and the scientific revolution*. Harper & Row, 1980.

Mies, Maria, and Vandana Shiva. *Ecofeminism*. Zed Books, 1993.

Mohanty, Chandra. Under Western eyes: Feminist scholarship and colonial discourses. *Feminist Review* 30, no. 1 (1988): 61–88.

Mollett, Sharlene, and Caroline Faria. Messing with gender in feminist political ecology. *Geoforum* 45 (2013): 116–125.

Mollett, Sharlene. Gender's critical edge: Feminist political ecology, postcolonial intersectionality, and the coupling of race and gender. In *Routledge handbook of gender and environment*, pp. 146–158. Routledge, 2017.

Moore, Jason W. The Capitalocene, Part I: On the nature and origins of our ecological crisis. *The Journal of Peasant Studies* 44, no. 3 (2017): 594–630.

New, Caroline. Man bad, woman good? Essentialisms and ecofeminisms. *New Left Review* (1996): 79–93.

Ojeda, Diana, Padini Nirmal, Dianne Rocheleau, and Jody Emel. Feminist ecologies. *Annual Review of Environment and Resources* 47 (2022): 149–171.

Ortner, Sherry B. Is female to male as nature is to culture? *Feminist Studies* 1, no. 2 (1972): 5–31.

Parreñas, Juno Salazar. *Decolonizing extinction: The work of care in orangutan rehabilitation*. Duke University Press, 2018.

Pellow, David Naguib. *What is critical environmental justice?*. John Wiley & Sons, 2017.

Plumwood, Val. *Feminism and the mastery of nature*. Routledge, 2002 [1993].

Plumwood, Val. Gender, eco-feminism and the environment. *Controversies in Environmental Sociology* 1 (2004): 43–60.

Ramamurthy, Priti. Material consumers, fabricating subjects: Perplexity, global connectivity discourses, and transnational feminist research. *Cultural Anthropology* 18, no. 4 (2003): 524–550.

Rao, Manisha. Ecofeminism at the crossroads in India: A review. *Dep* 20, no. 12 (2012): 124–142.

Rocheleau, Dianne. Maps, numbers, text, and context: Mixing methods in feminist political ecology. *The Professional Geographer* 47, no. 4 (1995): 458–466.

Rocheleau, Dianne, Barbara Thomas-Slayter, and Esther Wangari, eds. *Feminist political ecology: Global issues and local experience.* Routledge, 2013 [1996].

Rocheleau, Dianne, and Robin Roth. Rooted networks, relational webs and powers of connection: Rethinking human and political ecologies. *Geoforum* 38, no. 3 (2007): 433–437.

Rocheleau, Dianne, and Padini Nirmal. Feminist political ecologies: Grounded, networked and rooted on earth. (2015).

Salleh, A. Class, race, and gender discourse in the ecofeminism/deep ecology debate. *Environmental Ethics* 15, no. 3 (1993): 225–244.

Sandilands, Catriona. *The good-natured feminist: Ecofeminism and the quest for democracy.* University of Minnesota Press, 1999.

Sasser, Jade S. *On infertile ground.* New York University Press, 2018.

Saxena, Alder Keleman, Deepti Chatti, Katy Overstreet, and Michael R. Dove. From moral ecology to diverse ontologies: Relational values in human ecological research, past and present. *Current Opinion in Environmental Sustainability* 35 (2018): 54–60.

Sayre, Nathan F. Carrying capacity: Genesis, history and conceptual flaws. *Annals of the Association of American Geographers* 98, no. 1 (2008).

Seager, J. *Earth follies: Coming to feminist terms with the global environmental crisis.* Routledge, 1993.

Shiva, V. *Staying alive: Women, ecology, and development.* Zed Books, 1989.

Simon, Gregory L., Bryan Wee, Deepti Chatti, and Emily Anderson. Drawing on knowledge: Visual narrative analysis for critical environment and development research. *Environment and Planning E: Nature and Space* 5, no. 1 (2022): 293–317.

Sprague, Joey. *Feminist methodologies for critical researchers: Bridging differences.* Rowman & Littlefield, 2016.

Strathern, Marilyn, Jade S. Sasser, Adele Clarke, Ruha Benjamin, Kim Tallbear, Michelle Murphy, Donna Haraway, Yu-Ling Huang, and Chia-Ling Wu. Forum on making kin not population: Reconceiving generations. *Feminist Studies* 45, no. 1 (2019): 159–172.

Subramaniam, Banu. *"Overpopulation" is not the problem.* Public Books, 2018.

Sultana, Farhana. Suffering for water, suffering from water: Emotional geographies of resource access, control and conflict. *Geoforum* 42, no. 2 (2011): 163–172.

Sultana, Farhana. Gendering climate change: Geographical insights. *The Professional Geographer* 66, no. 3 (2014): 372–381.

Sultana, Farhana. Political ecology 1: From margins to center. *Progress in Human Geography* 45, no. 1 (2021): 156–165.

Sundberg, Juanita. Feminist political ecology. *International encyclopedia of geography: People, the earth, environment and technology* (2016): 1–12.

Traweek, Sharon. *Beamtimes and lifetimes.* Harvard University Press, 1988.

Truelove, Yaffa. (Re-)conceptualizing water inequality in Delhi, India through a feminist political ecology framework. *Geoforum* 42, no. 2 (2011): 143–152.

Visweswaran, K. *Fictions of feminist ethnography.* University of Minnesota Press, 1994.

Wolf, Diane L. *Feminist dilemmas in fieldwork.* Westview Press, 1996.

16 Racializing Nature

The Place of Race in Environmental Imaginaries and Histories

Alex A. Moulton

Introduction

In 1774 Edward Long published his three-volume tome *The History of Jamaica or, General Survey of the Antient and Modern State of that Island.* Among other things, Long reported that the abatement of the "putrid and malignant fevers" that plagued the early colonial West Indies was attributable "to the more extensive cultivation of the country, the cutting down of its thick woods in several parts, and melioration of its atmosphere" (Long 1774, 2: 554). The increased production of sugar was an added benefit. Long's history of Jamaica features documentation of the rich stores of flora and fauna. He details how white rule, Black enslavement, and Taino (Native Caribbean people) extermination had enabled the accumulation of great wealth to the diligent plantation owner. Writers like Long in the early years of English colonialism in Jamaica celebrated how environmental transformations had created "agreeable climates;" agreeable for sugarcane cultivation and for Europeans, who would otherwise purportedly degenerate into the kind of barbarism attributed to Black people, and others "native" to the torrid zone (Hulme 2008; Livingstone 2002).

Why begin a chapter on "racializing nature" with Long and his celebration of land clearance for plantations? Why note his suppositions about the relationship between land clearance and environmental health? How can all this change how political ecologists address the futurity of racialized natures? For starters, Long's comments allow us to contextualize the array of forest conservation policies that intensified a century later. Situating his comments on how bodily health and environmental conditions determine each other helps us understand the racial logics that inform later conservation discourse, which are steeped in anxieties about racial purity and environmental quality. If this is not all noticeably clear at first blush from Long's comments, it is because postulations about how race and nature come together in nature–society imaginaries, and how they are instantiated through regimes of socioecological governance are not always overt nor always explicitly racist. And, these postulations are not stable, changing according to shifting regimes of power, ideologies, and social, racial, and ecological contexts. Examining some of the major conceptual, theoretical, and empirical motifs in the changing conjunctions of race and nature helps show how racialized nature has always been a central dispositive of political ecology. Long helps us see how, in direct ways, our ecological crisis is a consequence of destruction and wholesale transformation of ecosystems to fuel capitalism. The human bondage and native genocide which provided initial subsidy for capitalism to become a world system was rationalized through racialization of nature and naturalization of racial constructions through theological, bio-evolutionary, and economic categorical systems (Murphy 2021; Wynter 1995).

DOI: 10.4324/9781003165477-21

Box 16.1 Key concept: the Plantationocene

The Plantationocene offers a robust critique of the limits of discussions of the global ecological crisis induced by anthropogenic climate change. Central to the notion of the plantationocene is the understanding that the slave plantation inaugurated novel forms of human brutality and exploitation, alongside widescale transformations of nature, and circulation of non-human species. The plantationocene insists on an understanding of the racially-mediated experience of "Man," as the default figure of human embodiment. The racial ordering practices of the plantation were rooted in racialization of nature, and a dichotomization of nature and culture. The wealth accumulated from the plantation drove European industrialization and provided the subsidies for the development of global capitalism. Any consideration of futures not foreclosed by the plantation and its organizing logics must attend to the persistence of intra-human inequalities and the forms of dehumanization still at work in our social and political systems.

Further Reading

Carney, J. A. (2021). Subsistence in the Plantationocene: Dooryard gardens, agrobiodiversity, and the subaltern economies of slavery. *The Journal of Peasant Studies*, 48(5), 1075–1099.

Davis, J., Moulton, A. A., Van Sant, L., and Williams, B. (2019). Anthropocene, capitalocene, ... plantationocene?: A manifesto for ecological justice in an age of global crises. *Geography Compass*, 13(5), e12438.

Murphy, M. W., and Schroering, C. (2020). Refiguring the Plantationocene. *Journal of World Systems Research*, 26(2), 400–415.

Unpacking the Racial Politics of Nature

A political ecology of racializing nature shows the historical role that race, and racialization have played in environmental outcomes, patterns of access to nature, perceptions of resource management, and visions of environmental justice. Given this fundamental role of race and nature as organizing concepts to modern society, scholars have insisted that we understand that:

1. Race is socially constructed. As a construct, race is differentiated by social and geographical context. By the social construction of race, we mean that race has no biological essence; it is not a static reality decided by bio-evolutionary forces (Omi and Winant 1993). While phenotypical differences are certainly visible across groups of people, it is our social (and thus cultural and political) determinations that ascribe meaning to those differences. The process of valorizing certain phenotypes vis-à-vis others lends itself to the creation hierarchies of phenotypic desirability. This process is the process of racialization, and it is what encodes meaning into otherwise biologically insignificant differences across the human species (Brahinsky et al. 2014; Miriti et al. 2023; Moore et al. 2003).

2. How people are racialized in places and across geographies requires attention to the specific histories, political economies, socio-cultural relations, and perceptions of place. This means that while we can discern patterns across diverse racialized socio-ecologies, frameworks explaining racialized nature cannot readily be picked up and dropped somewhere else without close attention to contextual differences. Therefore, this chapter cannot alone elucidate the workings of racialized nature, nor should it have to. Rather, the chapter offers a broad survey, which necessarily means some gaps.

While race and nature have a long association that has been restructured across multiple temporalities and spaces, not all scholarship on environmental politics, landscape history, and nature has attended to the logics that explain this race-nature relationship. Similarly, some scholarship has wholly neglected the processes accounting for the reproduction and recuperation of different race-nature dynamics. With increasing nuance political ecologists have challenged such elisions of race in the politics of nature; they have clarified how race and place are co-constituted and how conceptions of race and nature have been motivated by and mobilized towards projects of social and political discipline, economic dispossession, and environmental resource control.

Antecedents: Racialized Nature and the Figure of "Man"

For polygenists like Long, racist environmental theories provided easy explanations for variations among humans and differences in cultural groups, or "human civilizations." If differences were natural and self-evidently explainable by natural laws, then hierarchies were inevitable and beyond critique. Drawing on organismic analogies and biological metaphors, this early racializing of nature could see in the tropics, as Long did, opportunities to civilize the inferior races and produce commodities that would fuel superior races. The racialized character of nature in the racist colonial logic supposed that to survive in the tropics white people had to vanquish pestilential environmental conditions—even as enslaved Black and Indigenous bodies were put to work in their "natural" and geographically determined position (Huntington 1913, 1915, 1919; Semple 1901, 1911). Nature was God, and God had instituted racialized ecologies. Remarking on white people living in the mountains of Kentucky, Ellen Churchill Semple (1901: 594), a prominent cultural geographer of the early twentieth century, opined that "the inextinguishable excellence of the Anglo-Saxon race" was one such counteracting force enabling them to colonize and dominate virtually any environment, though she did warn of "deterioration" and "derangements" when white people are transferred to the tropics. Conversely, Black people suffered from "climatically conditioned exclusion," limiting them to tropical environments, in "the heat and moisture in which they thrive" (Semple 1911: 626).

For Ellsworth Huntington, a President of the Board of Directors of the American Eugenics Society from 1934 to 1938, physical geographical factors were even more determining than Semple supposed. As he put it in *World-Power and Evolution*, "it appears that today the distribution of civilization is almost in harmony with the degree to which the climate of the various parts of the world resembles that in which man's mind made its most rapid evolution" (Huntington 1919: 145). These early modern articulations of racialized nature in geographical sciences obfuscated the role of power and racist political ideology through naturalizing discourses of race. In the process, these environmental determinists explained phenotypical difference, interpreted as evidence of essential racial difference or non-homogeneity of the human species, as the outcome of physical geographic factors, such as climate and topography (Blaut 1993; Lavery 2022; Neumann 2014; Peet 1985; Radcliffe et al. 2010).

Determinists, eugenicists, and self-described naturalists—whether subscribing to regressive or progressive conceptions of human difference—asserted that features of the environment had a causative effect on intelligence, temperament, and physiognomy (Miriti et al. 2023; Wynter 2003). As David Livingstone (1992: 221) notes, such notions about climate, morality, and racial topographies would characterize nineteenth- early and early-twentieth-century geographical scholarship. Importantly, recent interventions such as by LaToya Eaves and Karen Al-Hindi (2023) remind us that contemporaries of Semple, such as Ida B. Wells-Barnett, offered analyses of race and space that more critically attended to the role of racism and epistemological domination in imaginaries and experience of place.

Importantly, arguments that the environment *determined* human races, separating *civilized "Man"* from the supposed Indian *savages* and Black subhuman, did not mean that the forces of nature could not be abated. Nature could be changed; transformed by Man. Indeed, as we have seen from Long, dominant groups modified the environment with dual interests in preserving racial purity or energy and securing economic profit to maintain ethno-class hegemony. These kinds of concerns about Man's adverse impact on nature animated the discourse and practices around racialized nature from the dawn of European colonization until they intensified in the mid-nineteenth to the twentieth century (Grove 1996). As I elaborate more clearly later, "Man" in these discourses should be understood as having been imagined in specific ethno-class terms that excluded non-white people and most non-male elite white humans (Wynter 2003).

George Perkins Marsh was among the leading commentators of the mid-1800s. A prominent environmentalist, Marsh is popularly regarded as the US's first environmentalist and was early to express concerns about Man's transformations of nature, and anticipated the notion of the Anthropocene:

> [Man's] destruction of the forests, the drainage of lakes and marshes, and the operations of rural husbandry and industrial art have tended to produce great changes in the hygrometric, thermometric, electric, and chemical condition of the atmosphere, though we are not yet able to measure the force of the different elements of disturbance, or to say how far they have been compensated by each other, or by still obscurer influences.
> (Marsh 2003: 13–14)

But Marsh and other early American environmentalists, including John James Audubon, John Muir, Henry David Thoreau, Theodore Roosevelt, and Gifford Pinchot were concerned with nature's alteration precisely because of how it was racialized (Merchant 2007; Mirzoeff 2022; Taylor 2016). Nature was white space; through their encounter with and transformation of nature, white men could reaffirm and enact their racial superiority and vigor. This thinking was given academic credence by Frederick Jackson Turner's (1894) *The Significance of the Frontier in American History* (Cronon 1987; Merchant 2004).

This "frontier thesis" reminds us that theory is never divorced from social consciousness; it is contingent on and mediated (in production and purpose) by social forces. And so it was that as the US frontier was vanishing, African Americans were gaining freedom, and immigration to the US was increasing—a conjunctural moment—race and racialization of nature were made to serve the political ends of "Man," the figure of the superior race. Conceived as at risk from racial pollution and environmental degradation, ostensibly from the same threat—non-white people—nature was cast as in need of protection. Since nature was racialized as white, conservation of nature became necessary for the conservation of whiteness itself. For white supremacists and xenophobes, white blood and soil had to be protected. The rhetoric of fear and risk in this era of conservation cannot escape the specter of the eugenics movement for white racial purity (Kosek 2004, 2006; Mirzoeff 2022). Out of this new rationality for racializing nature, one of the most insidious concepts of racialized nature would be created, "wilderness."

Clarifying this genealogy of "wilderness" as a racialized form of nature alerts us to the dangers of seeing the term as benign, rather than furthering racial ecological imaginaries. As William Cronon (1983, 1987, 1996) explains, this influence was possible because wilderness was sacralized, imbricated with Judeo-Christian conceptions of nature as a space of terror and temptation, romanticism of the sublime, and transcendentalism. This idea of wilderness propped up the *Yellowstone Model*—the suite of approaches to protected area management first associated with the establishment of the US Yellowstone National Park in 1872. In racializing nature as a space to be protected from the ravages of non-white people, the Yellowstone Model violently produced nature as white space. This violence was symbolic, epistemic, and bodily.

For Black people conservation reinscribed "nature" as a space of racial terror, rather than as a space for recreation and leisure (Finney 2014; Davis 2019). Under slavery, the policing of nature according to race meant Black people were punishable by death if they were seen in nature, out of their place on plantations. After slavery, this policing structured Black people's uneven exposure to environmental harms and confinement to polluted environments. To be sure, the ecologies of the Underground Railroad, figures like George Washington Carver, the maroons, or Black agrarians and environmental laborers offer environmental histories and ecological constellations that are not reducible to racialized environmental inequality and violence, these were wholly elided by white environmentalism (Ferdinand 2022; Hosbey and Roane 2021; Hyman 2021; Roane 2018). Black, Indigenous, and Latinx resistance to such "racially encoded" environmental relationships spurred the environmental justice movement (Bullard 1990; Bullard and Wright 2012; Bullard et al. 2007). For Indigenous peoples it became clear that nature was the new and colonial name for what they had only known as the multitude of more-than-human relations that sustained life. It was a cruel irony; wilderness, supposedly untrammeled land had to be purged of its long-term inhabitants to become declared wild (Adams and Hutton 2007; Eichler and Baumeister 2018; LaDuke 1999; Simpson 2001; Whyte 2018b).

Worryingly, the transit of wilderness across time-space has taken place "with troubling invisibility and stunning audacity" (Kosek 2004: 118). Some of this invisibility is the result of a sleight of hand in racializing nature not through overt racial discourses, but under the aegis of cultural sensibilities and class consciousness. The production and distribution of racialized political imaginaries of nature does not only create and reproduce symbolic racialized terrains, but it also results in materially different human-environmental relations. Racialized environments as socio-ecologies structured by race, are ecologies of differentiated access, displacement, dispossession, and domination (Grove 1996; Kosek 2006).

Territorializing Race: The Global Hegemony of Nature and Racialized Subject Formation

The character of racialized natures beyond the US underscore the diffusion, adoption and adaptation of distinctly US discourses, rationalizations, and practices of racial socio-environmental domination. There are of course also non-US racial discourses or politics of difference unique and emergent from within the localities they are contested in. However, the geographies of racialized natures globally, more importantly, highlight the ways in which racial regimes of governance are recuperated and adjusted through other-than-direct-colonial methods. To be sure, these colonial methods of imposing territories of racialized nature were crucial, especially in Africa, where the process of cordoning-off nature from Indigenous Africans communicated multiple messages about race and nature: Africans needed to be taken out of nature—the jungle—in order to be civilized; Africans could not appreciate nature and were destructive; nature had to be ordered and managed for whiteness to thrive; and nature's resources were for white use and enjoyment (Brockington 2002; Fairhead and Leach 1995; Jones 2006; Neumann 1998). However, today these other-than-direct-colonization methods constitute a new "scramble for Africa," whereby nature's value to Western conservation organization, "green" capitalists, and elite conservationists authorizes the enclosure of more of the African commons (Beymer-Farris and Bassett 2012; Ogada 2021; Ramutsindela et al. 2022; Sène 2022).

This land grab in the name of nature should remind us of plantation owners like Edward Long, for whom the racialization of nature and naturalization of race was not just about racism, but primitive accumulation—the acquisition of material wealth from colonies to colonizers as an initial down payment for capitalism. As Karl Marx put it:

> The discovery of gold and silver in America, the extirpation, enslavement and entombment in mines of the aboriginal population, the beginning of the conquest and looting of the East Indies, the turning of Africa into a warren for the commercial hunting of black-skins, signaled the rosy dawn of the era of capitalist production. These idyllic proceedings are the chief moments of primitive accumulation.
>
> (Marx 1990: 915)

For Marx, "this primitive accumulation plays in political economy about the same part as original sin in theology" (ibid.: 873). Observing this relationship between racialized nature and political economy, political ecology as a discipline early recognized the function of racism in governance of nature but moved too quickly to argue that it was capitalism, with its privatization and the commodification of nature that drove recent modern and ongoing environmental inequality.

As a consequence, class was and, in some ways, remains the axis of difference that political ecologists focus on. This to the detriment of not just racialization and racism, but gender. Such Marxian political ecology, anchored in historical materialism, took difference and difference making, except class, as emerging outside and distinct from the economy. To be sure, this traditional focus on class, did not wholly elide problematization of race. Race was implicit in the discussions of regimes of property, commodification, and privatization of nature (Bosworth 2021; Castree 2001; Mollett and Faria 2013; Robbins 2019). In Marx's framework, the theory of historical materialism, supposed that economic activities, structured by the dialectical relationship of material resources and modes of production, drove organization of society and the development of social institutions and culture (Blaikie and Brookfield 2015; Mann 2009; Prudham 2015). However, capitalism has always been racial capitalism, as Cedric Robinson explained in coining the phrase:

> The development, organization, and expansion of capitalist society pursued essentially racial directions, so too did social ideology. As a material force ... racialism would

> inevitably permeate the social structures emergent from capitalism. I have used the term "racial capitalism" to refer ... to the subsequent structure as a historical agency. (Robinson 2020: 2)

Put another way, capitalism required a structure of social difference, and racism provided it. Racialism remains central to ongoing capitalist economic relations, even as capitalism reinforces racialized differentiation (Gill 2021; Nishime and Williams 2018). Ruth Wilson Gilmore (2002: 16) offers a similar understanding, noting that:

> Racism functions as a limiting force that pushes disproportionate costs of participating in an increasingly monetized and profit-driven world onto those who, due to the frictions of political distance, cannot reach the variable levers of power that might relieve them of those costs.

This more nuanced language of racial capitalism, attentive to race not simply as a social reality, but to the processes of racialization came to political ecology by way of a broader post-structural turn in geography that gave way to greater engagement with Black Marxism and critical social theories of race. Challenging the thinking of structuralists, who were concerned with the role of transhistorical social structures and historical materialism in explanation of social problems, post-structuralist called attention to the way knowledge itself is produced. Post-structuralist questioned how the study of the world proceeds from presumptions about the very character of the world itself, and requires discursive analysis (Agrawal 2005; Escobar 1996; Forsyth 2008; Peet and Watts 1996). This post-structural turn, and the engagement with critical theories of race shone a light on how literary and visual representations, discourse and racial tropes, and theoretical frameworks are implicated in acts of social construction. These acts of social construction produce meanings that are not value neutral. This work has allowed political ecologists to see that it is not sufficient to notice race as a marker of difference that features in environmental governance and racism as cause of environmental inequality. Such an approach, in Michael Murphy's words, reproduces a "colonial unknowing" around race (see also Ferdinand 2021). As Murphy (2021: 123) explains, this means operating with "an epistemological orientation" that "renders unintelligible the entanglements of racialization and colonization, occluding the mutable historicity of colonial structures and attributing finality to events of conquest and dispossession." Relatedly, the colonial unknowing leaves unquestioned the ontological assumptions that "race is only environmentally relevant because it signifies which human populations are vulnerable to discrimination and exclusion and thus come to live in compromising ecological situations" (Murphy 2021: 123).

To the extent that political ecologists have taken on criticisms like Murphy's, their work has interrogated the lingering dualism between nature and society, as much as it cautions against viewing race as something that is exploited, rather than something inscribed. And inscribed through, among other things, environmental discourses, and practices! Race cannot just be treated as an additional variable for consideration. How race itself is reproduced, naturalized, and is imbricated in discourses and epistemologies needs to be examined (Ferdinand 2022). This offers a new map of racialized nature, beyond instantiations of racialized exclusion and expropriation of nature, and environmental injustice that manifests in racially differentiated vulnerability. Racialized nature are the socio-ecological arrangements of space in the afterlives of colonization, settler colonialism, slavery, and racial capitalism. Racialized natures are the harvest from the reproduction of whiteness and the integration of whiteness into the meshwork of modern life. I will return to the matter of whiteness later.

Racialized nature are socio-ecological configurations that are not limited to natural environments; they span the urban or built environment, the imagined and furtive environments, as well as literary and symbolic environments. The post-structural turn, therefore, has allowed political ecologists of racialized nature to show how the dynamic between political economy, state relations and bureaucracies, and socially politicized and constructed categories of differences structure subject formation in polyvalent ways that are inseparable from the construction of race and racialized subjects (Loftus 2020; Mollett 2011, 2021). Therefore, modes of governing nature and disciplining environmental relations are wrapped up in programs of racialized domination, which often tap into desires for improvement. Such imaginaries of improvement do often fall in the shadow of colonialism and the historical erosion of non-white people's sovereignty (Chao 2022a, 2022b; Li 2007; Moore 2005). Some emerging work engages with the environmental humanities to highlight how the development taxonomic order, classificatory systems, and scientific nomenclature—all supposedly confined to the realm of nature and having nothing to do with race—have race in mind, and place race and nature on the same grid of intelligibility as objects of knowledge (Moore et al. 2003; Yusoff 2018, 2020).

Contested Difference: Environmental Transformation, Race, and Resistance

So far, we have seen how racialization and racism have shaped the reproduction of ecologies of domination and exploitation. That story though is incomplete and dangerously misleading without accounts of resistance (Moore 1997; Peet and Watts 1996). The post-structuralist political ecologists insisted on a clear articulation of this, just as post-colonial theorists, feminist political ecologists (as well as ecofeminists), and subaltern studies make similar demands. These scholars argue that studies of colonial and imperial history, including their legacies and hegemonic logics that persist into the present, cannot just be understood from above, as imposed. They point to the histories from below, the counter-projects through which the governed express resistance to the impositions of racialized nature (Moore 1997). Some of these scholars call attention to the body as a material site of political ecologies (Doshi 2017; Mollett 2021; Vasudevan et al. 2022). By examining scale, political ecologists of racialized nature show what is missed by just a focus on territory. Racial ecologies, across multiple scales and temporalities show how difference and othering mediate environmental relationships in multifaceted ways. Racial ecologies texture environments from agrarian regimes to energy systems (Brand and Miller 2020; Luke 2022; Van Sant 2021; Van Sant et al. 2021a; Van Sant et al. 2021b; Williams 2020; Williams et al. 2020; Williams and Porter 2022). Importantly, this scholarship also calls for more attention to other social relations that intersect with race, as well as clarification of how discussions about population and ecological limits, disaster, food and housing (in)security, land grabbing, sustainable and international (un)development, or political ecologies of health must be understood as discussions about the political ecologies of race and difference (Fernando 2020; Garth and Reese 2020; Guthman and Mansfield 2013; Mansfield and Guthman 2015; Nishime and Williams 2018; Van Sant et al. 2021b). At the congruence of this intersectional and decolonial political ecology of race and nature are clearer mappings of place-based, environmental social movements, constituted by individuals with intersectional identities contesting environmental domination (Loftus 2020; Ojeda et al. 2022; Rocheleau et al. 1996).

An especially important avenue of research attending to these questions of the spatialization of race as it intersects with other axes of difference is being carried on by urban political ecologists (Burghardt et al. 2023; Heynen 2016; Heynen et al. 2006, 2018; Lawhon et al. 2014; Loughran 2017; Pulido 2016; Robbins 2012; Vasudevan 2021). This scholarship shows how the metabolism of the city, its relation to the rural and suburban is textured by racial imaginaries of belonging, racialized practices of surveillance and policing, and racialized mobility and labor.

As Matthew Gandy (2022: 26), notes this work on cities, also included attention to the "corporeal vulnerabilities, and affective dimensions to metropolitan nature" in racialized ecologies. Importantly, though, as Gandy points out, "The handling of race within the field of urban political ecology is clearly more than a question of enlarging the environmental justice framework but also extends to the epistemic critique of posthumanism."

The problematization of post-humanist futurity dovetails with epistemological and empirical decentering of North America (and more broadly, the "West"), to show how racialized projects of environmental management draw not only on race per se, but cognate constructs of difference (Asher 2009; Doshi 2013; Mollett 2011; Nightingale 2011). An especially noteworthy development is what can be called the Latin American school of political ecology (Alimonda 2019; Leff 2015) consolidating around the work of Kiran Asher, Anthony Bebbington, Arturo Escobar, Enrique Leff, Sharlene Mollett, and Diana Ojeda and others. Latin American political ecology analyzes the historical outcomes of ideologies of mestizaje and indigenismo, postcolonialism, multiculturalism, and the role of extractivism in contemporary state-building in Latin America (and the Hispanic Caribbean). This work highlights the deadly stakes of grassroots resistance of development qua extractivism and intensified racial ecologies (Bebbington and Bury 2013; Escobar 1996, 2008; Tetreault 2017). Importantly this critical consideration of what traditional political ecology—done by scholars from the so-called "First World" concerned with the "Third World"—has meant a political ecology attentive to environmental conflicts in the industrialized Global North countries (McCarthy 2002, 2005).

Box 16.2 Key methodological considerations: studying racialized natures

Racialized ecologies are often produced and maintained through terror. Resistance to the racialized hierarchies governing access to these territories of difference are, therefore, usually targeted by violence. Racialized natures are ecologies of trauma, traumatized ecologies. Studies of racialized natures require ethical engagements: political ecologists have to be attentive to how research can re-traumatize those at the center of our work—our encounters must be mindful of the psycho-affective toll our research can have when it is based on asking people to recount encounters with racism and environmental injustice; researchers have to be careful to not fetishize the violence that drives racialized ecologies—violence is not a metaphor or purely symbolic, but touches down in real bodies; while attentive to the difference that difference makes in racialized natures, scholars must be careful to not essentialize race as if it were biological fact, rather than socially constructed. If our work to clarify racialized ecologies does not also pursue liberation, abolition, repair, and futures beyond violence we should abandon them.

Further Reading

Tuck, E. 2009. Suspending damage: A letter to communities. *Harvard Educational Review*, 79(3), 409–428.

Moulton, A. A., Velednitsky, S., Harris, D. M., Cook, C. B., and Wheeler, B. L. (2021). On and beyond traumatic fallout: Unsettling political ecology in practice and scholarship. *Journal of Political Ecology*, 28, 678.

Woods, C. A. 2002. Life after death. *The Professional Geographer*, 54(1), 62–66.

Beyond the Ecologies of Whiteness

In this section, I want to focus on two subfields that have dramatically enriched the understandings of race in political ecology. These reflect my own interest, rather than their exclusivity in advancing broader understandings of racialization. Quite perceptively, W. E. B. Du Bois answered his own question on the nature of whiteness: "But what on earth is whiteness that one should so desire it? Then always, somehow, someway, I am given to understand that whiteness is the ownership of the earth, forever and ever" (Du Bois 1910: 30). Du Bois's question leads to a succinct, yet arresting definition of whiteness. Whiteness consists in the ownership. In much greater detail, Cheryl Harris (1993) shows how this relationship between racial identity and ownership renders whiteness as a set of property rights and as a property that authorizes rights. Racial ecologies can thus be newly understood as constructs of whiteness and its attempts to protect its mythologized claims to the "ownership of the earth" (Bhandar 2018; Bosworth 2021; Inwood and Bonds 2017; McCarthy and Hague 2004).

Racial ecologies are the socio-ecological outcomes of white European colonization and contemporary global political economy dominated by majority white countries whose domination has been subsidized by colonization and imperialism. As Sylvia Wynter and Michael Murphy show, the production of these racial ecologies in the image and likeness desired by whiteness reflects the way that the human was defined, and what that definition authorizes. Andrew Baldwin (2012) and Dylan Harris (2022) perceptively warn that we also contemplate how whiteness shapes our imaginaries of the future, and more-than or other-than-human worlds yet-to-come. Racialization props up socio-ecological realities and relationships as natural, and justifies the violence entailed in their maintenance (Nixon 2011). In this section, I highlight the work of Indigenous and Black political ecologies scholarship that denaturalizes the ecologies of whiteness.

Ecologies of Settler Colonialism: Indigenous Ecologies Against Racial-Ecological Violence

Indigenous peoples' cosmologies have challenged political ecologists' understanding of racialized natures in ways that begin with, but do not end at the insistence of naming the distinct regime of racialized nature and ecological political economy that constitutes settler colonialism. Settler colonialism, as a spatialization of a racialized socioecology, constitutes ecocide (Burow et al. 2018; De Leeuw and Hunt 2018; Tuck and Yang 2012). This by itself is an important corrective to traditional understandings of logics of racialized reproduction of nature. Beyond this, Indigenous political ecologies of racialized natures have also unsettled the understandings of nature's agency within the dualistic framework of "Anthropos" separated from and dominating "Nature." Indigenous political ecologies center nature as pedagogy and centers embodied ecological intelligence for repairing the socioecological disruptions of racialized "natural" resource extraction and ecological contamination (Goeman 2013; Liboiron 2021; Simpson 2014). This is rooted in and routed through more-than-human ontologies that unsettle the presumptions of political ecology's traditional historical materialism.

As such this scholarship highlights the Euro-American-centric fixation of what racializing nature does to the "native," as opposed to how racial regimes of settler domination, extraction, commodification, and transformation constitute ongoing violence that changes the very nature of more-than-human worlds more broadly (Curley 2021; Yazzie 2018). This work turns the gaze of US and Canadian political ecology inward, from case studies of oppression out there, to ongoing socio-ecological violence, extraction, and unsovereignty at home.

Indigenous political ecologies show that racial ecologies are violent (re)orderings of life that have fundamentally altered Indigenous worlds and futures. Indigenous political ecologies of racialized nature, therefore, alert us to how apocalypse has been a reality for Indigenous peoples, and not a new impending catastrophe related to anthropogenic climate change (Daigle 2018; Whyte 2018a). This reminder also challenges imaginaries of Earth's future subtended by settler logics, calling into question techno-scientific approaches such as geoengineering, reformist programs promoting ecosystem services as commodities. Such injunctions also highlight the dangers of post-humanist idealizations of multispecies kinship which are disconnected from discussions of power, difference, and ecological justice (Sundberg 2014; Collard et al. 2015; Fagan 2019). All this work demands ethical and politically intentional approaches that decenter whiteness and its vocabularies for "nature" and human–environmental relationships. Solutions cannot come in the ways political ecologists sometimes expect them; through political theories of recognition and political settlements with the state (Coulthard 2014; Morgensen 2011).

Ecologies of the Plantationocene: Black Ecologies Against Racist Environments and Despair

Black Geographies and Black Ecologies scholarship provides multiple openings for new approaches to political ecology concerned with the racialization of nature. Both bodies of work have distinct genealogies and analytical concerns that cannot be elaborated here (McKittrick and Woods 2007; Moulton and Salo 2022). Notwithstanding their difference, both scholarly traditions examine the afterlives and legacies of racialized forms of nature. In US focused literature, scholars have foregrounded the Black Civil Rights Movement, farm labor rights movement, Latinx community organizations, and Indigenous anticolonial movement in the emergence of the environmental justice movement (Pellow 2007; Pulido 2006, 2017; Pulido and De Lara 2018; Roane and Hosbey 2019). Much of this work centers the legacies of plantation ecologies ("the Plantationocene") and the spatialization of racialized socio-ecologies reproduced to continue the sequestration and exploitation of Black people. In this way, this work clarifies the links between contemporary spaces such as inner-cities and prisons (Gilmore 2007; Woods 2002), process such as displacement, gentrification, and systematic underdevelopment (Roane 2022; Vickers 2022; Wilson 2000; Woods 1998, 2002), and the plantation apparatus of racialized socio-natural domination. The afterlives of anti-Blackness are ecological as well as social (Bruno 2022; Williams and Porter 2022). But crucially, this work challenges the theoretical and empirical reduction of Black environmental histories to plantation and antiblack ecologies (Bledsoe and Wright 2019; Ferdinand 2021).

Black geographies and Black ecologies clarify and valorize the epistemologies of place and nature arising from Black experiences of and resistance to racialized nature and environmental racism (McKittrick 2011, 2013; Wright 2021). This scholarship, that is activist in orientation, shows the limits of traditional theoretical canons and methodologies for understanding Black practices of place-making (Allen et al. 2019; Allen 2020; McCutcheon 2019), marronage (Bledsoe 2017; Moulton 2022; Winston 2021; Wright 2020), and fugitivity in the face of racially politicized ecologies (Kelley 2021; Lewis 2020). This work thus shows how scholars have missed expressions of Black agency that reproduce antiracist ecologies of survival from the narrowest affordances in material and symbolic ecologies of racialized nature (Carney 2001; Carney and Rosomoff 2011; McKittrick 2006; Watkins 2021). This capacious project of mapping Black ecologies of livingness, resistance, repair, and abolition hold out much for political ecologists concerned with the futures beyond Black environmental debility (Freshour and Williams 2022; Heynen and Ybarra 2021).

Much of this futurity and prefigurative politics is radical because it challenges the figuration of "Man," the default human, in narrow terms that privilege knowledge, heterosexual masculinity, economic rationality, ecological visions, and politics of white supremacists (Davis et al. 2019; Wynter 2003). So that just as Indigenous Ecologies challenge negations of racialized power in post-human ecologies, Black Geographies and Ecologies challenge the denials of ongoing racism and the role of or race in imaginaries of the future and speculative political ecologies (Smith and Vasudevan 2017; Tuana 2019; Vasudevan et al. 2022). David Pellow (2016) urges us to be attentive to how "racial discourses of animality" become normalized in our vocabulary for racialized violence. Such a discourse works through comparisons between how nonhuman animals should be treated compared to humans, and how humans should act relative to nonhuman animals. But it relies on the logics of racialized nature and naturalization of race.

Future Directions

There are numerous new avenues for doing political ecologies concerned with racialized nature. These directions owe much to the insightful criticism of non-white scholars who have questioned the epistemological and methodological implications of doing political ecology that is overdetermined by North American, and particularly US frameworks. There is greater need for political ecologies of whiteness; political ecologies that clarify how whiteness is depoliticized and placed outside the understanding of race and racialization. Attention to the racializing ecological assemblages through which whiteness is articulated would no doubt result in a clearer picture of the global workings of whiteness and its afterlives, including the ways aspirations to whiteness, tenuous whiteness, and honorary whiteness function in the production of racial ecologies. Interrogating the racial ecologies of Western Europe vis-à-vis Eastern Europe, Eastern Europe vis-à-vis North Africa for example, might serve as a ground for such research. Such work would contribute to the efforts I noted above.

Africa and Asia remain spaces where political ecology is done, spaces where the effects of racialized ecologies are observed, rather than places from which to newly understand the organizing logics of racialized nature. If we are to gain truly global understandings of racial ecologies, Africa and Asia must be approached as spaces to think from, spaces that remark upon notions of race/difference, nature, and power in ways that challenge the hegemony of Western modernity. Southeast Asia, for example, urges us to notice that caste is not a synonym of race. The distinctions matter for how we are to understand ecologies of difference (Ranganathan 2022; Vandergeest and Roth 2016). Similarly, rather than remaining fixed on white impositions of racial ecologies and apartheid natures in Africa, political ecologist can take a cue from Black ecologies and examine Black livingness and agency, and how the pluralities of African ethnic politics, kinship, and spirituality relations mediate more-than-human ecologies. Black Geographies and Ecologies are already informing new approaches to African Studies and African diaspora studies, and ever more work is needed to disrupt the Western Atlantic fixity in examinations of Blackness and Black ecologies. The ways that a newly intensified plantation regime is driving transformation of Southeast Asia and Pacific ecologies should be examined in comparative terms to Black ecologies to ensure that plantations are not seen as new, but reorganized, redeployed, and rebranded apparatuses of socio-ecological discipline.

Racial ecologies structure racialized encounters and intimacies, shaping understandings of human affinities across differentiated embodiment and social positions (Lowe 2015). The racist intent behind racial ecologies means that these environments are not spaces of solidarity. However, emerging work at the intersections of Black, Latinx, Indigenous ecologies is disrupting the colonially inscribed lines of difference. This work charts new socio-ecological histories and futures beyond antagonisms and incommensurability. Thinking from borderlands and spaces of Black and Indigenous encounter, for example, Tiffany King and collaborators elucidate "otherwise worlds" that are only knowable through new theories and methodologies from political ecologies at the nexus of settler colonialism and anti-Blackness (King 2019; King et al. 2020). Work informed by such engagements should motivate new and critical inquiries into the racialized natures of the environmental humanities and continue to chart speculative ecologies beyond white supremacy, anti-Blackness, Indigenous genocide, anti-immigration, racialization of poverty, narrow (post)humanisms, and utopian imaginaries of the green new deal.

Rising eco-populism, the militarization of climate change adaptation, and growing interest in ocean resources and outer-space demand increased attention from political ecologists of racialized nature (Bennett 2019; McCarthy 2019). Greater engagement by political ecologists with environmental sociology, the environmental humanities, and ecologists should help these efforts. Against the destruction, denigration, death, and dying of racial ecologies, political ecologists must become concerned with anti-racist ecologies of repair, hope, and care (Moulton et al. 2021; Neely and Lopez 2022; Tuck 2009).

Conclusion

Edward Long (1774, 2: 554) reasoned, the "sickly island" of Jamaica was made healthier by racialized landscape transformation. Long though was not ignorant to the possible effects of the transformations of the landscape, he reasoned that there was "every reason to believe that the rains happen very differently now, both in time and quantity, in this island, from what they formerly did," and suggested that some of this be ascribed to "the clearing of woods in the mountainous parts" (Long 1774, 3: 646, 648, 651). He was not the first nor the last to come to that opinion. Yet the plantation machinery carried on the reproduction of a racialized ecology in the name of profit, and not until the end of the golden era of the plantation, as the formerly enslaved established free communities, did the adverse effects of this transformation get taken seriously. And then, only in the context of Black movement from enslavement to freedom. Forest degradation and environmental decline was framed as almost entirely the result of Black post-emancipation peasants. The prospects of Black natures, natures shaped by Black agrarianism, informed forestry as much as the growing awareness of earth and climate systems which were profoundly altered by the plantation system. We can end by returning to Long. Our current ecological crisis tied to climate change and capitalism is one manifestation of the afterlives of the racialization of nature and the naturalization of race, and imagining futures based on racial myopias will only reproduce the problem of plantation ecologies of anti-Blackness, Indigenous dispossession, and white supremacy.

References

Adams, W. M., and Hutton, J. (2007). People, parks and poverty: Political ecology and biodiversity conservation. *Conservation and Society*, 5(2), 147–183.

Agrawal, A. (2005). *Environmentality: Technologies of Government and the Making of Subjects*. Durham, NC: Duke University Press.

Alimonda, H. (2019). The coloniality of nature: An approach to Latin American political ecology. *Alternautas*, 6(1), 102–142.

Allen, D., Lawhon, M., and Pierce, J. (2019). Placing race: On the resonance of place with black geographies. *Progress in Human Geography*, 43(6), 1001–1019.

Allen, D. L. (2020). Black geographies of respite: Relief, recuperation, and resonance at Florida AandM University. *Antipode*, 52(6), 1563–1582.

Asher, K. (2009). *Black and Green: Afro-Colombians, Development, and Nature in the Pacific Lowlands*. Durham, NC: Duke University Press.

Baldwin, A. (2012). Whiteness and futurity: Towards a research agenda. *Progress in Human Geography*, 36(2), 172–187.

Bebbington, A., and Bury, J. (2013). *Subterranean Struggles: New Dynamics of Mining, Oil, and Gas in Latin America* (Vol. 8). University of Texas Press.

Bennett, N. J. (2019). In political seas: Engaging with political ecology in the ocean and coastal environment. *Coastal Management*, 47(1), 67–87.

Beymer-Farris, B. A., and Bassett, T. J. (2012). The REDD menace: Resurgent protectionism in Tanzania's mangrove forests. *Global Environmental Change*, 22(2), 332–341.

Bhandar, B. (2018). *Colonial Lives of Property: Law, Land, and Racial Regimes of Ownership*. Durham, NC: Duke University Press.

Blaikie, P., and Brookfield, H. (2015). *Land Degradation and Society*. Routledge.

Blaut, J. M. (1993). *The Colonizer's Model of the World: Geographical Diffusionism and Eurocentric History*. (Vol. 1). Guilford Press.

Bledsoe, A. (2017). Marronage as a past and present geography in the Americas. *Southeastern Geographer*, 57(1), 30–50.

Bledsoe, A., and Wright, W. J. (2019). The pluralities of black geographies. *Antipode*, 51(2), 419–437.

Bosworth, K. (2021). "They're treating us like Indians!": Political ecologies of property and race in North American pipeline populism. *Antipode*, 53(3), 665–685.

Brahinsky, R., Sasser, J., and Minkoff-Zern, L. A. (2014). Race, space, and nature: An introduction and critique. *Antipode*, 46(5), 1135–1152.

Brand, A. L., and Miller, C. (2020). Tomorrow I'll be at the table: Black geographies and urban planning: A review of the literature. *Journal of Planning Literature*, 35(4), 460–474.

Brockington, D., (2002). *Fortress Conservation: The Preservation of the Mkomazi Game Reserve, Tanzania*. Indiana University Press.

Bruno, T. (2022). Ecological memory in the biophysical afterlife of slavery. *Annals of the American Association of Geographers*, 1–11.

Bullard, R. (1990). *Dumping in Dixie: Race, Class, and Environmental Quality*. New York: Taylor & Francis.

Bullard, R., Mohai, P., Saha, R. and Wright, B. (2007). *Toxic Wastes and Race at Twenty, 1987–2007*. Cleveland, OH: United Church of Christ.
Bullard, R., and Wright, B. (2012). *The Wrong Complexion for Protection: How the Government Response to Disaster Endangers African American Communities*. NYU Press.
Burghardt, K. T., Avolio, M. L., Locke, D. H., Grove, J. M., Sonti, N. F., and Swan, C. M. (2023). Current street tree communities reflect race-based housing policy and modern attempts to remedy environmental injustice. *Ecology*, e3881.
Burow, P. B., Brock, S., and Dove, M. R. (2018). Unsettling the land: Indigeneity, ontology, and hybridity in settler colonialism. *Environment and Society*, 9(1), 57–74.
Carney, J. A. (2001). *Black Rice: The African Origins of Rice Cultivation in the Americas*. Harvard University Press.
Carney, J., and Rosomoff, R. N. (2011). *In the Shadow of Slavery: Africa's Botanical Legacy in the Atlantic World*. University of California Press.
Castree, N. (2001). Socializing nature: Theory, practice, and politics. In N. Castree and B. Braun (Eds.), *Social Nature: Theory, Practice, and Politics* (pp. 1–21). Malden, MA: Blackwell Publishers.
Chao, S. (2022a). *In the Shadow of the Palms: More-than-Human Becomings in West Papua*. Durham, NC: Duke University Press.
Chao, S. (2022b). Plantation. *Environmental Humanities*, 14(2), 361–366.
Collard, R. C., Dempsey, J., and Sundberg, J. (2015). A manifesto for abundant futures. *Annals of the Association of American Geographers* 105(2), 322–330.
Coulthard, G. S. (2014). *Red Skin, White Masks: Rejecting the Colonial Politics of Recognition*. Minneapolis, MN: University of Minnesota Press.
Cronon, W. (1983). *Changes in the Land: Indians. Colonists, and the Ecology of New England*. New York: Hill and Wang.
Cronon, W. (1987). Revisiting the vanishing frontier: The legacy of Frederick Jackson Turner. *The Western Historical Quarterly*, 18(2), 157–176.
Cronon, W. (1996). The trouble with wilderness: Or getting back to the wrong nature. *Environmental History*, 1(1), 7–28.
Curley, A. (2021). Unsettling Indian water settlements: The little Colorado river, the San Juan river, and colonial enclosures. *Antipode*, 53(3), 705–723.
Daigle, M. (2018). Resurging through Kishiichiwan. *Decolonization: Indigeneity, Education and Society*, 7(1), 159–172.
Davis, J. (2019). Black faces, black spaces: Rethinking African American underrepresentation in wildland spaces and outdoor recreation. *Environment and Planning E: Nature and Space*, 2(1), 89–109.
Davis, J., Moulton, A. A., Van Sant, L., and Williams, B. (2019). Anthropocene, capitalocene, … plantationocene?: A manifesto for ecological justice in an age of global crises. *Geography Compass*, 13(5), e12438.
De Leeuw, S., and Hunt, S. (2018). Unsettling decolonizing geographies. *Geography Compass*, 12(7), e12376.
Doshi, S. (2013). The politics of the evicted: Redevelopment, subjectivity, and difference in Mumbai's slum frontier. *Antipode*, 45(4), 844–865.
Doshi, S. (2017). Embodied urban political ecology: Five propositions. *Area*, 49(1):125–128.
Du Bois, W. E. B. (1910). *The Souls of White Folk*. New York: The Independent.
Eaves, L. E. and Al-Hindi, K. F. (2023). Intersectional sensibilities and the spatial analyses of Ida B. Wells-Barnett and Ellen Churchill Semple. *The Professional Geographer*, 75(4), 691–697.
Eichler, L., and Baumeister, D. (2018). Hunting for justice: An indigenous critique of the North American model of wildlife conservation. *Environment and Society*, 9(1), 75–90.
Escobar, A. (1996). Construction nature: Elements for a post-structuralist political ecology. *Futures*, 28(4), 325–343.
Escobar, A. (2008). *Territories of Difference: Place, Movements, Life, Redes*. Durham, NC: Duke University Press.
Fagan, M. (2019). On the dangers of an Anthropocene epoch: Geological time, political time and post-human politics. *Political Geography*, 70, 55–63.
Fairhead, J., and Leach, M. (1995). False forest history, complicit social analysis: Rethinking some West African environmental narratives. *World Development*, 23(6), 1023–1035.

Ferdinand, M. (2021). *Decolonial Ecology: Thinking from the Caribbean World*. John Wiley & Sons.

Ferdinand, M. (2022). Behind the colonial silence of wilderness: "In Marronage Lies the Search of a World". *Environmental Humanities*, 14(1), 182–201.

Fernando, J. L. (2020). The virocene epoch: The vulnerability nexus of viruses, capitalism and racism. *Journal of Political Ecology*, 27(1), 635–684.

Finney, C. (2014). *Black Faces, White Spaces: Reimagining the Relationship of African Americans to the Great Outdoors*. UNC Press Books.

Forsyth, T. (2008). Political ecology and the epistemology of social justice. *Geoforum*, 39(2), 756–764.

Freshour, C., and Williams, B. (2022). Toward "Total Freedom": Black ecologies of land, labor, and livelihoods in the Mississippi Delta. *Annals of the American Association of Geographers*, 1–10.

Gandy, M. (2022). Urban political ecology: A critical reconfiguration. *Progress in Human Geography*, 46(1), 21–43.

Garth, H., and Reese, A. M. (Eds.). (2020). *Black Food Matters: Racial Justice in the Wake of Food Justice*. Minneapolis, MN: University of Minnesota Press.

Gill, B. S. (2021). A world in reverse: The political ecology of racial capitalism. *Politics*, 0263395721994439.

Gilmore, R. W. (2002). Fatal couplings of power and difference: Notes on racism and geography. *The Professional Geographer*, 54(1), 15–24.

Gilmore, R. W. (2007). *Golden Gulag: Prisons, Surplus, Crisis, and Opposition in Globalizing California*. Berkeley: University of California Press.

Goeman, M. (2013). *Mark My Words: Native Women Mapping our Nations*. Minneapolis, MN: University of Minnesota Press.

Grove, R. H. (1996). *Green Imperialism: Colonial Expansion, Tropical Island Edens and the Origins of Environmentalism, 1600–1860*. Cambridge University Press.

Guthman, J., and Mansfield, B. (2013). The implications of environmental epigenetics: A new direction for geographic inquiry on health, space, and nature–society relations. *Progress in Human Geography*, 37(4), 486–504.

Harris, C. I. (1993). Whiteness as property. *Harvard Law Review*, 1707–1791.

Harris, D. M. (2022). The trouble with modeling the human into the future climate. *GeoHumanities*, 8(2), 382–398.

Heynen, N. (2016). Urban political ecology II: The abolitionist century. *Progress in Human Geography*, 40(6), 839–845.

Heynen, N., Aiello, D., Keegan, C., and Luke, N. (2018). The enduring struggle for social justice and the city. *Annals of the American Association of Geographers*, 108(2), 301–316.

Heynen, N., Perkins, H. A., and Roy, P. (2006). The political ecology of uneven urban green space: The impact of political economy on race and ethnicity in producing environmental inequality in Milwaukee. *Urban Affairs Review*, 42(1), 3–25.

Heynen, N., and Ybarra, M. (2021). On abolition ecologies and making "freedom as a place". *Antipode*, 53(1), 21–35.

Hosbey, J., and Roane, J. T. (2021). A totally different form of living: On the legacies of displacement and marronage as Black ecologies. *Southern Cultures*, 27(1), 68–73.

Hulme, M. (2008). The conquering of climate: Discourses of fear and their dissolution. *Geographical Journal*, 174(1), 5–16.

Huntington, E. (1913). Changes of climate and history. *American Historical Review*, 18(2), 213–232.

Huntington, E. (1915). *Civilization and Climate*. New Haven, CT: Yale University Press.

Huntington, E. (1919). *World-Power and Evolution*. New Haven, CT: Yale University Press.

Hyman, C. (2021). The Oak of Jerusalem: Flight, refuge, and reconnaissance in the Great Dismal Swamp Region. *A Digital Narrative with ArcGIS (A Story Map)*. www.arcgis.com/apps/Cascade/index.html?appid+f3a23e246b8ece52fb1463ce5d

Inwood, J., and Bonds, A. (2017). Property and whiteness: The Oregon standoff and the contradictions of the US settler state. *Space and Polity*, 21(3), 253–268.

Jones, S. (2006). A political ecology of wildlife conservation in Africa. *Review of African Political Economy*, 33(109), 483–495.

Kelley, E. (2021). "Follow the Tree Flowers": Fugitive mapping in beloved. *Antipode*, 53(1), 181–199.
King, T. L. (2019). *The Black Shoals: Offshore formations of Black and Native Studies*. Durham, NC: Duke University Press.
King, T. L., Navarro, J., and Smith, A. (Eds.). (2020). *Otherwise Worlds: Against Settler Colonialism and Anti-Blackness*. Durham, NC: Duke University Press.
Kosek, J. (2004). Purity and pollution: Racial degradation and environmental anxieties. In Peet, R. and Watts, M. (Eds) *Liberation Ecologies: Environment, Development and Social Movements* (pp. 115–152). London, UK: Routledge.
Kosek, J. (2006). *Understories: The Political Life of Forests in Northern New Mexico*. Durham, NC: Duke University Press.
LaDuke, W. (1999). *All our Relations: Native Struggles for Land and Life*. South End Press.
Lavery, C. (2022). The power of racial mapping: Ellsworth Huntington, immigration, and eugenics in the progressive era. *The Journal of the Gilded Age and Progressive Era* 21(4), 262–278.
Lawhon, M., Ernstson, H., and Silver, J. (2014). Provincializing urban political ecology: Towards a situated UPE through African urbanism. *Antipode*, 46(2), 497–516.
Leff, E. (2015). Political ecology: A Latin American perspective. *Desenvolv. Meio Ambiente*, 35(35), 29–64.
Lewis, J. S. (2020). *Scammer's Yard: The Crime of Black Repair in Jamaica*. Minneapolis, MN: University of Minnesota Press.
Li, T. M. (2007). *The Will to Improve: Governmentality, Development, and the Practice of Politics*. Durham, NC: Duke University Press.
Liboiron, M. (2021). *Pollution is Colonialism*. Durham, NC: Duke University Press.
Livingstone, D. N. (1992). *The Geographical Tradition: Episodes in the History of a Contested Enterprise*. Cambridge, MA: Blackwell.
Livingstone, D. N. (2002). Tropical hermeneutics and the climatic imagination. *Geographische Zeitschrift*, 65–88.
Loftus, A. (2020). Political ecology III: Who are "the people"? *Progress in Human Geography*, 44(5), 981–990.
Long, E. (1774). *The History of Jamaica: Or General Survey of the Antient and Modern State of the Island: With Reflections on Its Situation Settlements, Inhabitants, Climate, Products, Commerce, Laws, and Government*. (Vol. 1–3). T. Lowndes.
Loughran, K. (2017). Race and the construction of city and nature. *Environment and Planning A: Economy and Space*, 49(9), 1948–1967.
Lowe, L. (2015). *The Intimacies of Four Continents*. Durham, NC: Duke University Press.
Luke, N. (2022). Powering racial capitalism: Electricity, rate-making, and the uneven energy geographies of Atlanta. *Environment and Planning E: Nature and Space*, 5(4), 1765–1787.
Mann, G. (2009). Should political ecology be Marxist? A case for Gramsci's historical materialism. *Geoforum*, 40(3), 335–344.
Mansfield, B., and Guthman, J. (2015). Epigenetic life: Biological plasticity, abnormality, and new configurations of race and reproduction. *Cultural Geographies*, 22(1), 3–20.
Marsh, G. P. (2003). *Man and Nature*. University of Washington Press.
Marx, K. (1990). *Capital: A Critique of Political Economy. Vol. 1. 1867*. Trans. Ben Fowkes. London: Penguin.
McCarthy, J. (2002). First World political ecology: Lessons from the Wise Use Movement. *Environment and Planning A: Economy and Space*, 34(7), 1281–1302.
McCarthy, J. (2005). First World political ecology: Directions and challenges. *Environment and Planning A: Economy and Space*, 37(6), 953–958.
McCarthy, J. (2019). Authoritarianism, populism, and the environment: Comparative experiences, insights, and perspectives. *Annals of the American Association of Geographers*, 109(2), 301–313.
McCarthy, J., and Hague, E. (2004). Race, nation, and nature: The cultural politics of "Celtic" identification in the American West. *Annals of the Association of American Geographers*, 94(2), 387–408.
McCutcheon, P. (2019). Fannie Lou Hamer's freedom farms and black agrarian geographies. *Antipode*, 51(1), 207–224.

McKittrick, K. (2006). *Demonic Grounds: Black Women and the Cartographies of Struggle.* Minneapolis, MN: University of Minnesota Press.
McKittrick, K. (2011). On plantations, prisons, and a black sense of place. *Social and Cultural Geography*, 12 (8), 947–963.
McKittrick, K. (2013). Plantation futures. *Small Axe*, 17(3), 1–15.
McKittrick, K., and Woods, C. (Eds) (2007). *Black Geographies and the Politics of Place.* Toronto: Between the Lines.
Merchant, C. (2004). *Reinventing Eden: The Fate of Nature in Western Culture.* Routledge.
Merchant, C. (2007). *American Environmental History: An Introduction.* Columbia University Press.
Miriti, M. N., Rawson, A. J., and Mansfield, B. (2023). The history of natural history and race: Decolonizing human dimensions of ecology. *Ecological Applications*, 33(1), e2748.
Mirzoeff, N. (2022). The whiteness of birds. *Liquid Blackness*, 6(1), 120–137.
Mollett, S. (2011). Racial narratives: Miskito and Colono land struggles in the Honduran Mosquitia. *Cultural Geographies*, 18(1), 43–62.
Mollett, S. (2021). Hemispheric, relational, and intersectional political ecologies of race: Centring land-body entanglements in the Americas. *Antipode*, 53(3), 810–830.
Mollett, S., and Faria, C. (2013). Messing with gender in feminist political ecology. *Geoforum*, 45, 116–125.
Moore, D. S. (1997). Remapping resistance. In Pile, S., and Keith, M. (Eds) *Geographies of Resistance* (pp. 87–106). London Routledge.
Moore, D. (2005). *Suffering for Territory: Race, Place, and Power in Zimbabwe.* Durham, NC: Duke University Press.
Moore, D. S., Pandian, A., and Kosek, J. (2003). Introduction. The cultural politics of race and nature: Terrains of power and practice. In Moore, D. S., Pandian, A., and Kosek, J. (Eds.) *Race, Nature, and the Politics of Difference* (pp. 1–70). Durham, NC: Duke University Press.
Morgensen, S. L. (2011). The biopolitics of settler colonialism: Right here, right now. *Settler Colonial Studies*, 1(1), 52–76.
Moulton, A. A. (2022). Towards the arboreal side-effects of marronage: Black geographies and ecologies of the Jamaican forest. *Environment and Planning E: Nature and Space*, 25148486221103757.
Moulton, A. A., and Salo, I. (2022). Black geographies and Black ecologies as insurgent ecocriticism. *Environment and Society*, 13(1), 1–19.
Moulton, A. A., Velednitsky, S., Harris, D. M., Cook, C. B., and Wheeler, B. L. (2021). On and beyond traumatic fallout: Unsettling political ecology in practice and scholarship. *Journal of Political Ecology*, 28(1).
Murphy, M. W. (2021). Notes toward an anticolonial environmental sociology of race. *Environmental Sociology*, 7(2), 122–133.
Neely, A. H., and Lopez, P. J. (2022). Toward healthier futures in post-pandemic times: Political ecology, racial capitalism, and Black feminist approaches to care. *Geography Compass*, 16(2), e12609.
Neumann, R. P. (1998). *Imposing Wilderness: Struggles Over Livelihood and Nature Preservation in Africa* (Vol. 4). University of California Press.
Neumann, R. (2014). *Making Political Ecology.* Routledge.
Nightingale, A. J. (2011). Bounding difference: Intersectionality and the material production of gender, caste, class and environment in Nepal. *Geoforum*, 42(2), 153–162.
Nishime, L., and Williams, K. D. H. (Eds.) (2018). *Racial Ecologies.* University of Washington Press.
Nixon, R. (2011). *Slow Violence and the Environmentalism of the Poor.* Harvard University Press.
Ogada, M. (2021). The second Scramble for Africa. *Africa Is a Country.* https://africasacountry.com/2021/11/the-second-scramble-for-africa
Ojeda, D., Nirmal, P., Rocheleau, D., and Emel, J. (2022). Feminist ecologies. *Annual Review of Environment and Resources*, 47, 149–171.
Omi, M., and Winant, H. (1993). On the theoretical status of the concept of race. *Race, Identity and Representation in Education*, 3–10.
Peet, R. (1985). The social origins of environmental determinism. *Annals of the Association of American Geographers*, 75(3): 309–333.

Peet, R. and Watts, M. (Eds.) (1996). *Liberation Ecologies: Environment, Development and Social Movements*. London, UK: Routledge.

Pellow, D. (2007). *Resisting Global Toxics: Transnational Movements for Environmental Justice*. Cambridge, MA: MIT Press.

Pellow, D. N. (2016). Toward a critical environmental justice study: Black Lives Matter as an environmental justice challenge. *Du Bois Review: Social Science Research on Race*, 13(2), 221–236.

Prudham, S. (2015). Property and commodification. In Perreault, T. A., Bridge, G., and McCarthy, J. P. (Eds.) *The Routledge Handbook of Political Ecology* (pp. 430–445). Routledge.

Pulido, L. (2006). Geographies of race and ethnicity II: Environmental racism, racial capitalism and state-sanctioned violence. *Progress in Human Geography*, 41(4), 524–533.

Pulido, L. (2016). Flint, environmental racism, and racial capitalism. *Capitalism, Nature, Socialism*, 7(3):1–16.

Pulido, L. (2017). Geographies of race and ethnicity II: Environmental racism, racial capitalism and state-sanctioned violence. *Progress in Human Geography*, 41(4), 524–533.

Pulido, L., and De Lara, J. (2018). Reimagining "justice" in environmental justice: Radical ecologies, decolonial thought, and the Black Radical Tradition. *Environment and Planning E: Nature and Space*, 1(1–2), 76–98.

Radcliffe, S. A., Watson, E. E., Simmons, I., Fernández-Armesto, F., and Sluyter, A. (2010). Environmentalist thinking and/in geography. *Progress in Human Geography*, 34(1), 98–116.

Ramutsindela, M., Matose, F., and Mushonga, T. (Eds.) (2022). *The Violence of Conservation in Africa: State, Militarization and Alternatives*. Edward Elgar Publishing.

Ranganathan, M. (2022). Towards a political ecology of caste and the city. *Journal of Urban Technology*, 1–9.

Roane, J. T. (2018). Plotting the Black commons. *Souls*, 20(3), 239–266.

Roane, J. T. (2022). Black ecologies, subaquatic life, and the Jim Crow enclosure of the tidewater. *Journal of Rural Studies*, 94, 227–238.

Roane, J. T., and Hosbey, J. (2019). Mapping black ecologies. *Current Research in Digital History*, 2. https://doi.org/10.31835/crdh.2019.05

Robbins, P. (2012). *Lawn People: How Grasses, Weeds, and Chemicals Make Us Who We Are*. Philadelphia, PA: Temple University Press.

Robbins, P. (2019). *Political Ecology: A Critical Introduction*. John Wiley & Sons.

Robinson, C. J. (2020). *Black Marxism: The Making of the Black Radical Tradition*. UNC Press Books.

Rocheleau, D., Thomas-Slayter, B., and Wangari, E. (Eds.) (1996). *Feminist Political Ecology: Global Issues and Local Experiences*. Routledge.

Semple, E. C. (1901). The Anglo-Saxons of the Kentucky mountains: A study in anthropogeography. *The Geographical Journal*, 17(6), 588–623.

Semple, E. C. (1911). *Influences of Geographic Environment, on the Basis of Ratzel's System of Anthropo-geography*. H. Holt.

Sène, A. L. (2022). Land grabs and conservation propaganda. *Africa Is a Country*: https://africasacountry.com/2022/06/the-propaganda-of-biodiversity-conservation

Simpson, L. (2001). Aboriginal peoples and knowledge: Decolonizing our processes. *The Canadian Journal of Native Studies*, 21(1), 137–148.

Simpson, L. (2014). Land as pedagogy: Nishnaabeg intelligence and rebellious transformation. *Decolonization: Indigeneity, Education and Society* 3(3): 1–25.

Smith, S., and Vasudevan, P. (2017). Race, biopolitics, and the future: Introduction to the special section. *Environment and Planning D: Society and Space*, 35(2), 210–221.

Sundberg, J. (2014). Decolonizing posthumanist geographies. *Cultural Geographies*, 21(1), 33–47.

Taylor, D. E. (2016). *The Rise of the American Conservation Movement*. Durham, NC: Duke University Press.

Tetreault, D. (2017). Three forms of political ecology. *Ethics and the Environment*, 22(2), 1–23.

Tuana, N. (2019). Climate apartheid: The forgetting of race in the Anthropocene. *Critical Philosophy of Race*, 7(1), 1–31.

Tuck, E. (2009). Suspending damage: A letter to communities. *Harvard Educational Review*, 79(3), 409–428.

Tuck, E., and Yang, K.W. (2012). Decolonization is not a metaphor. *Decolonization: Indigeneity, Education, and Society* 1(1):1–40.
Turner, F. J. (1894). *The Significance of the Frontier in American History*. Madison: State Historical Society of Wisconsin.
Vandergeest, P., and Roth, R. (2016). A Southeast Asian political ecology. In Perreault, T. A., Bridge, G., and McCarthy, J. P. (Eds.) *The Routledge Handbook of Political Ecology* (pp. 82–98). London: Routledge.
Van Sant, L. (2021). "The long-time requirements of the nation": The US Cooperative Soil Survey and the Political Ecologies of Improvement. *Antipode*, 53(3), 686–704.
Van Sant, L., Hardy, D., and Nuse, B. (2021a). Conserving what? Conservation easements and environmental justice in the coastal US South. *Human Geography*, 14(1), 31–44.
Van Sant, L., Milligan, R., and Mollett, S. (2021b). Political ecologies of race: Settler colonialism and environmental racism in the United States and Canada. *Antipode*, 53(3), 629–642.
Vasudevan, P. (2021). An intimate inventory of race and waste. *Antipode*, 53(3), 770–790.
Vasudevan, P., Ramírez, M. M., Mendoza, Y. G., and Daigle, M. (2022). Storytelling earth and body. *Annals of the American Association of Geographers*, 1–17.
Vickers, M. P. (2022). On swampification: Black ecologies, moral geographies, and racialized swampland destruction. *Annals of the American Association of Geographers*, 1–8.
Watkins, C. (2021). *Palm Oil Diaspora: Afro-Brazilian Landscapes and Economies on Bahia's Dendê Coast*. Cambridge University Press.
Whyte, K. P. (2018a). Indigenous science (fiction) for the Anthropocene: Ancestral dystopias and fantasies of climate change crises. *Environment and Planning E: Nature and Space*, 1(12), 224–242.
Whyte, K. (2018b). Settler colonialism, ecology, and environmental injustice. *Environment and Society*, 9(1), 125–144.
Williams, B. (2020). "The fabric of our lives"? Cotton, pesticides, and agrarian racial regimes in the US South. *Annals of the American Association of Geographers*, 111(2), 422–439.
Williams, B., Van Sant, L., Moulton, A. A., and Davis, J. (2020). Race, land and freedom. In Domosh, M., Heffernan, M., and Withers, C. W. (Eds.). *The SAGE Handbook of Historical Geography* (Vol. 1) (pp. 179–198). Sage.
Williams, B., and Porter, J. M. (2022). Cotton, whiteness, and other poisons. *Environmental Humanities*, 14(3), 499–521.
Wilson, B. M. (2000). *Race and Place in Birmingham: The Civil Rights and Neighborhood Movements*. Rowman and Littlefield.
Winston, C. (2021). Maroon geographies. *Annals of the American Association of Geographers*, 111(7), 2185–2199.
Woods, C. (1998). *Development Arrested: The Blues and Plantation Power in the Mississippi Delta*. New York: Verso.
Woods, C. A. (2002). Life after death. *The Professional Geographer*, 54(1), 62–66.
Wright, W. J. (2020). The morphology of marronage. *Annals of the American Association of Geographers*, 110(4), 1134–1149.
Wright, W. J. (2021). As above, so below: Anti-Black violence as environmental racism. *Antipode*, 53(3), 791–809.
Wynter, S. (1995). "1492: A New World View." In Lawrence Hyatt, V., and Nettleford, R. (Eds.) *Race Discourse and the Origin of the Americas: A New World View* (pp. 5–57). Washington, DC: Smithsonian Institution.
Wynter, S. (2003). Unsettling the coloniality of being/power/truth/freedom: Towards the human, after man, its overrepresentation—An argument. *CR: The New Centennial Review*, 3(3), 257–337.
Yazzie, M. K. (2018) Decolonizing development in Diné Bikeyah: Resource extraction, anticapitalism, and relational futures. *Environment and Society*, 9(1), 25–39.
Yusoff, K. (2018). *A Billion Black Anthropocenes or None*. Minneapolis, MN: University of Minnesota Press.
Yusoff, K. (2020). The inhumanities. *Annals of the American Association of Geographers*, 111(3), 663–676.

17 Embodying Nature

De-centering and Re-centering Bodies as Socio-nature

Nari Senanayake

On the morning of October 14, 2018, Jayani Perera[1] died quietly and without a formal diagnosis of the ailments that ultimately ended her life. Her rapid deterioration and death unfolded against a backdrop of intense contestation about the source and toxicity of environmental contaminants in the place she called home. Jayani lived and labored in the agrarian settlement scheme of Padaviya, a locality of Sri Lanka that has now become synonymous with an epidemic of mysterious kidney disease that is widely believed to be tied to environmental toxins. Crucially, in this setting the causes of illness are uncertain, the consequences of disease are often deadly, and the possibilities of developing sickness in the future are both unpredictable and ever-present. Although a nationally funded mobile screening program promised to enact the mass diagnosis of asymptomatic individuals, at the height of the epidemic only 13% of the target population had been screened for kidney dysfunction through mobile clinics. Indeed, Jayani chose not to attend the mobile screening clinics that were hosted in Padaviya in the year prior to her death. Instead, she, and many others like her, preferred to live with uncertainty about their disease status, or to hover for extended periods of time between states of sickness and health.

In a region where cure is ephemeral and diagnosis is shot through with experiences of psychological distress and socio-economic fallout, the advantages of occupying the space between health and illness—at least in the short term and at the scale of the household—are by no means insignificant. By both embodying the potential to become ill at any time and actively resisting medical surveillance, Jayani's story requires us to develop much more *continuous* and less dichotomous understandings of health and illness (see also Hinchliffe et al. 2016; Senanayake 2021). Her experience is thus a window onto the contradictory, ambiguous, and liminal status of bodies in much of the world. And yet, awkward questions about Jayani's story remain for the field of political ecology. Her unruly body suggests that it may be necessary to rethink what it means to "embody nature" in the complex and novel environments of the Anthropocene. How do the socio-ecological hazards of our current moment unsettle distinctions between bodies and environments, as well as what "health" means and who is "deserving" of it? How might we tell stories that are more reflective of, and sympathetic to, the experiences of bodies which routinely transgress the health/illness dichotomy? How do we conceptualize bodies that refuse neat classification and who simultaneously index toxic and enabling relationships with nature? In taking such unruly bodies as sites of inquiry and sources of knowledge, I ask how political ecology might better account for their converged experiences of health, risk, and disease?

In this chapter, I review some of the ways that the body has been conceptualized and de- and re-centered in the field of political ecology to identify existing strengths but also,

DOI: 10.4324/9781003165477-22

new directions for future work. Building on several recent reviews (Abbots et al. 2020; Mansfield 2017, 2018; Nichols and Del Casino 2021; Senanayake and King 2019), I examine what counts as "political ecology of the body" and trace this literature's development over time and along three related strands. First, existing literature highlights porosity, plasticity, and body-ecologies as important conceptual vocabularies for describing the making and remaking of bodies, particularly in the dual contexts of late capitalism and the Anthropocene. Second, political ecologists have generated important theorizations of embodied materiality. Taking seriously the emotional, affective, and biochemical dimensions of bodies, this strand of work broadens our understanding of how health, harm, and biological variation materialize in everyday life. Third, by increasingly centering the relational, intersectional, and uneven entanglements of bodies and nature, scholars have productively extended the field's conceptualizations of difference, agency, and causality. Each of these thematic clusters contains tensions, overlaps and discontinuities. However, like much of the field writ large, political ecology's take on bodies and nature is tied together by a tradition of evolution and ambivalence (Forsyth 2003; Guthman 2011, 2012; Robbins 2015). In dialogue with this contradictory impulse to "advance and undermine explanation" (Robbins 2015: 91), I conclude by exploring potentials to "crip" existing literature through the concept of alterlife.

The Body as Socio-nature

Permeability: Environmental Change Does Not Stop at the Skin

A review of literature on nature and embodiment suggests that the concepts of porosity/permeability have been important catalysts for the development of new frameworks, many of which self-define as "political ecologies of the body" (Abbots et al. 2020; Guthman 2011; Hayes-Conroy and Hayes-Conroy 2013, 2015; Neely 2015; Nichols and Del Casino 2021). For many political ecologists, the task is surely to transgress categorical distinctions between bodies and the environment. Accordingly, scholars insist on experiences of permeability to disturb normative ideas of bodies as impervious, static and "bounded by their skin" (Bosworth 2017; Braun 2007; Carson 2002; Jackson and Neely 2014; Langston 2010; Nash 2006; Senanayake and King 2019).

A key intervention turns on scaling down the places or environments that "count" for political ecologists, moving beyond a focus on landscapes to also examine bodies as dynamic ecosystems that are themselves subjects and sources of environmental transformation (see Guthman 2012; Neely 2015). This ecological understanding of the body tracks the interactions between microbes, cells, chemicals, toxins, or viruses within the body as well as the dynamic flow of materials between bodies and environments (see for instance Guthman and Mansfield 2013, 2015; Neely 2015). In other words, the concepts of permeability and porosity have allowed political ecologists to expand our understanding of how environmental change shapes bodies in dynamic ways and gives rise to novel patterns of variability, health, and harm.

Unsurprisingly then, much of the empirical work on the porous and permeable boundaries of bodies has focused on how anthropogenic environments and structural inequalities reconfigure human biology and patterns of health. This includes a rich body of scholarship on the contingent, non-linear, and cross-scalar implications for human and planetary health, wrought by global warming or industrial pollutants (Baer and Singer 2016; Mansfield 2012a, 2018; Nyantakyi-Frimpong 2021; Senier et al. 2017; Shattuck

2019, 2020). Work in political ecology also takes pains to critique the role of political economic forces in shaping bodies and biologies, generating important theorizations of the body as a spatial fix (Guthman 2011, 2015) as well as nuanced accounts of embodied neoliberalism (Sweet et al. 2018), bioprecarity (Shildrick 2019), and bodily wastage (Pulido 2016; Stanley 2015; Wright 2013). In dialogue with larger debates on on the parasitic interplay between neoliberalism and the body, political ecologists craft intimate portrayals of how bodies are produced as capitalist property, waste, and sites of surplus value in the wake of ecological upheaval and change. The (re-)emergence of infectious diseases and new interspecies interactions—particularly those between humans, animals, and pathogens—have also drawn attention to the nonlinearity and porous relationships between bodies and nature, creating new shocks and stresses to livelihoods, often in socially differentiated ways (Hinchliffe et al. 2016). And as I discuss further below, the boundaries and thresholds of the body have been challenged and reconceptualized by theorizations of the microbiome, multispecies ecologies, and research on environmental epigenetics. Together, these interventions have helped recast bodies as both objects and agents of environmental change: collapsing clear distinctions between human and microbial bodies, heredity and development biology, as well as natural and human history (Meloni et al. 2021; Nash 2006). Crucially, by rendering visible diverse embodiments of social, industrial, and ecological hazards, political ecologists seek to re-spatialize responsibility for environmental harm away from neoliberal models of personal accountability and rational self-management. Instead, much of this work takes pains to reveals how structures, such as settler colonialism and racial capitalism, produce uneven possibilities for health and well-being in the first place. As a consequence, this work better accounts for the embodiment of "slow violence" through everyday pollution, racism, and inequalities (McManus 2021; Meloni et al. 2021; Pathak 2020; Shadaan and Murphy 2020; Vasudevan 2021) as well as how embodied experiences may "register, mediate, shape, reproduce, resist, or disrupt broader social, environmental, and economic dynamics and trajectories" (Kinkaid 2019: 47).

Plasticity: body-environments remade

Within the past decade, the empirical and theoretical scope of literature on embodied natures has widened to encompass interventions from the post-genomic sciences. Much of this engagement turns on the concept of bodily plasticity, which describes how biophysical, psychosocial, and nutritive environments rework cellular processes within bodies in ways that variously contribute to either disease, health, *or* environmentally induced—but nonpathogenic—biological differences (Guthman and Mansfield 2015; Mansfield 2018; Meloni et al. 2021; see also Mansfield 2012a). As a corrective to static models of gene-environment interaction, the emphasis on bodily malleability, fluidity, and plasticity has pushed political ecologists to not only consider how the "supposedly external environment actively enters, shapes, and becomes part of the body" but also, how the body is often "active in its own remaking... [and] responds anew to the environment in an ongoing, iterative relationship" (Guthman and Mansfield 2013: 497). As Mansfield (2018: 221) elaborates, "in showing the openness of the body—the radical specificity of each body to its time and place—[the concept of plasticity shows] ... the body to be heterogeneous and changing rather than categorical and fixed." This emphasis on contingency and "becoming" also allows space to consider diverse embodiments of nature—beyond binary categories of health and harm—that nevertheless, may be significant in people's lives. A similar line of reasoning upends stubborn myths of, and aspirations for, bodily purity or "pristine" (bodily) natures. This literature instead demonstrates how "human activity has altered earth systems in ways that produce multi-scalar outcomes which not only do not stop at the skin but enter into our genome to shape fundamental processes," such as phenotypic development, genomic regulation, neuronal functioning, and cellular activity (Mansfield 2018: 220; see also Meloni et al. 2021).

More recently, this work has examined how the emerging science of biological plasticity intersects with histories of systematic advantage and disadvantage in ways that reproduce deeply normalizing regimes of biopolitics (Barbour and Guthman 2018; Mansfield 2012b, 2017, 2018; Pentecost and Cousins 2017). This work taps into a rich tradition of scholarship that highlights how experiences of social stress and inequality are rendered biological through dynamic chemical changes within bodies as well as how social environments structure and reproduce biological vulnerabilities across space and through (intergenerational) time (Kuzawa and Sweet 2009; Bollati and Baccarelli 2010; Landecker and Panofsky 2013). Building on this work, political ecologists examine how theorizing bodies as permeable, plastic, and "open" also unleashes new forms of politics, inequality, and power. Warning against an uncritical celebration of bodily plasticity, Mansfield (2012b, 2018: 224) takes pains to demonstrate how "non-dualist and non-essentialist approaches can [also] be normalizing and biopolitical, even indexing new forms of eugenics." Put simply, by recasting the body as malleable, ideas of bodily plasticity multiply possibilities for intervention into what, until now, has been seen as relatively fixed, reshaping our understanding of the "interplay of plasticity and evolution" (Leatherman and Goodman 2020: 1). By doing so, ideas of plasticity can unevenly enroll bodies into new forms of improvement and transformation by proliferating opportunities to "identify and eliminate traits defined as 'abnormal'" (Mansfield 2018: 223). Cutting across political ecology's engagement with the post-genomic sciences is thus a steadfast and healthy skepticism that "acknowledges the power laden implications of any ... foundational [albeit non-dualistic] account" of embodying nature (Robbins 2015, 93). Emblematic of political ecology writ large, existing scholarship on bodies and nature attempts to have it "both ways," integrating concepts from cognate fields to advance explanations of how nature is embodied, while revealing problematic assumptions that underpin these concepts and how they in turn generate highly political effects.

Body-Ecologies and the Microbiome

Recent scholarship on embodying nature has re-conceptualized, challenged, and extended the boundaries of the body, adding to critical vocabularies of porosity and plasticity by emphasizing how "the human is always intermeshed with the more-than-human world" (Alaimo 2010: 2; Blackman 2016; Greenhough et al. 2020; Landecker 2016; Lorimer 2017). At the center of this analytical shift are the concepts of body ecologies (Murphy 2006), local/situated biologies (Lock 2017) and the microbiome (Relman 2015), all of which provide frameworks for understanding the role of non-human actors and objects in the constitution of bodies and health.

The very question of what counts as the body, for instance, is destabilized by political ecology's encounters with multi-species or more-than-human geographies. A growing theoretical interest in the body as a site of multi-species interaction has led recent work to decenter and disrupt the boundaries of the human body. Instead, this work foregrounds "networks of social relations," "disease situations," or "entanglements" between multiple bodies—human, animal, and microbial, as well as between bodies and complex environments (see Hinchliffe et al. 2016; Jackson and Neely 2014; Nading 2014). One of the central concepts to enter political ecology as a result of this engagement is the microbiome, which has been leveraged to re-spatialize the figure of the human to include the diverse forms of microbial life in, on, and around bodies (Greenhough et al. 2020; Lorimer 2017; Stallins et al. 2018). Precisely because "microbial cells outnumber human cells in the body and the number of microbial genes in the body are at least 100 times greater than in the human genome," political ecologists increasingly put forward unstable, geographically configured, and *compositionally diverse* conceptualizations of the body (Knight et al. 2017: 66). Theorizations of "*Homo microbis*" posit a new figure of the human who is not only made up of varying configurations of bacteria, viruses, fungi, and archaea but also, "whose life course microbial exposures come to configure identity, subjectivity and health outcomes" (Lorimer 2017: 11; also see Relman 2015). Central to the contributions of this work is a much more relational conceptualization of bodies, pathogens and microbes. As Hinchliffe et al. (2016: 16) elaborate:

> In this alternative [conceptualization of bodies], pathogens or the microbial world more generally, are very much a part and parcel of life wherever it may be. This is a world not so much threatened by the microbial outside but one where the manner in which lives are made, and the ways in which the inevitable entanglements between hosts, environments and microbes are handled, are key to any prospect for safe and indeed good life.

Owing significant debts to this early work on the more-than-human dimensions of bodies as well as how health (and its absence) emerges across species boundaries, more recent scholarship in the field of political ecology has expanded in several exciting directions to include studies of air, atmospheres, and breath (Allen 2020; Graham 2015; McManus 2021; Mostafanezhad 2021), one-health (Craddock and Hinchliffe 2015), indoor ecologies (Biehler 2013; Biehler and Simon 2011; Murphy 2006), multi-species health justice (Chao et al. 2022), and multi-species treatment/care assemblages (Hausermann 2015; Lunstrum et al. 2021). Cutting across this work is a focus on how "bodies are materialized 'otherwise', in relation to the myriad non-human actors around and within them" (Jackson and Neely 2014: 58), as well as how dynamic multi-species relationships intersect with larger "questions of environmental governance, social and spatial inequality and public engagement" to configure human health and possible health interventions (Greenhough et al. 2020: 1; also see: Hinchliffe et al. 2016).

Equally important are insights that flow from encounters with the conceptual apparatus of medical anthropology, including important theorizations of local or situated biologies (Lock 2017) and body ecologies (Murphy 2000, 2006). These core concepts have pushed forward our understandings of bodies and health as emergent properties that index multi-scalar and "continual interactions of biological and social processes across time and space" (Leatherman and Goodman 2020: 7; see also Lock 2017; Senier et al. 2017). Body ecologies, for example, mark a creative reimagination of bodily thresholds and boundaries. By rematerializing bodies at the molecular level, this concept has been used to better document how bodies are drawn into unsteady cycles of re-capacitation and debilitation through the transformation of home environments (see Senanayake 2022b). Central to this lens is how an individual's response to toxins or pathogens not only depends on internal ecologies, such as different quantities of enzymes, minerals, vitamins, t-cells, and yeast within the body, but also the complex material composition of the environment it inhabits. The molecular body thus extends to all objects in the home (or inhabited space) *and* within the human body—taking the form of a wider personal ecosystem. For instance, Michelle Murphy's work on "illegitimate" and highly contested environmental illnesses such as sick building syndrome (Murphy 2006) and multiple chemical sensitivity (Murphy 2000), sets aside the question of whether these conditions are real. Instead, Murphy powerfully illustrates how practices of self-care and the regulation of body-ecologies in space give rise to different and useful materializations of bodies outside their abjection from traditional biomedical models of disease. As Murphy (2000: 108) observes, these practices of self-care have the effect of displacing the site of disability from the body into the environment: "attention to space, [and] the structures and objects of the built environment become what render bodies abled or disabled, healthy or sick. By… changing the material composition of the built environment, the body can be rendered able again, the rebellious body tamed" (for similar analyses of how health/illness are materialized through care/home landscapes also see Brown 2003; Dyck 1998; Dyck et al. 2005; Milligan 2016; Parr 2001). Perhaps of particular value for political ecologists is the work that these concepts do to de-essentialize attributes such as pathogenicity or toxicity and instead illustrate how they emerge through interactions between multiple factors, from microbial populations, chemical exposures, and immune responses to locally varying environments and disease management practices (see also Senier et al. 2017 for a discussion of the socio-exposome). Pushing the body "beyond the skin" to encompass body ecologies thus also allows political ecologists to rethink relations of responsibility in ways that avoid an unhelpful and/or a myopic focus on individual behaviors and univalent relationships between health and the environment (Mollett 2021; Murphy 2017a, 2017b; Senanayake 2021; Shadaan and Murphy 2020).

Bodily Materiality

To date, much work that self-defines as political ecology of the body takes up questions of bodily materiality, drawing on diverse theoretical traditions to consider bodies as both "agent[s] and canvas[es] of embodied material practices" (Carney 2014: 4). Such studies have evolved from an initial concern with how "visceral" realms (senses, memory, feelings) inform environmental practices (Carney 2014; Ham 2020; Hayes-Conroy and Hayes-Conroy 2013, 2015; Kinkaid 2019) to interests in the biochemical dimensions of bodily materiality (Guthman and Mansfield 2015; Neely 2015), and most recently, the potential to develop integrated frameworks (Abbots et al. 2020; Nichols 2022b; Nichols and Del Casino 2021). For example, across case studies of the slow food movement and school garden and cooking programs, Jessica and Allison Hayes-Conroy (Hayes-Conroy and Hayes-Conroy 2013, 2015: 662) ask "how and why the physical, visceral body comes to desire, revolt or remain apathetic in the production of specific food–body [and by extension, health–body] connections." For these authors, sensory habits and visceral and affective relations are critical, though often overlooked, determinants of health as well as key forces shaping motivations to change (or not change) behaviors towards the environment (also see Sultana 2011; Truelove 2011). This work has since served as a springboard for research in multiple directions, helping to specify links between affective bodily process and the formation of political and ecological subjectivities (see: Kinkaid 2019; Carney 2014; Ham 2020) as well as generating new insights into the stories bodies reveal about the changing conditions of our existence—stories that do not always match people's stated accounts, and that are sometimes stories that people cannot or will not tell (Krieger 2005: 350; also see Holmes 2013; Petryna 2013; Parr 2010; Senanayake 2019).

Building on this work, studies that take up the biological, biochemical and/or ecological materiality of the body push political ecology in new and exciting directions (Guthman and Mansfield 2013, 2015; Jackson and Neely 2014; Neely 2015). While appreciating the contributions of existing (post)-phenomenological approaches to bodily materiality, this strand of work moves away from documenting the visceral and/or affective body to instead "explicitly engage the biophysical processes of ... poisoning or pollution ... [or] for that matter, the [biosocial] mechanisms that give rise to [and are also, deeply shaped by] affective, embodied responses" (Guthman and Mansfield 2015: 562). Across a series of publications, Julie Guthman and Becky Mansfield demonstrate how ecological, nutritional, and psychosocial stresses are transcribed (metabolically and phenotypically) onto bodies, rendering them unevenly vulnerable to the development of disease, health, or novel forms of biosocial differentiation. They argue that "current approaches to the political ecology of the body either neglect to engage the materiality of the body altogether or emphasize embodied experience at the expense of biological and chemical changes *within* the body and as they interact with diverse environments" (Guthman and Mansfield 2013, 2015: 566; Guthman 2011; Mansfield 2017, 2018). A key point of innovation in this literature is that it leverages work on the biological and biochemical materiality of the body to elaborate how political ecologists might better account for the materialization of environmentally induced human difference, as well as novel patterns of health and harm (also see: Lock and Nguyen 2018; Neely 2015).

As these various conceptualizations of bodily materiality take root in the field of political ecology, they have generated lively debate as well as new attempts at synthesis. For example, both Nichols and Del Casino (2021) and Abbots et al. (2020) push past the tendency in political ecology to either "conceptualize bodily materiality as ... complex socio-ecologies of 'blood, brains, and bones' *or* affective, visceral, and never fully representable" to develop integrated frameworks (Nichols and Del Casino 2021: 776). Specifically, Nichols and Del Casino (2021) chart a twofold path forward for political ecology. First, by holding in tension approaches to the body as both a "historical socio-biochemical assemblage and also an affective/non- representable happening," Nichols and Del Casino (2021: 786–787) highlight dynamic corporeal and affective interactions that shape how bodies experience (dis)ease and wellbeing across space and time. As they provocatively argue, "rather than looking at bodies as dis-eased or well... [this] approach allows us to see bodies—and their internal ecologies—as always in flux, actualizing moment-to-moment in differing states" (786; see also Allen 2020). A second important intervention of this approach is that it allows for scholars to more fully elaborate the myriad and reciprocal ways that "biophysical functions [shape and] are shaped by emotional and affective embodied experiences" (783; see Abbots et al. 2020 for an excellent discussion of how these reciprocal dynamics play out in the case of obesity). This reframing reveals new and surprising impacts on health and health practices by bringing into view the "webbed complexities... [and] entanglements among structures, materialities, visceralities, [physiological pathways] and bodies" (Abbots et al. 2020: 360). These innovations are important for countering a tendency in early work on bodily materiality to either bracket the affective or biochemical aspects of disease experience and/or to unfold within disciplinary silos. Moreover, such an approach allows space to consider how political ecologists may better participate in transdisciplinary conversations by integrating a political ecology of bodily materiality into current models of biomedical research and development policy (Abbots et al. 2020; Nichols 2022a, 2022b).

Bodily Entanglements: Revisiting Embodied Difference and Causality in Political Ecologies of Health

Intersectionality and Bodily Difference

Political ecologists have also brought relational, intersectional, and uneven entanglements of bodies and nature to bear on the field's conceptualizations of difference. For many this has involved exploring how ecological upheaval (re)-produces embodied disparities, or how social, ecological, and bodily differences are routinely co-produced through everyday practices of resource use. Promising to generate new insights, political ecologists leverage the concept of intersectionality to demonstrate how "changing environmental conditions *bring into existence* categories of social [and bodily] difference ... In other words, [difference] ... itself is re-inscribed in and through practices, policies and responses associated with changing environments and shifting modes of resource governance" (Elmhirst 2015: 523; see also Truelove 2011, 2019; Truelove and O'Reilly 2021). In taking up such ideas, political ecologists have revealed how "*particular* bodies bear the brunt of subsidizing and compensating" for ecological change and/or environmental interventions (Truelove 2011: 150), variously theorizing these outcomes as examples of "corporeal spatial precarity" (Mollett 2021), embodied "structural brutality" (Speed 2019), embodied disparity (Carney 2014), and the hidden embodied costs of natural resource management/transformation (Nichols 2020; Nightingale 2021; Nyantakyi-Frimpong 2021; Ranganathan 2022; Sultana 2011; Truelove 2011). Against neat or predetermined modes of categorization, this work approaches the question of bodily difference as a question of intersectional and deeply knotted socio-natural-corporeal relationships. Viewing the embodiment of nature as a process of co-production, where patterns of social inequality, environmental variability, and bodily difference are replicated, reconstituted, and intersect forces us to look more closely at how we (re)-make difference in the present, and how embodied disparities get reproduced over and over.

Schemes to promote kidney health in dry zone Sri Lanka, for example, set in motion interdependent and recursive processes of social, ecological, and bodily differentiation that rework the distinctions between what harms and what heals, as well as who is harmed and healed (Senanayake 2022a). Through long-term ethnographic fieldwork, my own work has demonstrated that the production of differentiated harms and uneven exposures to risk are, in fact, central to how well-intentioned disease mitigation strategies unfold on the ground. By analyzing these dynamics through the concept of intersectionality, polyvalent relationships between health and environment can be brought into sharp relief, including, in this case, how and why health improvement schemes generate new yet unevenly experienced harms by: (i) intensifying labor burdens and labor-related health conditions among economically disadvantaged women; and (ii) intensifying debt and dependence on agrochemical use among economically and ecologically resource-poor farmers. Two important contributions emerge from this expanded engagement with intersectionality, environmental change, and embodiment. First, by foreground intersecting, compounding, and sometimes mutually reinforcing differences (social, ecological, bodily), an intersectional framework can enrich explanations of how and why health inequalities can be so difficult to transcend, as well as how these knotted relationships create conditions for further disablement and degradation (see also Dillon and Sze 2016). Second, and conversely, intersectional approaches to embodiment have led political ecologists to investigate how the synergistic and interactive effects of social, ecological, and bodily difference produce "lateral and unexpected" consequences for health, including new openings for transgressing health exclusions and new spaces for collective action to improve socio-environmental conditions (see also Jampel 2018; Nightingale 2011: 155; Nightingale 2021).

Entanglement and Causality

Recent texts in the field of political ecology also illustrate what is at stake in rethinking body-environment interactions through the concept of entanglement (Hinchliffe et al. 2016; Neely 2021; Senanayake and King 2019). Specifically, work in this vein moves away from explanatory models which assume that agency can be ascribed to separate and independent factors (i.e., pathogen, parasite, or individual toxin), to instead document how context and the relations between entities fundamentally configure bodily and health outcomes. In practice, this focus on entanglements hinges on a key conceptual shift, signaled by the vocabulary of assemblages, topologies, and ecologies. These analytical tools move away from documenting conscious, controlled, and contained processes of causality to instead highlight distributed agency, synergistic effects, and overlapping yet distinct assemblages that might better illuminate the radical contingency of disease onset. As I argue elsewhere with Brian King (Senanayake and King 2019), these types of analytical moves provide a strong foundation for re-theorizing cause and effect relationships between the environment and human health. Beyond a focus on discrete causal relationships and the atomized body, these interventions highlight relational entanglements, emergent effects and the political and ecological processes that shape broader dynamics between health and environment. Doing so can reorient intervention away from treatment or eradicating "inherently dangerous" pathogens/toxins to investigating how "dangerous situations" arise, shape the intensities and scope of human-ecological interaction, and in turn amplify disease risk (Hinchliffe et al. 2016; King 2017; Neely 2015). It can also help recalibrate what counts as evidence of health risks to better address cumulative and synergistic exposures, disease bundles, ecological relations within and between bodies, and the structural determinants of disease. By extension, through its focus on relationships, the concept of entanglement helps reconfigure the norms that govern disease management to better reflect the complex socio-ecological situations that configure possibilities for health in the first place.

Abigail Neely's (2021) recent work leverages the concept of entanglement to further challenge assumptions about causality and agency in scholarship on bodies and nature. For Neely, existing frameworks cannot fully account for the embodiment of witchcraft illnesses. Like Murphy's (2006) work on contested environmental illnesses, Neely sets aside *prescriptive* conceptualizations of causality and agency to instead illustrate how witchcraft illnesses are embodied through dynamic entanglements of ancestors, healers, plant and animal parts, and incantations. In making this intervention, her approach "accounts for the relationships among all kinds of actors without a prescriptive account of *how* they relate." Instead, by theorizing the embodiment of nature as a more than human entanglement, her analysis "allows for the possibility that witchcraft makes a person sick without necessarily knowing the exact form relationships take or the direct causal pathways of the components of illness" (Neely 2021: 974). Moving beyond an obsession in biomedicine, and even existing approaches in political ecology with mapping discrete (albeit multiple) causal pathways, Neely leverages the analytical potential of entanglement to demonstrate how supra-natural beings like ancestors or incantations can materialize embodied states of being that matter to local communities, "even as it is not clear how the various components of the illness relate to each other" (ibid.: 972). Theorizing causality and agency as the "reconfiguration of bodily entanglements" also generates important normative implications. As Neely argues, not only does this framework provide valuable scaffolding for pushing forward nascent political ecologies of comorbidity, ecosyndemics, and the embodiment of multiple diseases/syndromes (Baer and Singer 2016; King and Rishworth 2022), but it also multiples possibilities for intervention and treatment. As she ultimately concludes, "if agency is the reconfiguration of entanglements, and entanglements are many things are once, then multiple entry points for (material) political change" and for healthier futures emerge (Neely 2021: 979).

Alterlife

> It is the story that makes the difference, because it is the story that hid my humanity from me.
>
> —Ursula Le Guin

While this robust conceptual vocabulary has generated important insights into bodies and nature, it does not fully grapple with the complex, contradictory, and fraught negotiation of embodied precarity unfolding within many communities: negotiations that simultaneously index acts of resignation and resistance, as well as states of debility and capacity, and health and harm. Only recently have scholars turned their attention to forms of embodiment that arguably *exceed* these kinds of binaries, instead taking pains to highlight how embodying nature is often polyvalent and deeply paradoxical. This includes important theorizations of slow observation (Davies 2018, 2023), resigned activism (Lora-Wainwright 2017), and wagering life (Valdivia 2018). Together this conceptual armory indexes forms of what Michelle Murphy (2017b: 5) calls *alterlife*: a concept that examines how life forged through ongoing structural, environmental, and chemical violence, is also "resurgent life, which asserts and continues nonetheless, and is [ultimately], life open to becoming something else" or to becoming "otherwise in the aftermath" (Murphy 2017b: 5–10). By highlighting embodiments of nature that *simultaneously* erode and enable life, work in this vein "allows us to both acknowledge and go beyond a concern with the inequities of power, which so strongly signal an expectation of negativity and lack of social justice, to ask how... precarious bodies might [also] signal a potential for [alternative modes of] communality," relationality and revitalization (Shildrick 2019: 595). As a consequence, the concept of alterlife is particularly useful for documenting complex and fraught negotiations of precarity beyond states of victimhood. Rejecting the notion that embodied precarity is only "a condition of negativity" or a state of abjection, this work "reimagine[s] how life might flourish in a different register" (Shildrick 2019: 610). Simply put, alterlife has the capacity to encompass forms of embodiment that exceed the health/illness binary, gesturing towards ways of being in common that are currently being invented.

Taking crip, queer, feminist, and decolonial critiques of "damage-based" frameworks/concepts seriously, the concept of alterlife also encourages political ecologists to be more reflexive about the stories that we tell about bodies and nature. As Michelle Murphy argues:

> despite often anti-racist intentions, this damage-based research has pernicious effects, placing the focus on chemical violence by virtue of rendering lives and landscapes as pathological. Such work tends to resuscitate racist, misogynist, and homophobic portraits of poor, Black, Indigenous, female, and queer lives as damaged and doomed, as inhabiting irreparable states that are not just unwanted but less than fully human.
>
> (Murphy 2017a: 496)

In short, the concept of alterlife contributes to a growing reappraisal of pathogenicity in existing literature, suggesting that we need to concurrently recognize the violence of "already being altered [through toxic exposures, ecological violence, racial capitalism etc.] *and* the struggle to become otherwise in the aftermath" (Murphy 2017b: 5). Put differently, it requires us to grapple with unruly bodies that routinely transgress health/illness, debility/capacity, and resistance/resignation dichotomies. In doing so, the concept of alterlife dovetails with interventions in feminist disability studies which "crip" or unsettle normative understandings of an ableist, atomistic body and a "successful" life "with its neoliberal dream of full capacitation and unimpeded agency" to instead theorize possibilities "for thinking [and living] different forms of sociality and flourishing" (Shildrick 2019: 607–610). As Shildrick argues, "crip theory does not claim to resolve loss, anxiety, or suffering. Rather it employs the risk strategy of embracing abject states of being and finding different meaning and resistance within them" (ibid.: 607).

In their work on everyday life in nuclear landscapes, for example, Thom Davies and Abel Polese (2015) identify important negotiations of abjection that are useful for understanding how residents of contaminated environments shape the ideas and politics of intervention to their own ends. They argue that for residents around the Chernobyl Exclusion Zone, subverting the post-nuclear everyday through environmentally risky food practices, informal economic activity, and social networks are central to survival in toxic environments and critically, create ways to "enact agency and subvert their post-disaster status as 'bare life'" (ibid.: 37). Embodied states of alterlife are thus indissociable from "both the quagmire we are in and from the times, places, and modes of being yet to come" (McRuer 2018: 147). In many regions of the world, life and its possibilities are molded by such polyvalent and paradoxical relationships between health, bodies, and environments. This polyvalence reminds us that, as Murphy argues,

> life forged in ongoing chemical violence is also life open to becoming something else… Alterlife acknowledges that one cannot simply get out, that this hurtful and deadly entanglement forms part of contemporary existence in this moment, in the ongoing aftermath. And yet the openness to alteration may also describe the potential to become something else, to defend and persist, to recompose relations to water and land, to becomes alter-wise in the aftermath.
>
> (Murphy 2017a: 500–501)

Alterlife, together with larger interventions in feminist disability studies, thus offers new potential avenues for political ecology to come.

Conclusion

The relationships between bodies and nature have generated substantial critical engagements in the field of political ecology. Spanning diverse empirical contexts, political ecologists leverage an ever-expanding conceptual toolkit to theorize uneven and power-laden embodiments of nature. Among the field's chief contributions include work which: (1) brings porosity and plasticity to bear on existing conceptualizations of the body; (2) provides alternative spatialization of bodies as dynamic, multi-species or local/situated/body ecologies; (3) expands the pathways of environmental change that count to better explain the patterning of health and disease over time and space; and (4) pushes forward our understanding of materiality, difference, and causality in ongoing debates about bodies and nature. Collectively, these contributions attest to the power and continued relevance of political ecology in conversations about embodying nature.

And yet, to return to the chapter's introduction, Jayani's story holds several lessons for pushing forward existing work in political ecology. Her experience suggests that many bodies are much more ambiguously configured by porosity, plasticity, local/situated biologies, and material bodily entanglements than accounted for in existing literature. Given that liminal bodily states are not only increasingly common but also profoundly mediate possibilities for health, how might political ecologists better foreground these types of experiences? How might we better account for instability, ambiguity, and contingency in our stories of bodies and their uneven entanglements in environments? Simply put, what possibilities exist for political ecologists to not only trace dynamic pathways of embodiment, but also to elaborate the often-paradoxical *experiences* of bodily states that are produced as a consequence?

To grapple with these questions in my own work, I have turned to Michelle Murphy's (2017b: 5) theorization of *alterlife* to theorize life that is "already recomposed, pained, and damaged [by environmental violence] but potentiality, nonetheless." Scholars such as Michelle Murphy, Eve Tuck, Margrit Shildrick, and Anna Tsing challenge political ecologists to learn with a "thick archive and dense present" of creative struggles for bodily re-capacitation, where communities devise strategies to maintain their families, assert attachments to land, affirm life, and reclaim dignity in a permanently polluted world. By foregrounding various moments of alterlife in the shadows of toxic pollution, colonialism, racial capitalism, and ecological rupture, political ecology can also begin to think beyond the binaries of health/harm, debility/capacity, and resistance/resignation in its accounts of embodied nature. Indeed, this concept helps us grapple with our knotted relationships to social and environmental toxicity as well as reflect on how we might meet our research collaborators, like Jayani, more commensurately in their accounts of everyday life, and especially in the value and potentiality that they continue to attach to their lives and bodies even under conditions of debility and slow death.

Note

1 To protect the privacy of my interlocutors, individual names are replaced with pseudonyms in this chapter.

References

Abbots E-J, Eli K and Ulijaszek S (2020) Toward an affective political ecology of obesity: Mediating biological and social aspects. *Public Culture* 16(3): 346–366.

Alaimo S (2010) *Bodily Natures: Science, Environment, and the Material Self*. Indiana University Press.

Allen IK (2020) Thinking with a feminist political ecology of air-and-breathing-bodies. *Body and Society* 26(2): 79–105. DOI: 10.1177/1357034X19900526

Baer H and Singer M (2016) *Global Warming and the Political Ecology of Health: Emerging Crises and Systemic Solutions*. New York: Routledge. DOI: 10.4324/9781315428017

Barbour M and Guthman J (2018) (En)gendering exposure: Pregnant farmworkers and the inadequacy of pesticide notification. *Journal of Political Ecology* 25(1): 332–349.

Biehler D (2013) *Pests in the City: Flies, Bedbugs, Cockroaches, and Rats*. University of Washington Press.

Biehler D and Simon G (2011) The great indoors: Research frontiers on indoor environments as active political-ecological spaces. *Progress in Human Geography* 35(2): 172–192.

Blackman L (2016) The new biologies: Epigenetics, the microbiome and immunities. *Body and Society* 22(4): 3–18.

Bollati V and Baccarelli A (2010) Environmental epigenetics. *Heredity* 105: 105–112.

Bosworth K (2017) Thinking permeable matter through feminist geophilosophy: Environmental knowledge controversy and the materiality of hydrogeologic processes. *Environment and Planning D: Society and Space* 35(1): 21–37. DOI: 10.1177/0263775816660353

Braun B (2007) Biopolitics and the molecularization of life. *Cultural Geographies* 14(1): 6–28.

Brown M (2003) Hospice and the spatial paradoxes of terminal care. *Environment and Planning A* 35(5): 833–851.

Carney MA (2014) The biopolitics of "food insecurity": Towards a critical political ecology of the body in studies of women's transnational migration. *Journal of Political Ecology* 21(1): 1–18.

Carson R (2002) *Silent Spring*. Boston: Houghton Mifflin.

Chao S, Bolender K and Kirksey E (2022) *The Promise of Multispecies Justice*. Duke University Press.

Craddock S and Hinchliffe S (2015) One world, one health? Social science engagements with the one health agenda. *Social Science and Medicine* 129: 1–4.

Davies T (2018) Toxic space and time: Slow violence, necropolitics, and petrochemical pollution. *Annals of the American Association of Geographers*: 1–17. DOI: 10.1080/24694452.2018.1470924

Davies T (2023) Slow observation: Witnessing long-term pollution and environmental racism in Cancer Alley. In: *Toxic Timescapes: Examining Toxicity across Time and Space*. Ohio University Press, pp. 50–71.

Davies T and Polese A (2015) Informality and survival in Ukraine's nuclear landscape: Living with the risks of Chernobyl. *Journal of Eurasian Studies* 6(1): 34–45.

Dillon L and Sze J (2016) Police power and particulate matters: Environmental justice and the spatialities of in/securities in US cities. *English Language Notes* 54(2): 13–23.

Dyck I (1998) Women with disabilities and everyday geographies: Home space and the contested body. In: R. Kearns and W. Gesler (eds) *Putting Health into Place: Landscape, Identity, and Well-Being*. Syracuse University Press, pp. 102–109.

Dyck I, Kontos P, Angus J, et al. (2005) The home as a site for long-term care: Meanings and management of bodies and spaces. *Health and Place* 11(2): 173–185. DOI: 10.1016/j.healthplace.2004.06.001

Elmhirst R (2015) Feminist political ecology. In: Perreault T, Bridge G, and McCarthy J (eds) *The Routledge Handbook of Political Ecology*. Routledge, pp. 519–530.

Forsyth T (2003) *Critical Political Ecology: The Politics of Environmental Science*. New York: Routledge.

Graham S (2015) Life support: The political ecology of urban air. *City* 19(2–3): 192–215. DOI: 10.1080/13604813.2015.1014710

Greenhough B, Read CJ, Lorimer J, et al. (2020) Setting the agenda for social science research on the human microbiome. *Palgrave Communications* 6(1): 1–11. DOI: 10.1057/s41599-020-0388-5

Guthman J (2011) *Weighing in: Obesity, Food Justice, and the Limits of Capitalism*. University of California Press.

Guthman J (2012) Opening up the black box of the body in geographical obesity research: Toward a critical political ecology of fat. *Annals of the Association of American Geographers* 102(5): 951–957.

Guthman J (2015) Binging and purging: Agrofood capitalism and the body as socioecological fix. *Environment and Planning A* 47(12): 2522–2536. DOI: 10.1068/a140005p

Guthman J and Mansfield B (2013) The implications of environmental epigenetics: A new direction for geographic inquiry on health, space, and nature-society relations. *Progress in Human Geography* 37(4): 486–504.

Guthman J and Mansfield B (2015) Nature, difference, and the body. In: Perreault T, Bridge G, and McCarthy J (eds) *The Routledge Handbook of Political Ecology*. Routledge, pp. 558–570.

Ham JR (2020) "Every day it's tuo zaafi": Considering food preference in a food insecure region of Ghana. *Agriculture and Human Values* 37(3): 907–917. DOI: 10.1007/s10460-020-10027-7

Hausermann HE (2015) "I could not be idle any longer": Buruli ulcer treatment assemblages in rural Ghana. *Environment and Planning A: Economy and Space* 47(10): 2204–2220. DOI: 10.1177/0308518X15599289

Hayes-Conroy A and Hayes-Conroy J (2015) Political ecology of the body: A visceral approach. In: Bryant RL (ed.) *The International Handbook of Political Ecology*. Edward Elgar Publishing, pp. 659–672.

Hayes-Conroy J and Hayes-Conroy A (2013) Veggies and visceralities: A political ecology of food and feeling. *Emotion, Space and Society* 6: 81–90.

Hinchliffe S, Bingham N, Allen J, et al. (2016) *Pathological Lives*. John Wiley & Sons.

Holmes S (2013) *Fresh Fruit, Broken Bodies: Migrant Farmworkers in the United States*. University of California Press.

Jackson P and Neely A (2014) Triangulating health: Toward a practice of a political ecology of health. *Progress in Human Geography* 39(1): 47–64.

Jampel C (2018) Intersections of disability justice, racial justice and environmental justice. *Environmental Sociology* 4(1): 122–135. DOI: 10.1080/23251042.2018.1424497

King B (2017) *States of Disease: Political Environments and Human Health*. Berkeley: University of California Press.

King B and Rishworth A (2022) Infectious addictions: Geographies of colliding epidemics. *Progress in Human Geography* 46(1): 139–155. DOI: 10.1177/03091325211052040

Kinkaid E (2019) Embodied political ecology: Sensing agrarian change in north India. *Geoforum* 107: 45–53. DOI: 10.1016/j.geoforum.2019.10.013

Knight R, Callewaert C, Marotz C, et al. (2017) The microbiome and human biology. *Annual Review of Genomics and Human Genetics* 18: 65–86.

Krieger N (2005) Embodiment: A conceptual glossary for epidemiology. *Journal of Epidemiology and Community Health* 59(5): 350–355.

Kuzawa CW and Sweet E. (2009) Epigenetics and the embodiment of race: Developmental origins of US racial disparities in cardiovascular health. *American Journal of Human Biology* 21(1): 2–15.

Landecker H (2016) Antibiotic resistance and the biology of history. *Body and Society* 22(4): 19–52.

Landecker H and Panofsky A (2013) From social structure to gene regulation, and back: A critical introduction to environmental epigenetics for sociology. *Annual Review of Sociology* 39: 333–357.

Langston N (2010) *Toxic Bodies: Hormone Disruptors and the Legacy of DES*. Yale University Press.

Leatherman T and Goodman A (2020) Building on the biocultural syntheses: 20 years and still expanding. *American Journal of Human Biology* 32(4). 1–14. DOI: 10.1002/ajhb.23360

Lock M (2017) Recovering the Body. *Annual Review of Anthropology* 46(1): 1–14. DOI: 10.1146/annurev-anthro-102116-041253

Lock M and Nguyen V-K (2018) *An Anthropology of Biomedicine*. John Wiley & Sons.

Lora Wainwright A (2017) *Resigned Activism: Living with Pollution in Rural China*. MIT Press.

Lorimer J (2017) Parasites, ghosts and mutualists: A relational geography of microbes for global health. *Transactions of the Institute of British Geographers*: 1–15.

Lunstrum E, Ahuja N, Braun B, et al. (2021) More-than-human and deeply human perspectives on COVID-19. *Antipode* 53(5): 1503–1525. DOI: 10.1111/anti.12730

Mansfield B (2012a) Gendered biopolitics of public health: Regulation and discipline in seafood consumption advisories. *Environment and Planning D: Society and Space* 30(4): 588–602. DOI: 10.1068/d11110

Mansfield B (2012b) Race and the new epigenetic biopolitics of environmental health. *BioSocieties* 7(4): 352–372.

Mansfield B (2017) Folded futurity: Epigenetic plasticity, temporality, and new thresholds of fetal life. *Science as Culture* 26(3): 355–379. DOI: 10.1080/09505431.2017.1294575

Mansfield B (2018) A new biopolitics of environmental health: Permeable bodies and the Anthropocene. In: *The SAGE Handbook of Nature: Three Volume Set*. Sage Publications, pp. 216–230. DOI: 10.4135/9781473983007

McManus P (2021) A more-than-urban political ecology of bushfire smoke in eastern Australia, 2019–2020. *Australian Geographer* 52(3): 243–256. DOI: 10.1080/00049182.2021.1946244

McRuer R (2018) *Crip Times: Disability, Globalization, and Resistance*. NYU Press.

Meloni M, Wakefield-Rann R and Mansfield B (2021) Bodies of the Anthropocene: On the interactive plasticity of earth systems and biological organisms. *The Anthropocene Review*: 20530196211001516. DOI: 10.1177/20530196211001517

Milligan C (2016) *There's No Place like Home: Place and Care in an Ageing Society*. Routledge.

Mollett S (2021) Hemispheric, relational, and intersectional political ecologies of race: Centring land-body entanglements in the Americas. *Antipode* 53(3): 810–830. DOI: 10.1111/anti.12696

Mostafanezhad M (2021) The materiality of air pollution: Urban political ecologies of tourism in Thailand. *Tourism Geographies* 23(4). 4: 855–872. DOI: 10.1080/14616688.2020.1801826

Murphy M (2000) The "elsewhere within here" and environmental illness; or, how to build yourself a body in a safe space. *Configurations* 8(1): 87–120. DOI: 10.1353/con.2000.0006

Murphy M (2006) *Sick Building Syndrome and the Problem of Uncertainty: Environmental Politics, Technoscience, and Women Workers*. Duke University Press.

Murphy M (2017a) Alterlife and decolonial chemical relations. *Cultural Anthropology* 32(4): 494–503.

Murphy M (2017b) What can't a body do? *Catalyst: Feminism, Theory, Technoscience* 3(1): 1–15.

Nading A (2014) *Mosquito Trails: Ecology, Health, and the Politics of Entanglement*. University of California Press.

Nash L (2006) *Inescapable Ecologies: A History of Environment, Disease, and Knowledge*. University of California Press.

Neely A (2015) Internal ecologies and the limits of local biologies: A political ecology of tuberculosis in the time of AIDS. *Annals of the Association of American Geographers* 105(4): 791–805.

Neely AH (2021) Entangled agencies: Rethinking causality and health in political-ecology. *Environment and Planning E: Nature and Space* 4(3): 966–984. DOI: 10.1177/2514848620943889

Nichols CE (2020) The wazan janch: The body-mass index and the socio-spatial politics of health promotion in rural India. *Social Science and Medicine* 258: 113071. DOI: 10.1016/j.socscimed.2020.113071

Nichols CE (2022a) Digesting agriculture development: Nutrition-oriented development and the political ecology of rice–body relations in India. *Agriculture and Human Values* 39(2): 757–771.

Nichols CE (2022b) Inflammatory agriculture: Political ecologies of health and fertilizers in India. *Environment and Planning E: Nature and Space*: 25148486221113556.

Nichols CE and Del Casino VJ (2021) Towards an integrated political ecology of health and bodies. *Progress in Human Geography* 45(4): 776–795. DOI: 10.1177/0309132520946489

Nightingale AJ (2011) Bounding difference: Intersectionality and the material production of gender, caste, class and environment in Nepal. *Geoforum* 42(2): 153–162.

Nightingale AJ (2021) Gender, nature, body. In: Harcourt W, A. H. Akram-Lodhi, Kristina Dietz, et al. (eds) *Handbook of Critical Agrarian Studies*. Edward Elgar Publishing, pp. 131–138. Available at: https://www.elgaronline.com/view/edcoll/9781788972451/9781788972451.00023.xml (accessed 4 January 2022).

Nyantakyi-Frimpong H (2021) Climate change, women's workload in smallholder agriculture, and embodied political ecologies of undernutrition in northern Ghana. *Health and Place* 68: 102536. DOI: 10.1016/j.healthplace.2021.102536

Parr J (2001) Notes for a more sensuous history of twentieth-century Canada: The timely, the tacit, and the material body. *The Canadian Historical Review* 82(4): 720–745.
Parr J (2010) *Sensing Changes: Technologies, Environments, and the Everyday, 1953–2003*. UBC Press.
Pathak G (2020) Polycystic ovary syndrome, medical semantics, and the political ecology of health in India. *Anthropology and Medicine* 27(1): 49–63. DOI: 10.1080/13648470.2018.1544606
Pentecost M and Cousins T (2017) Strata of the political: Epigenetic and microbial imaginaries in post-Apartheid Cape Town. *Antipode* 49(5): 1368–1384. DOI: 10.1111/anti.12315
Petryna A (2013) *Life Exposed: Biological Citizens after Chernobyl*. Princeton University Press.
Pulido L (2016) Flint, environmental racism, and racial capitalism. *Capitalism Nature Socialism*. Taylor & Francis Group.
Ranganathan M (2022) Towards a political ecology of caste and the city. *Journal of Urban Technology* 29(1): 135–143. DOI: 10.1080/10630732.2021.2007203
Relman DA (2015) The human microbiome and the future practice of medicine. *Journal of the American Medical Association* 314(11): 1127–1128.
Robbins P (2015) The trickster science. In: Perreault T, Bridge G, and McCarthy J (eds) *The Routledge Handbook of Political Ecology*. Routledge, pp. 89–101.
Senanayake N (2019) Tasting toxicity: Bodies, perplexity, and the fraught witnessing of environmental risk in Sri Lanka's dry zone. *Gender, Place and Culture*: 1–25. DOI: 10.1080/0966369X.2019.1693345
Senanayake N (2021) Theorising liminal states of health: A spatio-temporal analysis of undiagnosis and anticipatory diagnosis in the shadow of toxic pollution. *Transactions of the Institute of British Geographers*. Wiley Online Library.
Senanayake N (2022a) Towards a feminist political ecology of health: Mystery kidney disease and the co-production of social, environmental, and bodily difference. *Environment and Planning E: Nature and Space*: 25148486221113964. DOI: 10.1177/25148486221113963
Senanayake N (2022b) "We are the living dead"; or, the precarious stabilisation of liminal life in the presence of CKDu. *Antipode*. Wiley Online Library.
Senanayake N and King B (2019) Health-environment futures: Complexity, uncertainty, and bodies. *Progress in Human Geography* 43(4): 711–728. DOI: 10.1177/0309132517743322
Senier L, Brown P, Shostak S, et al. (2017) The socio-exposome: Advancing exposure science and environmental justice in a postgenomic era. *Environmental Sociology* 3(2): 107–121. DOI: 10.1080/23251042.2016.1220848
Shadaan R and Murphy M (2020) EDC's as industrial chemicals and settler colonial structures. *Catalyst: Feminism, Theory, Technoscience* 6(1).
Shattuck A (2019) Risky subjects: Embodiment and partial knowledges in the safe use of pesticide. *Geoforum*. Elsevier.
Shattuck A (2020) Toxic uncertainties and epistemic emergence: Understanding pesticides and health in Lao PDR. *Annals of the American Association of Geographers*: 1–15.
Shildrick M (2019) Neoliberalism and embodied precarity: Some crip responses. *South Atlantic Quarterly* 118(3): 595–613.
Speed S (2019) *Incarcerated Stories: Indigenous Women Migrants and Violence in the Settler-Capitalist State*. UNC Press Books.
Stallins JA, Law DM, Strosberg SA, et al. (2018) Geography and postgenomics: How space and place are the new DNA. *GeoJournal* 83(1): 153–168. DOI: 10.1007/s10708-016-9763-6
Stanley A (2015) Wasted life: Labour, liveliness, and the production of value. *Antipode* 47(3): 792–811. DOI: 10.1111/anti.12128
Sultana F (2011) Suffering for water, suffering from water: Emotional geographies of resource access, control and conflict. *Geoforum* 42(2): 163–172.
Sweet E, DuBois LZ and Stanley F (2018) Embodied neoliberalism: Epidemiology and the lived experience of consumer debt. *International Journal of Health Services* 48(3): 495–511.
Truelove Y (2011) (Re-)conceptualizing water inequality in Delhi, India through a feminist political ecology framework. *Geoforum* 42(2): 143–152.
Truelove Y (2019) Rethinking water insecurity, inequality and infrastructure through an embodied urban political ecology. *Wiley Interdisciplinary Reviews: Water* 6(3): e1342.

Truelove Y and O'Reilly K (2021) Making India's cleanest city: Sanitation, intersectionality, and infrastructural violence. *Environment and Planning E: Nature and Space* 4(3): 718–735. DOI: 10.1177/2514848620941521

Valdivia G (2018) "Wagering life" in the petro-city: Embodied ecologies of oil flow, capitalism, and justice in Esmeraldas, Ecuador. *Annals of the American Association of Geographers* 108(2): 549–557.

Vasudevan P (2021) An intimate inventory of race and waste. *Antipode* 53(3): 770–790. DOI: 10.1111/anti.12501

Wright M (2013) *Disposable Women and Other Myths of Global Capitalism*. Routledge.

18 Unruly Nature

Non-human Intractability and Multispecies Endurance

Rosemary Collard

Introduction

On a narrow trail through summer-yellow moss, fir needles and flakes of arbutus bark, the director of a Vancouver Island wildlife conservation organization stops ahead of me and peers down at a small mound of dirt. With a stick she nudges some dirt off the pile. Underneath, the prize: moldy-white cougar scat (Figure 18.1).

This is the closest most people come to a cougar, even on Vancouver Island, home to one quarter of the lethal and non-lethal cougar attacks in North America in the past century (Collard 2012), and to what is likely the continent's densest population of cougars. I say "likely" because cougars are hard to count—ecologists writing in *Nature* recently called them "wide-ranging, cryptic, and notoriously difficult to detect" (Murphy et al. 2019: 1). A cougar study from the early 2000s on Vancouver Island is illustrative. Scientists set up eleven motion and heat detector cameras and thirty hair snag sites along multiple transects in Pacific Rim National Park. They did not catch a single cougar image or bit of fur. A sniffer dog (recently returned from tracking jaguars in Argentina) did find some scat. But cougar scat is hard to DNA test, which means it cannot be individuated—ten scat deposits could be from ten different cougars or just one. My interviews with the scientists involved left me with a key take-home: cougars' bodies and behaviors are unruly within scientific projects.

Cougars' range once blanketed the Americas. Persecution and habitat destruction have reduced that range by two thirds (Gross 2008). The cougars on Vancouver Island today are descended from survivors of a century-long attempt to eradicate them. In the mid-nineteenth century, as the colonial government forced First Nations people onto reserves, and white settlers occupied First Nations land to "improve" it through European-style agriculture, cougars emerged as a threat—namely to sport hunters' preferred "game" (deer) and to settlers' imported domesticated animal property. In response, one of the earliest colonial legal acts designated cougars and other "noxious predators" as "vermin," subject to a bounty hunt from 1864–1957 (Loo 2006). The first Provincial Game and Forest Warden in BC declared of cougars that a "special effort should be made to exterminate them" (Williams 1913). An average of 135 cougars were killed per year for bounty (Collard 2012). This attempted eradication of "vermin" was consistent with if not a lynchpin of the wider project and imaginary of ecological imperialism, where "superior" domesticated animal imports would replace native wildlife (Crosby 1986; Hubbard 2014; Belcourt 2015; Rutherford 2022). But cougars have been wayward within the settler-colonial project, too. Not only have they survived, but also other

DOI: 10.4324/9781003165477-23

Figure 18.1 Cougar scat in Sooke Mountain Park, August 2021.

Source: Photograph by the author.

settler-colonial "improvements" like industrial forest harvest and proliferating residential flower gardens have even amplified cougars' unruliness by boosting the population of deer, cougars' preferred prey.

The office of the BC Provincial Game and Forest Warden eventually morphed into what is today called the Conservation Officer Service, which no longer pays bounties for dead cougars. But their office still oversees the elimination of "conflict" cougars—deemed to threaten human life or property. In past decade 127 conflict cougars have been killed per year in BC—not far off from the 135 annually under the bounty system (Figure 18.2).

The director and I are discussing this ongoing program of destruction on our walk. She, like many around the world, is working to establish gentler ways of living with unruly nature—nature with "capacity to transgress human expectations" (Govindrajan 2015: 37), nature that is "difficult to control or categorize" (Krishnan et al. 2015: 5), like cougars. Many political ecologists are interested in this question too, and in nature's disruptiveness more generally. From what conditions does unruliness emerge? How do such natures manifest? Who benefits and who loses from them? Why are they important? This chapter addresses these questions in turn.

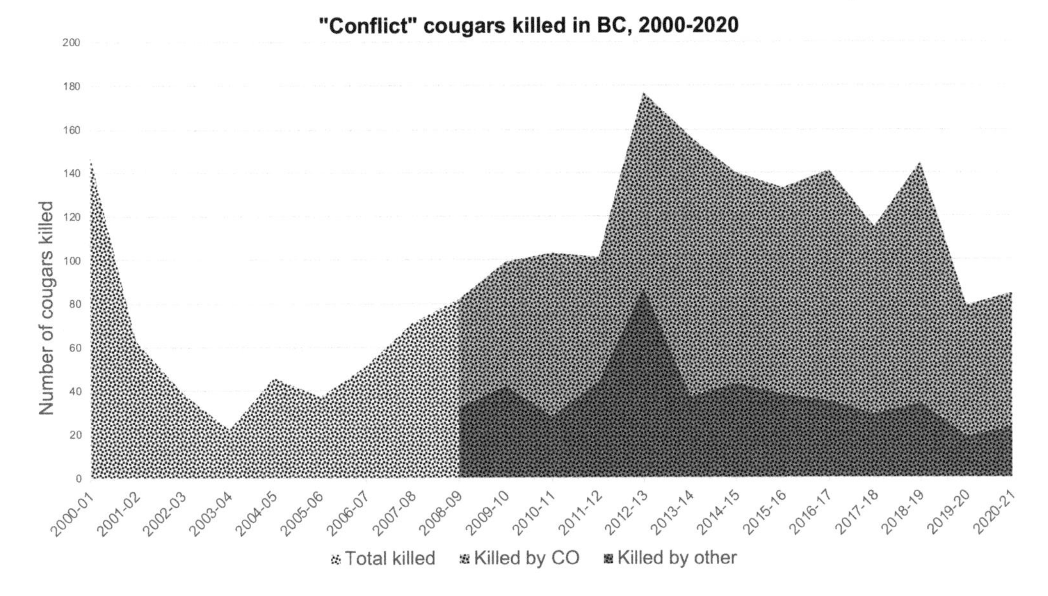

Figure 18.2 "Conflict" cougars killed annually in BC, 2000–2020.

Source: BC Conservation Officer Service, Provincial Cougar Conflict Statistics.

Nature Acts

This is the only chapter in the book that does not start with a verb. The other chapter titles invoke unspecified subjects acting on/with nature. This chapter is about nature acting back. Nature's agency—its capacity to act, to have influence—is by now a cornerstone of political ecological thought. From maple trees and moose (Simpson 2017) to parasites (Lorimer 2020), wildfires (Mauch 2015), fish (Todd 2016), and lawns (Robbins 2007), infinitely varied natures, from the microscopic to the cosmic (and including humans, of course), act with others to shape human lives, bodies, behaviors, capacities, desires, and the wider world. They do so through world-making actions as quick as a windstorm or as slow as glacial retreat, as spectacular as a cougar attack or as mundane as a slug infestation in the kale bed (see Ginn 2014).

These actions are not necessarily disruptive. Nature's unruliness is relational—it emerges at the intersection of two general forces: nature's agency, and humans' plans for and ideas about nature, which can and do fall apart when nature does not conform to expectation, whether by explicitly refusing to obey, or just going about its business unconcerned with human plans. Hugh Raffles's (2010: 4) description of insects could stand in for many natures, animal and not: "they are so busy, so indifferent, and so powerful. They'll almost never do what we tell them to do. They'll rarely be what we want them to be. They won't keep still. In every respect, they are really very complicated creatures." Unruliness, then, is not just about nature's recalcitrance; it is about nature's off script inventiveness and the "unintentional landscapes" and natures that result (Gandy 2016; Srinivasan 2019).

Political ecologists' understanding of nature's agency along these lines is indebted to Indigenous thought, in ways that are often unacknowledged (Watts 2013; Todd 2015, 2016; Sundberg 2014). While there is no singular Indigenous understanding of nature, around the world Indigenous and Black legal, intellectual and cultural systems have advanced an understanding of nature as not only agential but also in relations of kinship with humans (Estes 2019; McGregor 2018; Watts 2013; Todd 2018; Belcourt 2015; Simpson 2017; Bawaka Country et al. 2015; Bennett 2020). These ideas challenge still widespread and deep-rooted Western thinking that it is humans who act, and nature that is passively acted on.

Colonial and Capitalist "Improvement"

Colonial and capitalist development intent on "improving" human and nonhuman nature has generally revolved around an attempted separation from and mastery of nature. Within the liberal thought that emerged during and after the Enlightenment, becoming properly human or Man was accomplished by bending external nature and the body to one's rational mind (Anderson 1997; Wynter 2003). Accordingly, the settler-colonial project has involved attempts to transform ecologies into natures that are "disciplined," "tidy," and "orderly" (Qitsualik 2013: 27; Zahara and Hird 2016). For example, across the Americas, colonial states oversaw the replacement of diverse old growth forests with dense, homogenous stands. They suppressed forest fires, including banning Indigenous people's burning practices, casting them as "wild or wasteful" (Zahara 2020: 557; Pyne 2007; Vinyeta 2021), generally associating fire with deviance and racial "others" (Kosek 2006). (Most of these states now recognize that fire suppression combined with plantation-style stands creates a densely packed tinderbox that leads to megafires.) Altogether, as Julietta Singh (2018: 569) writes, "The colonial errand is an act of entering the 'wilderness,' in order to convert, to destroy, to civilize ... all that is 'wild' in the wilderness." British empire-building in particular involved "the concerted attempt to legitimate the supremacy of *Homo sapiens* (often as white and male bodies) through a symbolic and material mastery over animals across the many territories of imperial dominion" (Burton and Mawani 2020: 12–13).

These logics persist in contemporary development, which Krithika Srinivasan (2019: 383) describes as at base "about insulating the human from the vagaries of nature ... This ... translates into the desire for a sanitized living environment, one free of dirt, bugs, microbes and other risky creatures." In this sense, the modern human subject retains original Enlightenment ideals and fantasies of escaping and mastering nature, "securing the human through the control of unruly ecologies" (Lorimer 2017: 27). These ideals and plans are by no means universal—but they have been hegemonic and globalized within colonial, capitalist and modernist development projects over the past centuries. Despite their spread and force, these projects are everywhere incomplete, because of human resistance and also nature's nonconformity.

So, on one hand, nonhuman natures, like human ones, have will, agendas, and complex needs; they act accordingly and exert agency. On the other hand, capitalist and colonial systems are invested in selectively controlling, taming, deadening, siphoning off, accumulating and alienating those lively capacities, wills, and energies. Out of the meeting place of these, unruly natures are born. They are both an expression of nature's vitality and an artefact of certain human structural inclinations and interventions. For this reason it is perhaps more predictable than paradoxical that colonial and capitalist interventions to tame nature can actually engender intractability, and wayward nature often emerges "from inside the systems designed to control it" (Temple 2015: 14), as the following four manifestations of unruly natures show.

Spatial Transgressions

One of the most recurring and conspicuous manifestations of unruly nature is spatial: nature that upsets material and discursive borders around the nation-state, the city, the home, the body itself. Political ecologists looking to nature's spatial transgressions have revealed both how nonhuman nature is enrolled in border-making imaginaries and practices, and how nonhuman nature destabilizes boundary-making at all scales (see Sundberg 2011). For example, in the Israel-Palestine borderlands, Natalia Gutkowski (2021) tracks how the Israeli system of control in the West Bank operates through animal bodies—such as through Israel's seizure of Palestinians' donkeys, impeding their owners' ability to make a living—while also highlighting how pests, disease, and other spatially unpredictable natures disturb this system of control.

Urban animals are exemplar "out-of-place" natures. Whether chickens in Botswana (Hovorka 2008), coyotes in Toronto (Blue and Alexander 2015; Van Patter 2021) or snakes in Bangalore (Narayanan and Bindumadhav 2019: 402), the relative invisibility of mundane and spectacular animals in cities belies their centrality as "vital members of urban societies." Even animals who are familiar residents of cities—macaques in parts of Asia (Barua and Sinha), crows and raccoons (Heath Justice 2021) in North America—confound their human neighbors through spatial incursions into yards and homes, "abiding in the unruly borderlands of what many humans consider to be appropriate animal behaviour" (Heath Justice 2021: 7). In their multispecies ethnography at childcare centers in Vancouver, Veronica Pacini-Ketchabaw and Fikile Nxumalo (2015: 153) understand raccoons as "unruly subjects" who "radically disrupt colonial nature/culture and human/nonhuman divides" in spatial ways, by entering spaces wild animals are not supposed to be. In a particularly vivid account, raccoons enter the children's tent and settle in, rearranging the cushions.

Similarly, the movement of animals like possums (Power 2009) and insects like bedbugs (Biehler 2009, 2015) through homes challenge the perception of domestic space as autonomous and purely human. Dawn Biehler's (2009: 1018) history of pest management practices in US public housing documents how in the mid-twentieth century, chemicals like DDT promised to "not just control but obliterate nature in domestic space." But pest control practices inadvertently made homes more attractive to roaches, who became genetically and behaviorally resistant to pesticides. Biehler's political ecology of the home demonstrates the permeability of borders around not only home, but also the body itself. Other political ecologists have picked up here, highlighting how not only parasites (Lorimer 2017, 2020) but also air and breath highlight the body's "permeability, relationality, vulnerability and also excessive untameability" (Allen 2020).

Box 18.1 Methodological approaches and considerations

How do political ecologists research unruly nature? A common and arguably integral methodological starting point is the recognition that more-than-human nature is a lively force within the research itself—a co-producer of knowledge, not a passive object from which researchers extract data (Abrell and Gruen 2020). This recognition is core to the Bawaka Collective, an Indigenous and non-Indigenous, human-more-than-human research collective that includes Bawaka Country as a member and an author in its published work to honor how "non-humans—landscapes, seascapes, animals, winds, sun, moon, tides and spirits such as Bayini, a spirit woman of Bawaka—constantly shape and influence our research collaboration" (Lloyd et al. 2012: 1087). Nonhuman nature's agency, interests and capacities can shape all stages of the research, making research a relational achievement.

Because studying unruly nature generally involves attending to both more-than-human world- making and the human plans it transgresses, research often employs mixed methods. Researchers rely on a standard suite of qualitative methods, like interviews, surveys, and archival research, but some additionally draw on field science methods including wildlife tracking and ethology (e.g., Van Patter 2021), sometimes in collaboration with wildlife or animal behavior scientists (e.g., Barua and Sinha 2019). Observation is another commonly used research method, where researchers typically immerse themselves as participants or spectators in "more-than-human contact zones" (Isaacs and Otruba 2019; e.g., Gillespie 2018; Collard 2020), even conducting long-term "multispecies ethnographies" (Kirksey and Helmreich 2010).

Using field science and observation methods often involves up-close encounters between researchers and various natures. These encounters can inflict harm and rely on conditions of forced proximity, raising ethical concerns about cross-species communication and nonhuman consent, privacy, and well-being—issues researchers argue are absent from current institutional research ethics protocols and frameworks (Van Patter and Blattner 2020; Abrell and Gruen 2020). In response, some researchers are innovating their own supplementary ethical protocols, centering, inter-alia, "non-maleficence (vulnerability, confidentiality), beneficence (reciprocity, representation), and voluntary participation (mediated informed consent and ongoing embodied assent)" (Van Patter and Blattner 2020: 173; also see Abrell and Gruen 2020; Rubio-Ramon and Srinivasan 2022).

Reproductive Intractability

Unruly nature also manifests when nature's reproduction defies human expectation and intention. Institutions ranging from zoos to industrial agricultural firms have a long-standing preoccupation with nature's reproduction and how to control it. There are two sides of this coin: attempts to promote reproduction deemed precarious or low; and efforts to curb reproduction cast as excessive. Elizabeth Hennessy's (2013, 2019; also see Srinivasan 2014) meticulous tracing of captive breeding programs for giant tortoises in the Galapagos is illustrative of the first side of the coin. She follows decades of experiments by scientists, zookeepers and conservationists to marshal what she calls "assemblages of reproduction" toward keeping tortoises alive and ensuring they "would produce new life to carry the species forward" (Hennessy 2013: 74). But most tortoises died and would not reproduce. Hennessy details the ways "the keepers tested various 'tricks' with 'vibrant matter'—mud, adobe, sprinklers—to address reproductive problems" (ibid.: 76). Even when there were finally successes, "the process was far from routine. Practices that worked once did not necessarily work again; reproduction remained tricky" (ibid.: 76). Here, nature—tortoises—is unruly in that its reproduction is not guaranteed, because the ingredients of reproduction are delicate and complicated, challenging to replicate in "assemblages of reproduction."

On the flip side of the coin are animals variously cast as feral, invasive, and pestilent—animals deemed reproductively excessive. Various governments and scientists have employed a dizzying array of methods to curb these nature's reproduction, from toxic pest management practices (Biehler 2015) to various examples of "killing for conservation" (Atchison 2015; Hodgetts 2017; Perkins 2020) One dramatic case from Australia—the Pelorus experiment—represents a redoubled attempt to control reproductive capacities. Seeking to eradicate goats on Pelorus Island off the coast of Queensland, scientists and municipal councilors released two dingoes onto the island to kill the goats—dingoes implanted with an experimental slow-release toxic implant that would kill them after 600 days, before they themselves could reproduce. The dingoes did not die as expected, and the experiment was met with loud public backlash (Probyn-Rapsey and Lennox 2020). Pest management measures do often fail, and can even introduce new disruptions. Ryan Galt (2010) points to van den Bosch's concept of the "pesticide treadmill" to explain that because pesticides often kill pest species' natural enemies, and because pests often become resistant to the pesticide (as for Biehler's roaches), pesticides eventually backfire. Arguably there is no stronger example of this than the "superweeds" that have "developed ... strategies to resist glyphosate" and other herbicides (see Werner et al. 2022: 20). Interventions breed the need for more interventions; work to stabilize chemical and pest is only ever temporary (see Guthman 2019). This is especially evident when it comes to the third manifestation of unruly nature examined here: nature that does not yield to capital.

Natures Less Subjugated to Capital

Nonhuman nature forms two primary conditions of possibility for capitalist production: it provides inputs in the form of raw materials and absorbs outputs (e.g., waste). Capitalists seek profit in part by minimizing costs associated with both. This has amounted to a many-centuries long attempt to "improve" nature through techniques ranging from mechanization to genetic manipulation to the application of chemicals—essentially attempting to make nature a more efficient and predictable—and therefore cheaper—raw material producer and waste absorber. Despite these interventions, nature remains a "fictitious commodity"—one that is not produced in the first instance for sale in the market (Polanyi 1944). Markets (and capitalists) do not alone create the natures on which they rely—and cannot perfectly control them. Unruly nature—nature "less fully subjugated to capital" (Malm 2018: 3)—shapes, escapes and resists capitalist intervention. Richard White's (1995: 112) description of the Columbia river holds for trees (Scott 1998; Prudham 2003, 2005), chickens (Boyd 2003; Wadiwel 2018), and arguably all nature: it "has purposes of its own which do not readily yield to maximize profit." This tension between capitalism's reliance on nature and its inability to fully control it engenders fundamental, ongoing structural crises within capitalism (Polanyi 1944; O'Connor 1988; Scott 1998; Prudham 2005; Fraser 2021).

For example, capitalists have long sought—often in alliance with states and universities—to intervene in nonhuman nature's biological temporal rhythms, extending seasonal crop cycles and speeding up plant and animal growth periods. These efforts sometimes fail and even when they succeed, can introduce new problems. Scott Prudham (2003, 2005: 117) documents the challenges forest companies and managers have faced in their attempts to hasten slow- growing Douglas firs on the west coast of North America, not least because these trees' 60- to 80-year growth period means experiments in their "improvement" can take decades. Broiler chickens provide a spectacular example of temporal "improvements" that have "succeeded" but introduced new unruliness. Today's broiler chickens mature in half the time their ancestors took. Dinesh Wadiwel (2018: 529) theorizes this transformation as a quickening less of raw material production than of the "labor time required from animals to produce themselves as commodities within the production process." For Wadiwel, doing so involves finding the most efficient way to counter animals' resistance, including their escape, flight, and general insubordination (also see Hribal 2003; Gillespie 2016). But genetic, nutritional and spatial measures to quicken chickens' biological time have also tended to create more disturbance—like disease outbreaks (Boyd 2003). Similarly, Paul Robbins and April Luginbuhl (2005) frame the outbreak of disease at elk game farms in the US as an example of how elk, like other "fugitive resources" such as fish and water, "resist enclosure," leaving efforts to enclose and privatize wildlife incomplete.

Unknowable Natures

Part of what makes "fugitive resources" resistant to profit-seeking intervention is that nature can be intractable within knowledge projects, especially modern science—this is the fourth and final manifestation of unruly natures examined here. Since Michel Callon's (1986) pathsetting study of scallops in St Brieuc Bay, France, natures that are difficult to know have been of particular interest to scholars working at the intersection of science and technology studies and political ecology. Callon (1986) follows scientists studying and seeking to cultivate scallops, or "domesticate" them, as he says. This requires locking the scallops into place—in this case, onto testing devices—within the alliance of actors that made up the study. But parasites, ocean currents, and multiple other factors interfere; the scallops do not anchor themselves to the testing devices, preventing the alliance from forming. Callon's study informed decades of future work that understands science as performed within networks or collectives of actors, including nonhumans.

The will to know nonhuman nature through science has been a driving force within projects of colonial and capitalist improvement. From their beginnings, colonial exploration, conquest and imperialism have been enmeshed with disciplines of natural history, ecology, medicine and genetics (Pratt 1992; Schiebinger 2007; Chakrabarti 2013; Subramaniam 2014), particularly seeking to extricate infinitely varied life forms from "the tangled threads of their life surroundings" and re-weave them "into European-based patterns of global unity and order," like taxonomy (Pratt 1992: 31; also see Johnson and Murton 2007). But as Callon's scallops, cougars and Galapagos tortoises suggest, nature is often an unruly scientific test subject. Young Rae Choi's (2021) study of South Korea's tidal flats, called *getbol*, provides a compelling example. Historically, getbol were seen "as barren and empty wastelands that had to be turned into something useful" (Choi 2021: 353). In the twentieth century, thousands of square kilometers of getbol were removed in a wider wave of "modernization"—first by Japanese colonizers seeking to maximize rice production and then by the South Korean developmental state, pursuing economic growth. More recently, conservationists have sought to reconceptualize getbol as teeming with life—"live clams, crabs, mudskippers, and baby octopuses" (ibid.: 353). In parallel, the South Korean state has become more interested in knowing *getbol* as part of its wider mission of "mapping, monitoring, and forecasting" to serve "various users including oceanographers, fishers, seafarers, and tourists" (ibid.: 346). But in constantly changing form from water to land, tidal flats prove to be "slippery" objects of knowledge; their "in-betweenness gives them particular materialities that constantly frustrate our efforts to know them … or place them into modern knowledge systems" (ibid.: 340).

For Choi, getbol's unknowability has been partly what led to their removal and what frustrates attempts to conserve them today. This has high stakes for the diverse nonhuman beings who live there, and for the local fishers who know the flats well enough to make a living there. But the presence of these fishers speaks to other ways of inhabiting unruly nature. As Anna Tsing (2015: 161) says, "no single standard for assessing disturbance is possible; *disturbance matters in relation to how we live.*" The previous sections have shown how systems of thought and development seeking to master nature have largely encountered nature's agency as a problem. But in many ways for the same reason, there are communities who for whom unruly nature has been a condition of possibility for freedom and cultural survival.

Refuge and Recovery

Natures that have defied "taming" and "improvement" have at times served as refuge and even a wellspring of individual and collective memory and social life, especially for racialized and Indigenous people escaping and resisting enslavement, genocide, oppression, and exploitation. Willie Wright (2020: 1134) argues that the very "ability of select fugitive groups to obtain forms of spatial autonomy is reliant on their ability to seek, find, and settle within difficult and seemingly uninhabitable landscapes." Wright specifically looks to marronage—fugitive slaves' formation of societies "beyond bondage and exploitation" (ibid.: 1135). He demonstrates how the landscapes "onto which freedom dreams are sutured" (ibid.: 1134) in these societies tend to be unruly ones: in North and Central America, enslaved African people escaped to swampland that was "illegible within the spatial imaginary of the participants of the plantocracy" (ibid.: 1140); Indigenous people in the Zapatista Army for National Liberation have organized an alternative way of life in "forgotten … rugged mountains" in Chiapas, Mexico (ibid.: 1141); Vietnamese "families, battalions, and communities" found protection and communal life in resilient subterranean tunnels dug in soil composed of clay and iron (ibid.: 1142). For Wright (2020: 1142; also see Malm 2018; Gandy 2021), these examples call for a reconceptualization of "defiant environments": not as deficient or worthless but rather as offering space and sustenance for people struggling for "self-determination, spatial autonomy, and just futures."

Unruly nature that survives or escapes capitalist and colonial intervention can also hold cultural memory and enable the recovery and continuance of cultural practices essential to Indigenous people's identity, social life and claims to land. Ruba Salih and Olaf Corry's (2021) study of settler-colonialism in Israel–Palestine is beautifully illustrative. The Israeli state's attempts to enclose Palestinian lands and "settle" them have included efforts to replace some cacti and vegetation. But the "unruly work of cacti, fruit trees and groves" (ibid.: 12) has meant that these natures have persisted around destroyed Palestinian villages. The presence of these resilient natures "incites refugees' practices of return" (ibid.: 12) to harvest almonds trees and care for grape groves, to have family picnics "sitting under, smelling, and eating the fruits and vegetables of the land which the settlers … tried, unsuccessfully, to eradicate" (ibid.: 14). Natures persisting in the ruins "not only bring to light what has been destroyed, allowing the recovery of traces of a previous life, but also most crucially have an afterlife" (Salih and Corry 2021: 16). Palestinian olive groves and almond trees unsettle Israeli narratives—used to dampen Indigenous claims to land—that cast Palestinian landscapes as barren wastelands to be recovered and improved through Israeli control.

Humility and Creative Openings

Unruly natures have fascinated political ecologists for decades. In part, this is because they spotlight nature's agency and thus invite humility in the recognition that humans do not construct the world alone; all natures are, like Raffles (2010: 100) describes insects, "not merely the opportunity to culture but its co-authors." This co-construction is not necessarily harmonious. Some natures are more inconveniencing and threatening than others. As this chapter has shown, the capitalist, colonial response to these kinds of natures has reliably been aggression (to extend Malm 2018 slightly). There is a tension here, then, between acknowledging nature's agency without denying the violent forces that have consistently been applied to quash it. As Antionette Burton and Renisa Mawani (2020: 11) suggest, an "emphasis on the disruptive and disorderly force of nonhuman animals" should not be read "as a triumphalist case for animal resistance. Given the violence to which animals were subject and the extinction rates under global imperial regimes, 'agency' must be carefully calibrated." Unruly natures are an ideal site for walking this line. Studying them involves acknowledging the power and consequence of capitalist and colonial interventions into nature while also reminding ourselves "of the limits of environmental control in an era of technological and institutional hubris" (Krishnan et al. 2015: 5). Looking to nature's nonconformity helps us understand "the operations, *but also the instability* of colonial power" (Salih and Corry 2021: 6, emphasis added), issuing "a challenge to histories of untrammeled species mastery that continue to shape grand narratives of modern British imperialism" (Burton and Mawani 2020: 5; also see Smalley 2017).

For some political ecologists, these natures beyond control and comprehension are also alive with creative political and theoretical possibility. Places like "the water's edge—that unsettled boundary between the solid and the liquid—[are] a potent site of theoretical possibility, potential, and dreaming" (Boon, Butler, and Jefferies 2018: 1). Such natures hint at the possibility of "a world where capital is not the master-builder, where things come into and go out of existence on their own, where the curse of exchange-value has been lifted and all sorts of other generative forces are given free reign" (Malm 2018: 28). Unruly nature in its waywardness holds both new and old wisdom, pasts and futures—on the one hand, it can form what Saidiya Hartmann (2019: 227–228) describes as "insurgent ground that enables new possibilities and new vocabularies" for rebellious subjects; on the other hand, by virtue of its survival, unruly nature, like the persistent cacti in razed Palestinian villages, can be a participant in and co- holder of longstanding cultural memory and practices.

Conclusion

Last spring I moved back to my hometown in T'Sou-ke Nation territory on Vancouver Island and rented a 1960s prefab cabin on an orderly oceanfront property where an elderly widow hired my dad to build a house. Deer wander the property, attracted by colorful beds of dahlias, roses, and rhododendrons with neatly trimmed edges.

On the fireplace mantel in the cabin are two small trophies from the Sooke Fall Fair for "Flowers" in 1974 and 1979. The lawn is mowed Mondays. One day a cougar disrupted this orderliness and routine, setting the neighborhood abuzz by walking across everyone's lawns. I did not see it. And when a rotting stench began emanating from the bushes along the short path to the beach, I obliviously held my breath when I passed. But when the widow's son visited, his small dogs quickly uncovered the smell's source. Placed in a gap in the bush on the shore's edge was a dismembered deer carcass—entrails and limbs piled, fur dotting the ground: a cougar cache. The widow's son, a wildlife biologist, threw the deer parts into the ocean. The next day, they reappeared on the pebbly tidal flats. He threw them back in, but the tide kept bringing them back.

Cougars are classic intractable natures. Their waywardness is born at the intersection of their powerful, life-altering, world-making agency, and centuries-long attempts by (colonial) states and subjects to eradicate or at least constrain them. Unruliness is always relational in this way—it emerges at the intersection of different forces: namely, nature's agency and human design. Cougars also exemplify how attempts to quash unruliness can actually breed more of it—the "settled" landscapes of lawns and flower gardens attracting deer who feed more cougars. This is not to say that in the absence of flower gardens or cougar eradication campaigns, cougars and people would live together harmoniously. Cougars are dangerous, elusive and they require a lot of space. What would it mean to leave more room for them, to be better neighbors? The director of the Vancouver Island wildlife conservation organization I introduced at the beginning of this chapter has ideas small and big. Her organization is running a pilot program testing fencing that prevents cougars from preying on farm animals (which usually leads to cougars being killed). A bigger issue is the seemingly unstoppable push of residential development into cougar habitat, suggesting the need to integrate cougar habitat models in municipal planning. These ideas both speak to a way of negotiating unruliness that is less about eliminating a specific nature and more about living with it in such a way that it becomes less troublesome. As long as there are attempts to bend nature to one's will, to structural will, there will be unruly natures. Perhaps the trick is to bend with nature instead.

References

Abrell, Elan and Lori Gruen. 2020. *Ethics and Animal Ethnography Working Paper*. Middletown, CT: Wesleyan Animal Studies and Brooks Institute for Animal Rights Law and Policy. www.wesleyan.edu/animalstudies/WASEvents/FinalEthicsandAnimalEthnographyDraft

Allen, Irma Kinga. 2020. Thinking with a feminist political ecology of air-and-breathing-bodies. *Body and Society* 26 (2): 79–105.

Anderson, Kay. 1997. A walk on the wild side: A critical geography of domestication. *Progress in Human Geography* 1 (4): 463–485.

Atchison, J. 2015. Experiments in co-existence: The science and practices of biocontrol in invasive species management. *Environment and Planning A: Economy and Space* 47 (8): 1697–1712.

Barua, Maan and Anindya Sinha. 2019. Animating the urban: An ethological and geographical conversation. *Social and Cultural Geography* 20 (8): 1160–1180.

Bawaka, Country, Sarah Wright, Sandie Suchet-Pearson, Kate Lloyd, Laklak Burarrwanga, Ritjilili Ganambarr, Merrkiyawuy Ganambarr-Stubbs, Banbapuy Ganambarr, and Djawundil Maymuru. 2015. Working with and learning from country: Decentring human authority. *Cultural geographies* 22 (2): 269–283.

Belcourt, Billy-Ray. 2015. Animal bodies, colonial subjects: (Re)locating animality in decolonial thought. *Societies* 5: 1–11.

Bennett, Joshua. 2020. *Being Property Once Myself: Blackness and the End of Man*. Cambridge, MA: Harvard University Press.

Biehler, Dawn. 2009. Permeable homes: A historical political ecology of insects and pesticides in US public housing. *Geoforum* 40: 1014–1023.
Biehler, Dawn. 2015. *Pests in the City: Flies, Bedbugs, Cockroaches, and Rats*. Seattle: University of Washington Press.
Blue, Gewndolyn and Shelley Alexander. 2015. Coyotes in the city: Gastro-ethical encounters in a more-than-human world. In *Critical Animal Geographies*, eds K Gillespie and R-C Collard, 149–163. London: Routledge.
Boon, Sonja, Lesley Butler and Daze Jefferies. 2018. *Autoethnography and Feminist Theory at the Water's Edge: Unsettled Islands*. Palgrave Macmillan.
Boyd, William. 2003. Making meat: Science, technology and American poultry production. *Technology and Culture* 42 (4): 631–664.
Burton, Antionette and Renisa Mawani, editors. 2020. *Animalia: An Anti-Imperial Bestiary for Our Times*. Durham: Duke University Press.
Callon, Michel. 1986. Some elements of a sociology of translation: Domestication of the scallops and the fisherman of St. Brieux Bay. In *Power, Action and Belief: A New Sociology of Knowledge?*, edited by J. Law, 196–229. London: Routledge.
Chakrabarti, Pratik. 2013. *Medicine and Empire, 1600–1960*. Palgrave Macmillan.
Choi, Young Rae. 2021. Slippery ontologies of tidal flats. *Environment and Planning E: Nature and Space* 5 (1): 340–361.
Collard, Rosemary-Claire. 2012. Cougar-human entanglements and the un/making of safe space. *Environment and Planning D: Society and Space* 30 (1).
Collard, Rosemary-Claire. 2020. *Animal Traffic: Lively Capital in the Global Exotic Pet Trade*. Durham, NC: Duke University Press.
Crosby, Alfred. 1986. *Ecological Imperialism: The Biological Expansion of Europe, 900–1900*. Cambridge University Press.
Estes, Nick. 2019. *Our History is the Future: Standing Rock versus the Dakota Access Pipeline, and the Long Tradition of Indigenous Resistance*. Verso.
Fraser, Nancy. 2021. Climates of capital. *New Left Review* 127: https://newleftreview.org/issues/ii127/articles/nancy-fraser-climates-of-capital
Galt. Ryan. 2010. Scaling up political ecology: The case of illegal pesticides on fresh vegetables imported into the United States, 1996–2006. *Annals of the Association of American Geographers* 100 (2): 327–355.
Gandy, Matthew. 2016. Unintentional landscapes. *Landscape Research* 41: 433–440.
Gandy, Matthew. 2021. An Arkansas parable for the Anthropocene. *Annals of the Association of American Geographers*, 112 (2): 368–386.
Gillespie, Kathryn. 2016. Nonhuman animal resistance and the improprieties of live property. In *Animals, Biopolitics, Law: Lively Legalities*, edited by Irus Braverman, 137–154. London: Routledge.
Gillespie, Kathryn. 2018. *The Cow with Ear Tag #1389*. University of Chicago Press.
Ginn, Franklin. 2014. Sticky lives: Slugs, detachment and more-than-human ethics in the garden. *Transactions of the Institute of British Geographers* 39 (4): 532–544.
Govindrajan, Radhika. 2015. The man-eater sent by God: Unruly interspecies intimacies in India's central Himalayas. In *Unruly Environments*, edited by Siddhartha Krishnan, Christopher Pastore and Samuel Temple, 33–37. Munich: Rachel Carson Centre Perspectives.
Gross, Liza. 2008. No place for predators? *PLoS Biology* 6 (2): 199–204.
Guthman, Julie. 2019. *Wilted: Pathogens, Chemicals, and the Fragile Future of the Strawberry Industry*. Berkeley: University of California Press.
Gutkowski, Natalie. 2021. Bodies that count: Administering multispecies in Palestine/Israel's borderlands. *Environment and Planning E: Nature and Space* 4(1): 135–157.
Hartmann, Saidiya. 2019. *Wayward Lives, Beautiful Experiments: Intimate Histories of Riotous Black Girls, Troublesome Women, and Queer Radicals*. New York: Norton.
Heath Justice, Daniel. 2021. *Raccoon*. London: Reaktion.
Hennessy, Elizabeth. 2013. Producing "prehistoric" life: Conservation breeding and the remaking of wildlife genealogies. *Geoforum* 49: 71–80.
Hennessy, Elizabeth. 2019. *On the Backs of Tortoises: Darwin, the Galapagos, and the Fate of an Evolutionary Eden*. New Haven: Yale University Press.

Hodgetts, Timothy. 2017. Wildlife conservation, multiple biopolitics and animal subjectification: Three mammals' tales. *Geoforum* 79: 17–25.

Hovorka, Alice. 2008. Transspecies urban theory: Chickens in an African city. *Cultural Geographies* 15 (1): 95–117.

Hribal, Jason. 2003. "Animals are part of the working class": A challenge to labor history. *Labour History* 44 (4): 435–453.

Hubbard, Tasha. 2014. Buffalo genocide in nineteenth-century North America: "Kill, skin, and sell". In *Colonial Genocide in Indigenous North America*, edited by Alexander Laban Hinton, Andrew Woolford and Jeff Benvenuto, 292–305. Durham, NC: Duke University Press.

Isaacs, Jenny and Ariel Otruba. 2019. Guest introduction: More-than-human contact zones. *Environment and Planning. E, Nature and Space* 2 (4): 697–711.

Johnson, Jay and Brian Murton. 2007. Re/placing native science: Indigenous voices in contemporary constructions of nature. *Geographical Research* 45 (2): 121–129.

Kirksey, Eben and Stefan Helmreich. 2010. The emergence of multispecies ethnography. *Cultural Anthropology* 24 (4): 545–576.

Kosek, Jake. 2006. *Understories: The Political Life of Forests in Northern New Mexico*. Durham, NC: Duke University Press.

Krishnan, Siddhartha, Christopher Pastore and Samuel Temple. 2015. Introduction: Unruly matters. In *Unruly Environments*, edited by Siddhartha Krishnan, Christopher Pastore and Samuel Temple, 5–7. Munich: Rachel Carson Centre Perspectives.

Loo, Tina. 2006. *States of Nature: Conserving Canada's Wildlife in the Twentieth Century*. Vancouver: UBC Press.

Lorimer, Jamie. 2017. Probiotic environmentalities: Rewilding with wolves and worms. *Theory, Culture and Society* 34: 27–48.

Lorimer, Jamie. 2020. *The Probiotic Planet: Using Life to Manage Life*. Minneapolis: University of Minnesota Press.

Lloyd, Kate, Sarah Wright, Sandie Suchet-Pearson, Laklak Burarrwanga, and Bawaka Country. 2012. Reframing development through collaboration: Towards a relational ontology of connection in Bawaka, North East Arnhem Land. *Third World Quarterly* 33 (6): 1075–1094.

McGregor, Deborah. 2018. Mino-mnaamodzawin: Achieving Indigenous environmental justice in Canada. *Environment and Society* 9 (1): 7–24.

Malm, Andreas. 2018. In wildness is the liberation of the world: On Maroon ecology and partisan nature. *Historical Materialism* 26 (3): 3–37.

Mauch, Christof. 2015. Unruly paradise—nature and culture in Malibu, California. In *Unruly Environments*, edited by Siddhartha Krishnan, Christopher Pastore and Samuel Temple, 45–54. Munich: Rachel Carson Centre Perspectives.

Murphy, Sean, David Wilckens, Ben Augustine, Mark Peyton, and Glenn Harper. 2019. Improving estimation of puma (*Puma concolor*) population density: Clustered camera-trapping, telemetry data, and generalized spatial mark-resight models. *Scientific Reports* 9 (1): 1–13.

Narayanan, Yamini and Sumanth Bindumadhav. 2019. "Posthuman cosmopolitanism" for the Anthropocene in India: Urbanism and human-snake relations in the Kali Yuga. *Geoforum* 106: 402–410.

O'Connor, James. 1988. Capitalism, nature, socialism: A theoretical introduction. *Capitalism, Nature, Socialism* 1 (1): 11–38.

Pacini-Ketchabaw, Veronica and Fikile Nxumalo. 2015. Unruly raccoons and troubled educators: Nature/culture divides in a childcare centre. *Environmental Humanities* 7: 151–168.

Perkins, Harold A. 2020. Killing one trout to save another: A hegemonic political ecology with its biopolitical basis in Yellowstone's native fish conservation plan. *Annals of the American Association of Geographers* 110 (5): 1559–1576.

Polanyi, Karl. 1944. *The Great Transformation: The Political and Economic Origins of Our Time*. Boston: Beacon Press.

Power, Emma. 2009. Border-processes and homemaking: Encounters with possums in suburban Australian homes. *Cultural Geographies* 16 (1): 29–54.

Pratt, Mary Louise. 1992. *Imperial Eyes: Travel Writing and Transculturation*. London: Routledge.

Probyn-Rapsey, Fiona, and Rowena Lennox. 2020. Feral violence: The Pelorus experiment. *Environment and Planning E: Nature and Space* 5 (1): 362–380.

Prudham, S. 2003. Taming trees: Capital, science, and nature in Pacific Slope tree improvement. *Annals of the Association of American Geographers* 93 (3): 636–656.
Prudham, S. 2005. *Knock on Wood: Nature as Commodity in Douglas-Fir Country*. New York: Routledge.
Pyne, Stephen. 2007. *Awful Splendour: A History of Fire in Canada*. Vancouver: UBC Press.
Qitsualik, Rachel. 2013. Innummarik: Self-sovereignty in classic Inuit thought. In *Nilliajut: Inuit Perspectives on Security, Patriotism and Sovereignty*, edited by Scot Nickels, Karen Kelley, Carrie Grable, Martin Lougheed, and James Kuptana, 23–34. Ottawa, ON: Inuit Tapiriit Kanatami.
Raffles, Hugh, 2010. *Insectopedia*. New York: Vintage Books.
Robbins, Paul. 2007. *Lawn People: How Grasses, Weeds, and Chemicals Make Us Who We Are*. Philadelphia: Temple University Press.
Robbins, Paul and April Luginbuhl. 2005. The last enclosure: Resisting privatization of wildlife in the Western United States. *Capitalism Nature Socialism* 16 (1): 45–61.
Rubio-Ramon, Guillem and Krithika Srinivasan. 2022. Methodologies for animal geographies: Approaches within and beyond the human. In *The Routledge Handbook of Methodologies in Human Geography*, eds Rosenberg M, Lovell S, Coen S. London: Routledge.
Rutherford, Stephanie. 2022. *Vermin, Villain, Icon, Kin. Wolves and the Making of Canada*. McGill-Queens Press.
Salih, Ruba and Olaf Corry. 2021. Displacing the Anthropocene: Colonisation, extinction and the unruliness of nature in Palestine. *Environment and Planning E: Nature and Space* 5 (1): 381–400.
Schiebinger, Londa. 2007. *Plants and Empire: Colonial Bioprospecting in the Atlantic World*. Harvard University Press.
Scott, James. 1998. *Seeing like a State: How Certain Schemes to Improve the Human Condition Have Failed*. Yale University Press.
Simpson, Leanne Betasamosake. 2017. *As We have Always Done: Indigenous Freedom through Radical Resistance*. Minneapolis: University of Minnesota Press.
Singh, Julietta. 2018. Errands for the wild. *South Atlantic Quarterly* 117 (3): 567–580.
Smalley, Andrea. 2017. *Wild by Nature: North American Animals Confront Colonization*. Johns Hopkins University Press.
Srinivasan, Krithika. 2014. Caring for the collective: Biopower and agential subjectification in wildlife conservation. *Environment and Planning D: Society and Space* 32 (3): 501–517.
Srinivasan, Krithika. 2019. Remaking more-than-human society: Thought experiments on street dogs as "nature". *Transactions of the Institute of British Geographers* 44 (2): 376–391.
Subramaniam, Banu. 2014. *Ghost Stories for Darwin: The Science of Variation and the Politics of Diversity*. Champaign: University of Illinois Press.
Sundberg, Juanita. 2011. Diabolic caminos in the desert and cat fights on the Río: A posthumanist political ecology of boundary enforcement in the United States–Mexico borderlands. *Annals of the Association of American Geographers* 101 (2): 318–336.
Sundberg, Juanita. 2014. Decolonizing posthumanist geographies. *Cultural Geographies* 21(1): 33–47.
Temple, Samuel. 2015. Unruly marshes: Obstacles or agents of empire in French North Africa? In *Unruly Environments*, edited by Siddhartha Krishnan, Christopher Pastore and Samuel Temple, 11–18. Munich: Rachel Carson Centre Perspectives.
Todd, Zoe. 2015. Indigenizing the Anthropocene. *Art in the Anthropocene: Encounters among aesthetics, politics, environments and epistemologies*, eds H Davis and E Turpin, 241–254. London: Open Humanities Press.
Todd, Zoe. 2016. An Indigenous feminist's take on the ontological turn: "Ontology" is just another word for colonialism. *Journal of Historical Sociology* 29(1): 4–22.
Todd, Zoe. 2018. Refracting the state through human-fish relations. *Decolonization: Indigeneity, Education and Society* 7 (1): 60–75.
Tsing, Anna. 2015. *The Mushroom at the End of the World*. Princeton University Press.
Van Patter, Lauren. 2021. Individual animal geographies for the more-than-human city: Storying synanthropy and cynanthropy with urban coyotes. *Environment and Planning E: Nature and Space* 5 (4): 2216–2239.
Van Patter, Lauren and Charlotte Blattner. 2020. Advancing ethical principles for non-invasive, respectful research with nonhuman animal participants. *Society and Animals* 28: 171–190.

Vinyeta, Kirsten. 2021. Under the guise of science: How the US Forest Service deployed settler colonial and racist logics to advance an unsubstantiated fire suppression agenda. *Environmental Sociology* 8 (2): 134–148.

Wadiwel, Dinesh. 2018. Chicken harvesting machine: Animal labor, resistance, and the time of production. *The South Atlantic Quarterly* 117 (3): 527–549.

Watts, Vanessa. 2013. Indigenous place-thought and agency amongst humans and non-humans (First Woman and Sky Woman go on a European world tour!). *Decolonization: Indigeneity, Education and Society* 2 (1): 20–34.

Werner, Marion, Christian Berndt and Becky Mansfield. 2022. The glyphosate assemblage: Herbicides, uneven development, and chemical geographies of ubiquity. *Annals of the Association of American Geographers* 112 (1): 19–35.

White, Richard, 1995. *The Organic Machine: The Remaking of the Columbia River*. New York: Hill and Wang.

Williams, A. 1913. Report for 1912. Victoria: Government of BC. BC Archives GR- 0446—Provincial game warden records, Box 38 File 12.

Wright, Willie Jamaal. 2020. The morphology of marronage. *Annals of the American Association of Geographers* 110 (4): 1134–1149.

Wynter, Sylvia. 2003. Unsettling the coloniality of being/power/truth/freedom: Towards the human, after man, its overrepresentation—an argument. *The New Centennial Review* 3 (3): 257–337.

Zahara, Alexander. 2020. Breathing fire into landscapes that burn: Wildfire management in a time of alterlife. *Engaging Science, Technology, and Society* 6: 555–585.

Zahara, Alexander and Myra Hird. 2016. Raven, dog, human: Inhuman colonialism and unsettling cosmologies. *Environmental Humanities* 7 (1): 169–190.

Index

Page numbers in *italics* indicate an illustration, **bold** a table

T - #0237 - 111024 - C0 - 246/174/15 - PB - 9780367760953 - Matt Lamination